PLANETARY LANDERS AND ENTRY PROBES

T0185736

This book provides a concise but broad overview of the engineering, science and flight history of planetary landers and atmospheric entry probes – vehicles designed to explore the atmospheres and surfaces of other worlds. It covers engineering aspects specific to such vehicles, such as landing systems, parachutes, planetary protection and entry shields, which are not usually treated in traditional spacecraft engineering texts. Examples are drawn from over thirty different lander and entry probe designs that have been used for lunar and planetary missions since the early 1960s. The authors provide detailed illustrations of many vehicle designs from space programmes worldwide, and give basic information on their missions and payloads, irrespective of the mission's success or failure. Several missions are discussed in more detail, in order to demonstrate the broad range of the challenges involved and the solutions implemented. *Planetary Landers and Entry Probes* will form an important reference for professionals, academic researchers and graduate students involved in planetary science, aerospace engineering and space mission development.

ANDREW BALL is a Postdoctoral Research Fellow at the Planetary and Space Sciences Research Institute at The Open University, Milton Keynes, UK. He is a Fellow of the Royal Astronomical Society and the British Interplanetary Society. He has twelve years of experience on European planetary missions including Rosetta and Huygens.

JAMES GARRY is a Postdoctoral Research Fellow in the School of Engineering Sciences at the University of Southampton, UK, and a Fellow of the Royal Astronomical Society. He has worked on ESA planetary missions for over ten years and has illustrated several space-related books.

RALPH LORENZ is a Scientist at the Johns Hopkins University Applied Physics Laboratory, USA. He is a fellow of the Royal Astronomical Society and the British Interplanetary Society. He has fifteen years of experience in NASA and ESA spaceflight projects and has authored several space books.

VIKTOR KERZHANOVICH is a Principal Member of Technical Staff of the Mobility and Robotic Systems Section of the Autonomous Systems Division, NASA Jet Propulsion Laboratory, USA. He was a participant in all Soviet planetary Venus and Mars entry probe programmes.

PLANETARY LANDERS AND ENTRY PROBES

ANDREW J. BALL
The Open University

JAMES R. C. GARRY
Southampton University

RALPH D. LORENZ
Johns Hopkins University Applied Physics Laboratory

VIKTOR V. KERZHANOVICH
NASA Jet Propulsion Laboratory

CAMBRIDGE
UNIVERSITY PRESS

CAMBRIDGE UNIVERSITY PRESS
Cambridge, New York, Melbourne, Madrid, Cape Town, Singapore,
São Paulo, Delhi, Dubai, Tokyo

Cambridge University Press
The Edinburgh Building, Cambridge CB2 8RU, UK

Published in the United States of America by Cambridge University Press, New York

www.cambridge.org
Information on this title: www.cambridge.org/9780521129589

First published 2007
This digitally printed version 2009

A catalogue record for this publication is available from the British Library

ISBN 978-0-521-82002-8 Hardback
ISBN 978-0-521-12958-9 Paperback

Contents

v

Preface

This book is intended as a concise but broad overview of the engineering, science and flight history of planetary landers and atmospheric probes. Such vehicles are subject to a wide range of design and operational issues that are not experienced by 'ordinary' spacecraft such as Earth-orbiting satellites, or even by interplanetary flyby or orbital craft. Such issues deserve special attention, and we have attempted to bring together in one place brief discussions of many of these aspects, providing pointers to more detailed (but dispersed) coverage in the wider published literature. This volume also draws heavily on real examples of landers and probes launched (or, at least, where the launch vehicle's engines were started with that intention!).

More than 45 years have passed since the first vehicles of this type were designed. To a certain extent some past missions, of which there are over one hundred, may now be considered irrelevant from a scientific point of view, outdated from an engineering point of view and perhaps mere footnotes in the broader history of planetary exploration achievements. However, we believe they all have a place in the cultural and technical history of such endeavours, serving to illustrate the evolving technical approaches and requirements as well as lessons learned along the way. They stand as testament to the efforts of those involved in their conception and implementation.

Part one of the book addresses the major engineering issues that are specific to the vehicles considered, namely atmospheric entry probes, landers and penetrators for other worlds. For material common to spacecraft in general we would refer the reader to other, existing sources. Part II aims to collect together in one place some key information on previous vehicles and their missions, with reference to the main sources of more detailed information. Part III covers some of these missions in further detail as 'case studies'.

January, 2006

Acknowledgements

The authors wish to thank Susan Francis and her colleagues at Cambridge University Press for their encouragement and patience. Many colleagues and contacts have helped with specific queries, including: Dave Atkinson, Aleksandr T. Basilevsky, Jens Biele, Jacques Blamont, Peter Bond, Jim Burke, Ed Chester, Chad Edwards, Alex Ellery, Bernard Foing, Sven Grahn, Aleksandr Gurshtein, Leonid Gurvits, Ari-Matti Harri, Mat Irvine, Bobby Kazeminejad, Oleg Khavroshkin, Vladimir Kurt, Bernard Laub, Mikhail Ya. Marov, Serguei Matrossov, Michel Menvielle, Don P. Mitchell, Dave Northey, Colin T. Pillinger, Sergei Pogrebenko, Jean-Pierre Pommereau, Lutz Richter, Andy Salmon, Mark Sims, Oleg A. Sorokhtin, Yuri A. Surkov, Fred W. Taylor, Stephan Ulamec, Paolo Ulivi, David Williams, Andrew Wilson, Ian P. Wright, Hajime Yano, and Olga Zhdanovich. We would also like to thank Professor John Zarnecki and the staff at the Open University Library. The diagrams that populate Part II were drawn using information gleaned from a variety of sources. While researching specific details for spacecraft, the authors were glad to receive help from the following people: Charles Sobeck, Bernard Bienstock, Corby Waste, Pat Flannery, Marty Tomasko, Marcie Smith, Dan Maas, Doug Lombardi, Debra Lueb, Martin Towner, Mark Leese, Steve Lingard and John Underwood.

List of acronyms and abbreviations

ACP	Aerosol Collector/Pyrolyser
ADS	Active Descent System
AFM	Atomic Force Microscope
AIAA	American Institute of Aeronautics and Astronautics
ALSEP	Apollo Lunar Surface Experiments Package
AMICA	Asteroid Multiband Imaging Camera
AMTEC	Alkali Metal Thermionic Emission Technology
ANC	ANChor
APEX	Athena Precursor EXperiment
APX	Alpha-Proton X-ray spectrometer OR Alpha Particle X-ray spectrometer
APXS	Alpha-Proton X-ray Spectrometer OR Alpha Particle X-ray Spectrometer
ARAD	Analog Resistance Ablation Detector
ARES	Atmospheric Relaxation and Electric field Sensor
ASAP	Ariane Structure for Auxiliary Payloads
ASI	Atmospheric Structure Instrument
ATMIS	ATmospheric structure and Meteorological Instrument System
AU	Astronomical Unit
AXS	Alpha-X-ray Spectrometer
AZ (АЗ)	Aerostatic Zond (Аэростатный Зонд)
BER	Bit Error Rate
BOL	Beginning-Of-Life
BPSK	Binary Phase-Shift Keying
CASSE	Cometary Acoustic Surface Sounding Experiment
CCD	Charge-Coupled Device

CD	Compact Disk
CDMS	Command and Data Management System
CDMU	Command and Data Management Unit
CFRP	Carbon Fibre Reinforced Plastic
CHARGE	CHemical Analysis of Released Gas Experiment
CIRCLE	Champollion Infrared and Camera Lander Experiment
ÇIVA	Comet nucleus Infrared and Visible Analyser
CNES	Centre National d'Études Spatiales
CNP	Comet Nucleus Penetrator
CNSR	Comet Nucleus Sample Return
CoM	Centre of Mass
CONSERT	COmet Nucleus Sounding Experiment by Radiowave Transmission
COSAC	COmetary Sampling And Composition experiment
COSPAR	COmmittee on SPAce Research
CPPP	Comet Physical Properties Package
CR	Cosmic Ray
CRAF	Comet Rendezvous/Asteroid Flyby
CSM	Command and Service Module
DAS (ДАС)	Long-lived Autonomous Station (Долгоживущую Автономную Станцию)
DC	Direct Current
DCP	Data and Command Processor
DESCAM	DEScent CAMera
DGB	Disc-Gap-Band
DIM	Dust Impact Monitor
DIMES	Descent Image Motion Estimation System
DISR	Descent Imager – Spectral Radiometer
DLBI	Differential Long Baseline Interferometer
DLR	Deutsches Zentrum für Luft- und Raumfahrt (German Aerospace Centre)
DNA	Deoxyribonucleic Acid
DoD	Depth of Discharge
DPI	Descent Phase Instrument
DSC	Differential Scanning Calorimeter
DSN	Deep Space Network
DS-2	Deep Space 2
DTE	Direct To Earth
DVLBI	Differential Very Long Baseline Interferometry
DWE	Doppler Wind Experiment

EADS	European Aeronautic, Defence and Space Company
EASEP	Early Apollo Surface Experiments Package
EDI	Entry, Descent and Inflation
EDL	Entry, Descent and Landing
EDLS	Entry, Descent and Landing System
EEPROM	Electrically-Erasable Programmable Read-Only Memory
EM	ElectroMagnetic
EPDM	Ethylene Propylene Diene, Modified
EPI	Energetic Particles Instrument
ESA	European Space Agency
ESS	Environmental Sensors Suite
ETR	Eastern Test Range
FBC	Faster, Better, Cheaper
FBS	Fan-Beam Sensor
FMCW	Frequency Modulated Continuous Wave
FRCI	Fibrous Refractory Composite Insulation
FSK	Frequency-Shift Keying
GAP	Gas Analysis Package
GCMS	Gas Chromatograph/Mass Spectrometer
GCR	Galactic Cosmic Ray
GPR	Ground-Penetrating Radar
GRAM	Global Reference Atmospheric Model
GZU (ГЗУ)	Ground Sampling Device (ГрунтоЗаборное Устройство)
HAD	Helium Abundance Detector
HASI	Huygens Atmospheric Structure Instrument
HGA	High Gain Antenna
IDD	Instrument Deployment Device
IDL	Interactive Data Language
IF	Intermediate Frequency
IKI (ИКИ)	Institute for Space Research
	(Институт Космических Исследований)
IMP	Imager for Mars Pathfinder
IMU	Inertial Momentum Unit
IPIU	Instrument Power Interface Unit
IR	InfraRed
ISAS	Institute of Space and Astronautical Science
ISEE-3	International Sun – Earth Explorer 3
ISIS	In Situ Imaging System
ITU	International Telecommunication Union
IUS	Inertial Upper Stage

JPL	Jet Propulsion Laboratory
KEP	Kinetic Energy Penetrator
KhM-VD (ХМ-ВД)	Running Model – Wind Engine (Ходовых Макет – ВетроДвигатель)
KSC	Kennedy Space Center
LAS	Large Atmospheric Structure
LCPS	Large Cloud Particle Size Spectrometer
LET	Linear Energy Transfer
LGA	Low Gain Antenna
LGC	Large Gas Chromatograph
LIDAR	LIght Detection And Ranging
LIR	Large Infrared Radiometer
LK (ЛК)	Lunar Ship (Лунный Корабль)
LM	Lunar Module
LN	Large Nephelometer
LNMS	Large Neutral Mass Spectrometer
LRD	Lightning and Radio emissions Detector
LRF	Laser Range-Finder
LRV	Lunar Roving Vehicle
LS (ЛС)	Lunar Seismometer (Лунный Сейсмометр)
LSFR	Large Solar Flux Radiometer
LTA	Lighter Than Air
MAE	Materials Adherance Experiment
MAG	Magnetometer
MAGNET	Magnetometer for NetLander
MAHLI	MArs HandLens Imager
MAPEX	Microelectronics And Photonics EXperiment
MARDI	MARs Descent Imager
MARIE	MArtian Radiation envIronment Experiment
MB	MössBauer spectrometer
MBS	MössBauer Spectrometer
MECA	Mars Environmental Compatibility Assessment
MECA	Microscopy, Electrochemistry and Conductivity Analyser
MEDLI	MSL Entry, Descent and Landing Instrumentation
MEEC	Mars Experiment on Electrostatic Charging
MEKOM (МЕКОМ)	Meteorological Complex (Метеокомплекс)
MER	Mars Exploration Rover
MESUR	Mars Environmental SURvey
MET	Meteorological Package

MET	Modular Equipment Transporter
MEx	Mars Express
MFEX	Microrover Flight EXperiment
MGS	Mars Global Surveyor
MI	Microscopic Imager
MIC	Mars Microphone
MIC	Microscope
MINERVA	MIcro/Nano Experimental Robot Vehicle for Asteroid
Mini-TES	Miniature Thermal Emission Spectrometer
MIP	Mars In situ Propellant production precursor
MIS	Meteorology Instrument System
MLI	Multi-Layer Insulation
MMRTG	Mutli-Mission Radioisotope Thermoelectric Generator
MNTK (МНТК)	International Scientific and Technical Committee (Международный Научный Технически Комитет)
MOx	Mars Oxidant experiment
MPAe	Max-Planck-Institut für Aeronomie
MPL	Mars Polar Lander
MPRO	atMospheric PROpagation
MSB (МСБ)	Small Solar Battery (Малая Солнечная Батарея)
MSL	Mars Science Laboratory (previously Mars Smart Lander)
MUPUS	MUlti-PUrpose Sensors for surface and sub-surface science
MUSES-C	MU Space Engineering Spacecraft C
MUSES-CN	MUSES-C Nanorover
MTUR	atMospheric TURbulence
MVACS	Mars Volatiles And Climate Surveyor
MWIN	atMospheric WINd
NASA	National Aeronautics and Space Administration
NEAR	Near-Earth Asteroid Rendezvous
NEIGE	NEtlander Ionospheric and Geodesic Experiment
NEO	Near-Earth Object
NEP	Nephelometer
NFR	Net Flux Radiometer
NII (НИИ)	Scientific Research Institute (Научно-Исследовательский Институт)
NII PDS (НИИ ПДС)	Scientific Research Institute for Parachute Landing Service (Научно-Исследовательский Институт Парашютно-Десантной Службы)
NIRS	Near-InfraRed Spectrometer
NMS	Neutral Mass Spectrometer

NPO (НПО)	Scientific Production Association (Научно-Производственное Объединение)
NTS	NEC Toshiba Space Systems
ODS	Optical Depth Sensor
ODT	Orbiter Delay Time
OKB (ОКБ)	Experimental Design Bureau (Опытно-Конструкторское Бюро)
ONC	Optical Navigation Camera
OPTIMISM	Observatoire PlanéTologIque: MagnétIsme et Sismologie sur Mars
PANCAM	PANoramic CAMera
PAW	Position Adjustable Workbench
PBO	Polybenzoxazole
PC	Personal Computer
PCM	Pulse Code Modulation
PCU	Pyro Control Unit
PEN	PENetrator
PI	Principal Investigator
PLL	Phase-Locked Loop
PLUTO	PLanetary Underground TOol
PM	Phase Modulation
PP	Permittivity Probe
PROM	Programmable Read-Only Memory
PrOP (ПрОП)	Instrument for the Evaluation of Passability (Прибор Оценки Проходимости)
PROP-F (ПРОП-Ф)	Mobile Robot for the Evaluation of the Surface of Phobos (Подвижной Робот Оценки Поверхности Фобоса)
PROP-M (ПРОП-М)	Mobile Robot for the Evaluation of the Surface of Mars (Подвижной Робот Оценки Поверхности Марса)
PrOP-V (ПрОП-В)	Instrument for the Evaluation of the Surface of Venus (Прибор Оценки Поверхности Венера)
PSE	Probe Support Equipment
PSK	Phase-Shift Keying
PTFE	PolyTetraFluoroEthylene
PTUW	Pressure, Temperature, hUmidity and Wind
PV	PhotoVoltaic
RA	Robotic Arm
RAATS	Robotic Arm Atmospheric Temperature Sensor
RAC	Robotic Arm Camera
RAD	Radiation Assessment Detector

RAD	Rocket-Assisted Descent
RADVS	Radar Altimeter & Doppler Velocity Sensor
RAM	Random Access Memory
RAT	Rock Abrasion Tool
RF	Radio Frequency
RHU	Radioisotope Heater Unit
RIFMA (РИФМА)	Roentgen Isotopic Fluorescence Method of Analysis (Рентген Изотопное Флуоресцирование Метод Анализа)
RKK (РКК)	Rocket-Space Corporation (Ракетно-Космическая Корпорация)
RMS	Root-Mean-Square
RNII (РНИИ)	Russian Scientific Research Institute (Российский Научно-Исследовательский Институт)
RNII KP	Russian Scientific Research Institute for Space Device Engineering
(РНИИ КП)	(Российский Научно-Исследовательский Институт Космического Приборостроения)
ROLIS	ROsetta Lander Imaging System
ROMAP	ROsetta lander Magnetometer And Plasma monitor
RPA	Retarding Potential Analyser
RTG	Radioisotope Thermoelectric Generator
RX	Receiving
SAA	South Atlantic Anomaly
SAM	Sample Analysis at Mars
SAMPLL	Simplified Analytical Model of Penetration with Lateral Loading
SAS	Small Atmospheric Structure
SCS	Stereo Camera System
SD2	Sampling, Drilling and Distribution system
SEIS	SEISmometer
SESAME	Surface Electrical, Seismic and Acoustic Monitoring Experiments
SEU	Single Event Upset
SI	Système Internationale
SINDA	Systems Improved Numerical Differencing Analyzer
SIRCA-SPLIT	Silicone-Impregnated Reusable Ceramic Ablator – Secondary Polymer Layer-Impregnated Technique
SIS	SISmomètre
SLA	Super-Lightweight Ablator
SMSS	Soil Mechanics Surface Sampler

SN	Small Nephelometer
SNFR	Small Net Flux Radiometer
SNR	Signal-to-Noise Ratio
SPICE	Soil Properties: thermal Inertia and Cohesion Experiment
SPIU	System Power Interface Unit
SSB	Space Studies Board
SSI	Surface Stereo Imager
SSP	Surface Science Package
SSV	Small Science Vehicle OR Small Separable Vehicle
STP	Soil Temperature Probe
TDL	Tunable Diode Laser
TECP	Thermal and Electrical Conductivity Probe
TEGA	Thermal and Evolved Gas Analyzer
TIRS	Transverse Impulse Rocket System
TM	Thermal Mapper
TNO	Trans-Neptunian Object
TPS	Thermal Protection System
TsUP (ЦУП)	Mission Control Centre (Центр Управления Полетами)
TV	Television
TX	Transmission
UDMH	Unsymmetrical DiMethyl Hydrazine
UHF	Ultra High Frequency
UIU (УИУ)	Acceleration Measuring Device (Устройство Измерения Ускорения)
UK	United Kingdom
US	United States
USA	United States of America
USO	Ultra-Stable Oscillator
UV	UltraViolet
VCO	Voltage-Controlled Oscillator
VeGa (ВеГа)	Venus-Halley (Венера-Галлей)
VHF	Very High Frequency
VLBI	Very Long Baseline Interferometry
VNIITransMash (ВНИИТрансМаш)	All-Russian Scientific Research Institute of Transport Machine-Building (Всероссийский Научно-Исследовательский Институт Транспортного Машиностроения)
WAE	Wheel Abrasion Experiment
WCL	Wet Chemistry Laboratory
WEB	Warm Electronics Box

WW2	World War 2
XRD	X-Ray Diffraction
XRF	X-Ray Fluorescence
XRFS	X-Ray Fluorescence Spectrometer
XRS	X-Ray Spectrometer
2MV (2MB)	2nd generation Mars/Venus (2 Марс/Венера)
3-DL	3-Dimensional Laminate
3MV (3MB)	3rd generation Mars/Venus (3 Марс/Венера)

Part I

Engineering issues specific to entry probes, landers or penetrators

This part of the book is intended to act as a guide to the basic technological principles that are specific to landers, penetrators and atmospheric-entry probes, and to act as a pointer towards more detailed technical works. The chapters of this part aim to give the reader an overview of the problems and solutions associated with each sub-system/flight phase, without going into the minutiae.

1

Mission goals and system engineering

Before journeying through the various specific engineering aspects, it is worth examining two important subjects that have a bearing on many more specific activities later on. First we consider systems engineering as the means to integrate the diverse constraints on a project into a functioning whole. We then look at the choice of landing site for a mission, a decision often based on a combination of scientific and technical criteria, and one that usually has a bearing on the design of several sub-systems including thermal, power and communications.

1.1 Systems engineering

Engineering has been frivolously but not inaptly defined as 'the art of building for one dollar that which any damn fool can build for two'. Most technical problems have solutions, if adequate resources are available. Invariably, they are not, and thus skill and ingenuity are required to meet the goals of a project within the imposed constraints, or to achieve some optimum in performance.

Systems engineering may be defined as

the art and science of developing an operable system capable of meeting mission requirements within imposed constraints including (but not limited to) mass, cost and schedule.

The modern discipline of systems engineering owes itself to the development of large projects, primarily in the USA, in the 1950s and 1960s when projects of growing scale and complexity were undertaken. Many of the tools and approaches derive from operational research, the quantitative analysis of performance developed in the UK during World War II.

Engineering up to that epoch had been confined to projects of sufficiently limited complexity that a single individual or a team of engineers in a dominant discipline could develop and implement the vision of a project. As systems

3

became more sophisticated, involving hundreds of subcontractors, the more abstract art of managing the interfaces of many components became crucial in itself.

A general feature is that of satisfying some set of requirements, usually in some optimal manner. To attain this optimal solution, or at least to satisfy as many as possible of the imposed requirements, usually requires tradeoffs between individual elements or systems. To mediate these tradeoffs requires an engineering familiarity and literacy, if not outright talent, with all of the systems and engineering disciplines involved. Spacecraft represent particularly broad challenges, in that a wide range of disciplines is involved – communications, power, thermal control, propulsion and so on. Arguably, planetary probes are even more broad, in that all the usual spacecraft disciplines are involved, plus several aspects related to delivery to and operation in planetary environments, such as aerothermodynamics, soil mechanics and so on.

While engineers usually like to plough into technical detail as soon as their task is defined, it is important to examine a broad range of options to meet the goals. As a simple example, a requirement might be to destroy a certain type of missile silo. This in turn requires the delivery of a certain overpressure onto the target. This could be achieved, for example, by the use of a massive nuclear warhead on a big, dumb missile. Or one might attain the same result with a much smaller warhead, but delivered with precision, requiring a much more sophisticated guidance system. Clearly, these are two very different, but equally valid, solutions.

It is crucial that the requirements be articulated in a manner that adequately captures the intent of the 'customer'. To this end, it is usual that early design studies are performed to scope out what is feasible. These usually take the form of an assessment study, followed by a Phase A study and, if selected, the mission proceeds to Phases B and C/D for development, launch and operation. During the early study phases a mission-analysis approach is used prior to the more detailed systems engineering activity. Mission analysis examines quantitatively the top-level parameters of launch options, transfer trajectories and overall mass budget (propellant, platform and payload), without regard to the details of subsystems.

In the case of a planetary probe, the usual mission is to deliver and service an instrument payload for some particular length of time, where the services may include the provision of power, a benign thermal environment, pointing and communications back to Earth.

The details of the payload itself are likely to be simply assumed at the earliest stages, by similarity with previous missions. Such broad resource requirements as data rate/volume, power and mass will be defined for the payload as a whole. These allow the design of the engineering system to proceed, from selecting among a broad choice of architectures (e.g. multiple small probes, or a single mobile one) through the basic specification of the various subsystems.

The design and construction of the system then proceeds, usually in parallel with the scientific payload (which is often, but not always, developed in institutions other than that which leads the system development), perhaps requiring adaptation in response to revised mission objectives, cost constraints, etc. Changes to a design become progressively more difficult and expensive to implement.

1.1.1 The project team

The development team will include a number of specialists dedicated to various aspects of the project, throughout its development. In many organizations, additional expertise will additionally be co-opted on particular occasions (e.g. for design reviews, or particularly tight schedules).

The project will be led by a project manager, who must maintain the vision of the project throughout. The project manager is the single individual whose efforts are identified with the success or otherwise of the project. The job entails wide (rather than deep) technical expertise, in order to gauge the weight or validity of the opinions or reports of various subsystem engineers or others and to make interdisciplinary tradeoffs. The job requires management skills, in that it is the efforts of the team and contractors that ultimately make things happen – areas where members of the team may variously need to be motivated, supported with additional manpower, or fired. Meetings may need to be held, or prevented from digressing too far. And this demanding job requires political skill, to tread the compromise path between constraints imposed on the project, and the capabilities required or desired of it.

A broadly similar array of abilities, weighted somewhat towards the technical expertise, is required of the systems engineer, usually a nominal deputy to the project manager. A major job for the systems engineer is the resolution of technical tradeoffs as the project progresses. Mass growth, for example, is a typical feature of a project development – mass can often be saved by using lighter materials (e.g. beryllium rather than aluminium), but at the cost of a longer construction schedule or higher development cost.

A team of engineers devoted to various aspects of the project, from a handful to hundreds, will perform the detailed design, construction and testing. The latter task may involve individuals dedicated to arranging the test facilities and the proper verification of system performance. Where industrial teams are involved, various staff may be needed to administer the contractual aspects. Usually the amount of documentation generated is such as to require staff dedicated to the maintenance of records, especially once the project proceeds to a level termed 'configuration control', wherein interfaces between various parts of the project are frozen and should not be changed without an intensive, formal review process.

In addition to the hardware and software engineers involved in the probe system itself, several other technical areas may be represented to a greater or lesser extent. Operations engineers may be involved in the specification, design, build and operation of ground equipment needed to monitor or command the spacecraft, and handle the data it transmits. There may be specialists in astro-dynamics or navigation. Finally, usually held somewhat independently from the rest of the team, are quality-assurance experts to verify that appropriate levels of reliability and safety are built into the project, and that standards are being followed.

In scientific projects there will be a project scientist, a position not applicable for applications such as communications satellites. This individual is the liaison between the scientific community and the project. In addition to mediating the interface between providers of the scientific payload and the engineering side of the project, the project scientist will also coordinate, for example, the generation or revision of environmental models that may drive the spacecraft design.

The scientific community usually provides the instruments to a probe. The lead scientist behind an instrument, the principal investigator (PI), will be the individual who is responsible for the success of the investigation. Usually this means pro-curing adequate equipment and support to analyse and interpret the data, as well as providing the actual hardware and software. An instrument essentially acts as a mini project-within-a-project, with its own engineering team, project manager, etc.

For the last decade or so, NASA has embraced so-called PI-led missions, under the Discovery programme. Here a scientist is the originator and authority (in theory) for the whole mission, guiding a team including agency and industry partners, not just one experiment. This PI-led approach has led to some highly efficient missions (Discovery missions have typically cost around $300M, com-parable with the ESA's 'Medium' missions) although there have also been some notable failures, as with any programme. The PI-led mission concept has been extended to more expensive missions in the New Frontiers line, and for Discovery-class missions in the Mars programme, called 'Mars Scout'.

A further class of mission deserves mention, namely the technology-devel-opment or technology-validation mission. These are intended primarily to demonstrate and gain experience with a new technology, and as such may involve a higher level of technical risk than one might tolerate on a science-driven mission. Some missions (such as those under NASA's New Millenium pro-gramme, notably the DS-2 penetrators) are exclusively driven by technology goals, with a minimal science payload (although often substantial science can be accomplished even with only engineering sensors). In some other cases, the science/technology borderline is very blurred – one example is the Japanese Hayabusa asteroid sample return: this mission offers a formidable scientific return, yet was originally termed MUSES-C (Mu-launched space-engineering satellite).

Whatever the political definitions and the origin of the mission requirements, it must be recognized that there is both engineering challenge and science value in any spacecraft measurement performed in a planetary environment.

A dynamic tension usually exists in a project, somewhat mediated by the project scientist. Principal investigators generally care only about their instrument, and realizing its maximum scientific return, regardless of the cost of the system needed to support it. The project manager is usually confronted with an already overconstrained problem – a budget or schedule that may be inadequate and contractors who would prefer to deliver hardware as late as possible while extorting as much money out of the project as possible. One tempting way out is to descope the mission, to reduce the requirements on, or expectations of, the scientific return. Taken to the extreme, however, there is no point in building the system at all. Or a project that runs too late may miss a launch window and therefore never happen; a project that threatens to overrun its cost target by too far may be cancelled. So the project must steer a middle path, aided by judgement and experience as well as purely technical analysis – hence the definition of systems engineering as an art.

1.2 Choice of landing site

Technical constraints are likely to exist on both the delivery of the probe or lander, and on its long-term operation. First we consider the more usual case where the probe is delivered from a hyperbolic approach trajectory, rather than a closed orbit around the target.

The astrodynamic aspects of arrival usually specify an arrival direction, which cannot be changed without involving a large delivered-mass penalty. The arrival speed, and the latitude of the incoming velocity vector (the 'asymptote', or V_∞, unperturbed by the target's gravity, is usually considered) are hence fixed. Usually the arrival time can be adjusted somewhat, which may allow the longitude of the asymptote to be selected for sites of particular interest, or to ensure the landing site is visible from a specific ground station. Occasionally this is fixed too, as in the case of Luna 9 where the descent systems would not permit any horizontal velocity component – the arrival asymptote would only be vertical at near-equatorial landing sites around 64°W.

The target body is often viewed in the planning process from this incoming V_∞: the plane going through the centre of the target body orthogonal to that vector is often called the 'B-plane'. The target point may be specified by two parameters. The most important is often called the 'impact parameter', the distance in the B-plane between the centre and the target point. For a given target body radius (either the surface radius, or sometimes an arbitrary 'entry interface' above which aerodynamic effects can be ignored) a given impact parameter will correspond to a flight-path angle, the angle between the spacecraft trajectory and the local horizon at that altitude. This may often be termed an entry angle.

The entry angle is usually limited to a narrow range because of the aero-thermodynamics of entry. Too high an angle (too steep) – corresponding to a small impact parameter, an entry point close to the centre of the target body – and the peak heating rate, or the peak deceleration loads, may be too high. Too

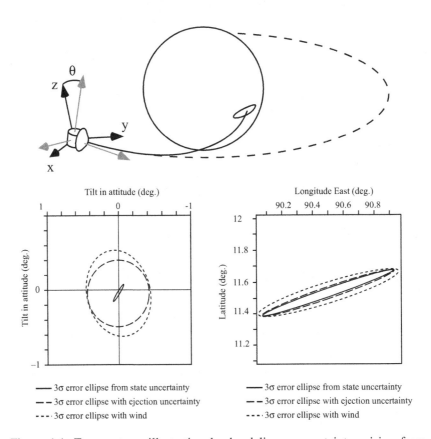

Figure 1.1. Top: cartoon illustrating lander-delivery uncertainty arising from uncertainties in the state vector at deployment. Bottom: attitude and landing-error ellipses for Beagle 2 (adapted from Bauske, 2004).

shallow an angle may result in a large total heat load; in the limiting case of a large impact parameter, the vehicle may not be adequately decelerated or may miss the target altogether.

The entry protection performance may also introduce constraints other than simple entry angle. For the extremely challenging case of entry into Jupiter's atmosphere, the $\sim$12.5 km s^{-1} equatorial rotation speed is a significant increment on the entry speed of $\sim$50 km s^{-1}. Heat loads vary as the cube of speed, and thus by aiming at the receding edge of Jupiter (i.e. the evening terminator, if coming from the Sun) the entry loads are reduced by a factor $(50+12.5)^3/(50-12.5)^3 = 4.6$, a most significant amelioration.

The second parameter is the angle relative to the target body equator (specifically where the equatorial plane crosses the B-plane) of the impact parameter. A B-plane angle of zero is on the equator; 90° means the entry point falls on the central meridian as seen from the incoming vector.

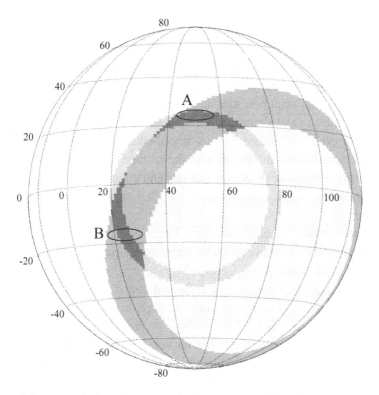

Figure 1.2. View of Titan from the arrival asymptote of the Huygens probe, with overlapping annuli reflecting the constraints on entry angle (light grey) and solar elevation (darker grey). Of the choice of two target locations where the regions overlap (A and B), only A accommodates the probe's delivery ellipse.

Other constraints include the communication geometry – if a delivery vehicle is being used as a relay spacecraft, it may be that there are external constraints on the relay's subsequent trajectory (such as a tour around the Saturnian system) which specify its target point in the B-plane. Targeting the flyby spacecraft on the opposite side of the body from the entry probe may limit the duration of the communication window. Current NASA missions after the Mars mission failures in 1999 now require mission-critical events to occur while in communication with the Earth: thus entry and landing must occur on the Earth-facing hemisphere of their target body.

Another constraint is solar. The entry may need solar illumination for attitude determination by a Sun sensor (or no illumination to allow determination by star sensor!), or a certain amount of illumination at the landing site for the hours following landing to recharge batteries. These aspects may influence the arrival time and/or the B-plane angle.

So far, the considerations invoked have been purely technical. Scientific considerations may also apply. Optical sensing, either of atmospheric properties, or surface imaging, may place constraints on the Sun angle during entry and descent. Altitude goals for science measurements may also drive the entry angle (since this determines the altitude at which the incoming vehicle has been decelerated to parachute-deployment altitude where entry protection – which usually interferes with scientific measurements – can be jettisoned).

The entry location (and therefore 'landing site') for the Huygens probe was largely determined by the considerations described so far (at the time the mission was designed there was no information on the surface anyway). The combination of the incoming asymptote direction and the entry angle defined an annulus of locations admissible to the entry system. The Sun angles required for scientific measurements of light scattering in the atmosphere, and desired shadowing of surface features defined another annulus. These two annuli intersected in two regions, with the choice between them being made partly on communications grounds.

There may be scientific desires and technical constraints on latitude. Latitude may be directly associated with communication geometry and/or (e.g. in the case of Jupiter), entry speed. For Mars landers in particular, the insolation as a function of latitude and season is a crucial consideration, both for temperature control and for solar power. Many Mars lander missions are restricted to 'tropical' landing sites in order to secure enough power.

So far, the planet has been considered only as a featureless geometric sphere. There may be scientific grounds for selecting a particular landing point, on the basis of geological features of interest (or sites with particular geochemistry such as polar ice or hydrated minerals), and, depending on the project specification, these may be the overriding factor (driving even the interplanetary delivery trajectory).

A subtle geographical effect applies on Mars, where there is extreme topographical variation – of order one atmospheric scale height. Thus selecting a high-altitude landing site would require either a larger parachute (to limit descent rate in the thinner atmosphere), or require that the landing system tolerate higher impact speed.

The landing sites for the Mars Exploration Rover missions (MER-A and -B) were discussed extensively (Kerr, 2002). The not-infrequent tradeoff between scientific interest and technical risk came to the fore. As with the Viking landing site selection, the most scientifically interesting regions are not the featureless plains preferred by spacecraft engineers. The situation is complicated by the incomplete and imperfect knowledge of the landing environment.

One constraint was that the area must have 20% coverage or less by rocks 0.5 m across or larger that could tear the airbags at landing. Rock distributions can be estimated from radar techniques, together with geological context from Mars orbit (while rocks cannot be seen directly, geological structures can – rocks are unlikely to be present on sand dunes, for example), and thermal inertia data.

Although there are no direct wind measurements near the surface in these areas, models of Martian winds are reaching reasonable levels of fidelity, and these models are being used to predict the windspeeds at the candidate landing sites. Winds of course vary with season (e.g. the Martian dust-storm season, peaking soon after solar longitude $L_s = 220°$, is best avoided!) and time of day.

The Pathfinder lander, for example, landed before dawn, at 3 a.m. local solar time, when the atmosphere was at its most stable. The MER had an imperative (following on in turn from the Mars Polar Lander failure) that it must be in communication with the Earth during its descent and landing. This requires that it land in the afternoon instead – when winds are strongest! Here, perversely, a politically driven engineering uncertainty introduces a deterministic (i.e. certain!) increase in risk.

On Earth, a handful of landing sites are used. The US manned missions in the 1960s and 1970s relied upon water landing; the mechanical properties of the ocean are well understood and uniform over some 60% of the globe, with the only variable being uncertain winds and sea state. Other missions (unmanned capsules and Russian manned missions) have landed on large flat areas, notably the Kazakhstan steppe, and Utah was used for the Genesis solar wind and Stardust comet-sample-return missions. A significant factor in the choice of landing is the accuracy with which the capsule can be targeted (oceans may be less desirable landing sites, but they are hard to miss) and whether a particularly rapid retrieval (e.g. for frozen comet samples) is required.

A final possibility, and a good example of systems engineering in action, is to avoid the landing problem altogether by retrieving the payload during its parachute descent. This approach was used for the film capsules in US Discoverer reconnaissance satellites, which were recovered by snaring the parachute lines with a frame suspended from a transport aircraft. The choice of this system may have been dictated partly by strategic concerns, rather than an EDL optimization from purely mass–performance considerations, but it nevertheless remains an option.

One way of reducing the importance of the landing-site selection problem is to provide mobility. This may pertain both to the landing itself, and operations after landing.

In terms of landing, the scale of feature that poses a hazard is comparable with that of the vehicle itself – landing on a half-metre sharp rock could dent a structure, puncture an airbag, or cause a tilt on a lander that might cripple its ability to generate power or communicate with Earth.

However, such small features cannot currently be imaged from orbit, nor can an unguided entry and descent system be assured of missing it. Such precision landing requires closed-loop control during the descent. Such guidance may require imaging or other sensing (a simple form of on-board image analysis was performed on the Mars Exploration Rovers, in order to determine the sideways drift due to winds, and apply a rocket impulse just prior to landing in order to suppress the sideways motion and the resultant loads on the airbag landing system). A technique that has been explored for Mars precision landing, and landing on small bodies such as asteroids, is LIDAR or laser ranging. This is able to produce a local high-resolution topographic map around the immediate landing area. The actuation involved in such precision landing may involve small thrusters, or conceivably steerable parachutes.

Clearly, if the goal is to analyse a rock with some instrument, the designer may equip a lander with a long, powerful arm that can bring the rock to the instrument. Or, the instrument may be brought to the rock, perhaps on a mobile vehicle (see Chapter 12).

Whether an arm is used, or a rover, their positions need to be controlled, and their positions (and that of the rock) need to be known. In general, ground-based analysis of image data is used for these tasks. However, goniometry (the measurement of arm position by recording joint rotations) and dead-reckoning (measuring the number of turns of a rover wheel) can permit some on-board autonomy. The latter suffers, especially on steep slopes of loose material, from wheel slippage – the wheel may turn without moving the vehicle forward.

Closed-loop navigation using on-board analysis of image data is beginning to find a role here. Additionally, crude hazard identification can be performed with structured light – such as a pattern of laser lines on the scene, which allows the ready identification of rocks or holes.

A cartoon, uppermost in Figure 1.1, shows how the body axes of a spacecraft are offset from a fixed inertial frame, and known to varying levels of accuracy. The landing site of an entry craft will vary as a result of uncertainties in the location of the combined spacecraft prior to release, and the path taken by the landing craft after ejection. This is illustrated in the lower two charts of Figure 1.1. The left-hand plot shows how the uncertainty in attitude of a Mars entry craft at a nominal altitude of 100 km varies as a result of different factors. In the right-hand plot, the landing footprint of the craft takes on an elliptical form, with the major axis of this error ellipse being dictated by uncertainties in ejection speed, cruise time from ejection to impact, and variability in aeroshell drag, amongst other effects.

A similar (although numerically different) problem confronted the Huygens probe. The choice of entry location is driven by several considerations: Figure 1.2 shows Titan's globe as seen from an incoming asymptote – in this case centred at 5°S, 50°E longitude. First, the entry angle must lie in a specified range of for example 60°–65°, denoted by the lightest grey circle.

Second, certain optical instruments require the Sun (here at 80° longitude, 24°S) to lie between 30° and 45° elevation as seen from the descending probe – this locus is denoted by the intermediate grey ring. The intersection of the two is shown by the dark grey areas – thus there is a choice of two target regions.

The delivery ellipse is shown centred on the two target locations. In general, the ellipse is narrow, one direction (often that associated with time of arrival) being typically larger than the other. This corresponds somewhat to the uncertainty of the spacecraft or target ephemeris and thus here the long axis of the ellipse is E–W.

It can be seen that only one of the two sites (A) is acceptable. At (B) the delivery uncertainty is such as to allow an unacceptable probability that the entry angle corridor would be violated. At (A) the long axis of the delivery ellipse is aligned with the long axis of the acceptable entry region and thus success is assured.

2

Accommodation, launch, cruise and arrival from orbit or interplanetary trajectory

The challenges involved in designing optimal trajectories for planetary landers or atmospheric probes are shared by many other types of spacecraft projects. Spacecraft, at least for the foreseeable future, have to be launched from the Earth's surface and then placed on a path that intersects the orbit of the target body. How this is achieved depends on the mission of the spacecraft and its associated cost and design details.

2.1 The launch environment

Spacecraft have been delivered to space on a wide variety of launchers, all of which subject their payloads to different acoustic, dynamic and thermal regimes. These parameters vary with the size and nature of the launcher, yet the complex launch vehicle industry often makes it difficult to isolate a preferred launcher type for a given mission. In Table 2.1 pertinent features of current launch vehicles are shown with data taken from their user manuals; the launcher market currently has over a dozen vehicles capable of lifting interplanetary payloads. Costs are not listed as many of the vehicles offer dual manifest capability, or other partial-occupancy accommodation (such as Ariane's ASAP) which can make heavy launchers and their capability available to even modestly funded missions.

Of particular interest are the mass values shown for the parameter C3. This quantity is the square of the hyperbolic escape speed; the speed an object would have upon leaving the influence of a gravitating body. Paths with a C3 greater than $0\,\mathrm{km}^2\,\mathrm{s}^{-2}$ are trajectories which never return to their origin. Trajectories originating at Earth's orbit, with a C3 of $10\,\mathrm{km}^2\,\mathrm{s}^{-2}$, are characteristic of Mars missions, with a C3 of 50 through $100\,\mathrm{km}^2\,\mathrm{s}^{-2}$ being representative of direct transfers to the outer planets. Realistic missions to such distant targets would endeavour to use more energy-efficient routes by the use of gravitational assists, and so the payload figures in Table 2.1 for such large C3 values are notional only.

Table 2.1. *Parameters of some current launch vehicles, a '/' is used where the value is not known from official sources*

Launcher	Peak axial 'g'	Deliverable payload mass (kg)			Notes
		C3 = 0 km^2 s^{-2}	C3 = 10 km^2 s^{-2}	C3 = 100 km^2 s^{-2}	
Ariane 5	4.2	6600	>3190	/	5.2° N site, multiple occupancy via ASAP
Delta IV Heavy	5.4	9588	8000	/	28.5° N site, dual launch capable
Long March 3C	6.1	2300	1700	/	Two sites: 37.5° N, 28.3° N
Taurus	7.2	329	263	35	37.5° N site
Atlas V (551)	5	6500	5500	850	Two sites: 28.7° N, 34.7° N
Proton	4	4838	4279	1061	45.6° N site
Soyuz	4.3	1600	1220	1220	62.8° N site

Missions to the Moon generally, by definition, have negative C3 values (around -2 km^2 s^{-2}), as hyperbolic escape never occurs. It can be appreciated that for greater speed changes, and larger C3 values, heavier and more expensive launchers are needed to deliver a given mass; this is shown concretely in the rocket equation described below. This is the first major tradeoff in a mission's design process as money is often the key finite resource in a mission, and so it is necessary to consider how a spacecraft or its mission could be resized so that a cheaper, and usually less flexible or less powerful, launcher can be used.

2.2 Transfer-trajectory choice

Rocket propulsion is the sole present means of producing the large speed changes associated with interplanetary travel. Although the technologies used in generating thrust can vary considerably, a rocket causes a change in speed, ΔV, that depends on the fractional amount of mass that is ejected and the rocket's efficiency, the specific impulse, I_{sp}. Usually, the word 'specific' refers to a unit of mass and so I_{sp} should be the impulse (N s) per unit mass (kg^{-1}) and have units of speed (m s^{-1}). For historical reasons I_{sp} refers to a unit weight of propellant, and so the preceding definition has its value divided by the gravitational acceleration, g_0, at the Earth's surface[1] to give dimensions of time (s). In the following equation m_i and m_f are the initial and final masses of the rocket.

---

[1] Thus, an I_{sp} of 2943 N s kg^{-1}, is equal to an 'I_{sp}' (by weight) of 300 s. This last formulation is widely used.

$$\Delta V = g_0 I_{sp} \ln \left[\frac{m_i}{m_f} \right] \qquad (2.1)$$

Typical I_{sp} values for propulsion systems that have been used to inject spacecraft into interplanetary trajectories range from 300 s to 340 s for liquid bipropellant stages,[2] to values of several thousand seconds for electric propulsion systems such as those used on the Deep Space 1 and Hayabusa craft. These two classes of rocket engine, chemical and electrical, have very different operating profiles. Chemical motors and engines are easily scaled to give very high thrusts with little impact on other spacecraft systems such as power generation. To give a certain impulse a chemical engine therefore needs to burn for a relatively brief period, unlike an electric propulsion system. Drives in this category, broadly, have I_{sp} values ten times those of chemical drives, but cannot be scaled to give high thrusts without the need for commensurately large and heavy power-raising equipment. Thus, to provide the same ΔV as chemical engines, electric drives are operated for much of the journey to the target body.

2.2.1 Transfer trajectories: impulsive

The high thrust levels delivered by chemical propulsion systems result in manoeuvres that last for a short fraction of the total transfer-trajectory duration. The burns needed at the start and end of the transfer path can be treated as being impulsive and of infinitesimal duration. With the exception of aerocapture or impact missions, the spacecraft executes at least two manoeuvres after being launched. To make best use of the Earth's orbital speed around the Sun, the first burn results in the craft leaving the Earth's orbit at a tangent, and moving along a trajectory that is part of a conic section. That trajectory is chosen so that it intersects the target body, with ϕ defined as the angle between the craft's velocity vector and the target planet's orbit. This arrangement is shown in Figure 2.1 for a craft being launched at a distance r_p from the Sun. For transfer A, $\phi \neq 0$ and the path intersects only one orbit at a tangent: that of the Earth. The hodograph for the arrival point of trajectory A is shown to the right of Figure 2.1. There are infinitely many one-tangent paths between two planets, with the transfer duration and required velocity changes fixed by the major axis of the elliptical path. Two paths A and B are shown in Figure 2.1, with different aphelia; $r_a(A)$ and $r_a(B)$.

The aphelion of path A, $r_a(A)$ in Figure 2.1 does not intersect the target planet. However, for path B the aphelion distance, $r_a(B)$, intersects the orbit of the target

---

[2] Such as mono-methyl hydrazine and nitrogen tetroxide, as used on the restartable Fregat transfer stage.

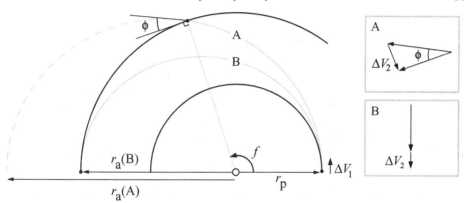

Figure 2.1. Showing two types of transfers between planets.

planet and this trajectory is known as a Hohmann transfer.[3] A Hohmann trajectory has the smallest propulsive requirement of any two-burn transfer and consequently has the longest duration. Other, more energetic transfers will be faster than a Hohmann transfer, and paths that have a change in polar angle, f, less than 180° are termed type 1 paths, such as route A shown in Figure 2.1. Craft on type 2 paths (not shown) meet their targets after aphelion, and so the polar angle changes by more than 180°.

In Table 2.2 the total ΔV values needed to perform various transfers between the orbits of the Earth and Mars are listed, with the Hohmann transfer in the first row. These values can be readily calculated from the algorithms available in many texts on orbital mechanics.

The data in Table 2.2 do not include the manoeuvre needed to leave the gravitational influence of the Earth, nor the course change required to brake into orbit about the target body. These quantities can be calculated easily from simple energy considerations. To escape a planet of mass M, a spacecraft must have a greater-than-zero hyperbolic escape speed, V. If the craft starts its journey in a circular parking orbit of radius R, a speed change at that point of size ΔV results in a hyperbolic escape speed given by:

$$V^2 = 2GM\left(\frac{1}{r} - \frac{1}{R}\right) + \left(\Delta V + \sqrt{\frac{GM}{R}}\right)^2 \tag{2.2}$$

Here r is the distance to the point at which the planet's gravity ceases to be significant compared with that of the Sun, and G is the universal constant of gravitation. This equation can be inverted to find the magnitude of the braking manoeuvre needed to convert a hyperbolic path to a closed circular path upon arrival at another planet, bearing in mind that the craft will enter the destination's

---

[3] Named after the German rocket scientist Walter Hohmann (1880–1945)

Table 2.2. *The ΔV requirements, transfer orbit semi-major axis, and duration for various Earth orbit to Mars orbit trajectories*

ΔV_1 (m s^{-1})	ΔV_2 (m s^{-1})	Semi-major axis (Gm)	Transfer duration (days)
2945	2650	188	258 (Hohmann)
3000	2975	189	230
3500	5105	199	177
5000	8980	235	129
7500	13 480	345	80
10 000	17 180	692	69

sphere of gravitational influence with a speed different from its hyperbolic escape speed at departure.

Furthermore, the preceding figures were calculated for the idealized case of Earth and Mars having coplanar and non-eccentric orbits. This simplification is useful only as a guide, and in reality the launch ΔV value can vary widely depending on the relative positions of the planets in their orbits. Plane-change manoeuvres are typically costly for interplanetary missions, and for small rotations of the orbit plane by Δi (radians), the ΔV requirement scales as $V\Delta i$. Generally plane changes are performed either as part of a launch sequence, or, later in a mission, to alter mapping coverage of an orbiter. Such manoeuvres rarely occur at interplanetary speeds, and when such alterations are needed non-impulsive techniques such as gravity assists can be used to rotate orbit planes significantly; the classic example being the purely ballistic Jupiter flyby of the Ulysses probe that led to a near-80° rotation of the craft's orbit plane.

In Figure 2.2 data for minimum, average and maximum duration Hohmann transfers are plotted for travel from the Earth to the other planets of the Solar System. The large eccentricity of Mercury's orbit reveals itself by the wide spread in spacecraft departure ΔV values, although the tabulated values represent an unreasonable span as they do take into account the actual values of the arguments of perihelion of each planet.

Clearly, for a given launch date there is a continuum of arrival dates. For realistic mission plans that account for planetary inclination and orbital eccentricity, the total ΔV changes as a function of the launch and arrival dates. A commonly encountered method of representing this information is the so-called 'pork-chop' plot. Such a plot is a contour map that shows some aspect of the launch energy (C3 or hyperbolic escape speed) with launch date and arrival date as the axes. An example of such a plot is shown in Figure 2.3, which shows the C3 value in km^2 s^{-2} for a transfer trajectory to Mars. The parallel grey lines are isochrones, contours of constant transfer

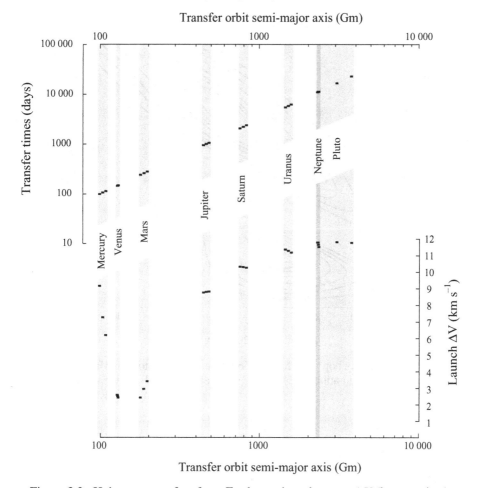

Figure 2.2. Hohmann transfers from Earth to other planets – ΔV (lower points) and durations (upper points) shown for shortest, average and longest trajectories.

duration. Superimposed on that grid are contours of constant C3 values, allowing a mission designer to balance the duration of a transfer path with the energy needed to achieve it. It should be noted that there is usually more than one minimum for the C3 needed. In this case the shorter of the two transfers, with a C3 of around $16 \, \text{km}^2 \, \text{s}^{-2}$, has a duration of ∼175 days and is a type 1 path. The second transfer, a type 2 path, lasts almost twice as long but saves $0.5 \, \text{km}^2 \, \text{s}^{-2}$ in C3, which corresponds to a difference in speed at Earth departure of ∼$6 \, \text{m} \, \text{s}^{-1}$.

2.2.2 Transfer orbits: gravity assists

By making a close pass to another planet, a spacecraft's velocity can be changed both in direction and magnitude. From the frame of reference of the planet being

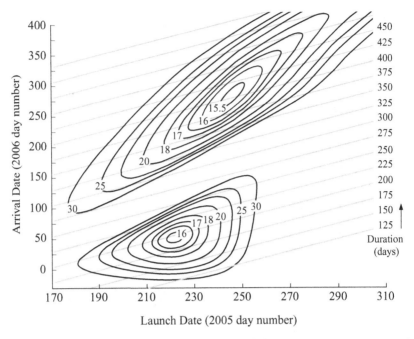

Figure 2.3. A 'pork-chop' plot for a notional transfer trajectory to Mars, showing contours of cruise duration (grey) and C3 values (black), the latter showing two minima.

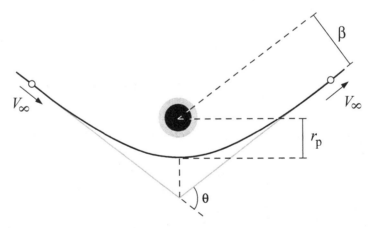

Figure 2.4. The geometry of a close flyby path in the planetocentric inertial frame.

used in this way, the spacecraft enters and leaves its influence with the same speed. However, the planet is in orbit about the Sun and from a heliocentric point of view a fraction of the planet's orbital speed can be given or taken away from the spacecraft, depending on the geometry of the flyby. Many texts deal with the derivation of this in detail and an overview is given here. In Figure 2.4, a craft makes a close pass of a planet in planetocentric coordinates, viewed perpendicular to the B-plane.[4] From the definition of a hyperbola, consideration of the angular momentum of the craft allows the periapsis of the craft to be found from its inbound trajectory.

For the impact parameter β, the periapsis of the swingby path, r_p, is given by

$$\frac{r_p}{\beta} = -\left(\frac{GM}{\beta V_\infty^2}\right) + \sqrt{1 + \left(\frac{GM}{\beta V_\infty^2}\right)} \qquad (2.3)$$

Note also the dependence of r_p on the inverse square of the approach speed. Uncertainties in the value of the spacecraft's speed therefore give rise to a commensurately larger scatter in periapsis distance. If the spacecraft makes a safe passage then the departure and arrival asymptotes are no longer colinear and the probe's path will have been bent through an angle θ, where

$$\sin\left(\frac{\theta}{2}\right) = \left(1 + \frac{r_p V_\infty^2}{GM}\right)^{-1} \qquad (2.4)$$

Gravity assists require precise navigational support when executed, and despite the additional complexity in the planning stage they can make otherwise impossible missions viable. Examples of multiple flyby missions were those of the Pioneer and Voyager spacecraft, and more recently the Cassini and Rosetta missions. What are not shown in Figure 2.5 are the speed boosts gained at each planetary encounter, which in total were equivalent to a heliocentric speed increase of $21.4\,\mathrm{km\,s^{-1}}$.

The present low cost of computing power allows the interested reader to examine such transfer schemes easily, and visualization/programming languages such as MATLAB or IDL are well-suited to such work.

2.2.3 Transfer orbits: continuous thrust

Spacecraft that use low-thrust propulsion systems to initiate interplanetary transfer necessarily must run their drives for long periods, and their trajectories cannot be modelled with the foregoing simple analysis. However, for co-planar transfers approximations exist (Fearn and Martin, 1995) that allow at least a first-cut to be made of ΔV requirements. For continuous thrust spiral orbits about a body of mass M, a craft starting with an initial mass of m_i has a final mass m_f and develops an exhaust of speed v_e. Here a_f and a_i are the semi-major axes of the final and initial orbits.

---

[4] The plane intersecting the target body's centre which is perpendicular to the inbound asymptote of the craft's path.

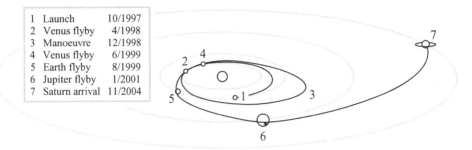

1	Launch	10/1997
2	Venus flyby	4/1998
3	Manoeuvre	12/1998
4	Venus flyby	6/1999
5	Earth flyby	8/1999
6	Jupiter flyby	1/2001
7	Saturn arrival	11/2004

Figure 2.5. A cartoon of the Cassini spacecraft's path to Saturn.

$$\frac{m_f}{m_i} = \exp\left[\frac{\sqrt{GM}}{v_e}\left(\frac{1}{\sqrt{a_f}} - \frac{1}{\sqrt{a_i}}\right)\right] \qquad (2.5)$$

Low, near-continuous thrust trajectories can have significant knock-on consequences for the mission as a result of the protracted transfer. Electric propulsion systems vary in their I_{sp} and in the mechanisms employed to heat the working fluid, which is generally a compressed gas. For example, the Mu10 engines of the Hayabusa craft delivered an I_{sp} of 4000 s, and used a microwave system to heat and ionize xenon gas, but gave a peak thrust of only 20 mN. A measure of efficiency for an electric propulsion system is its specific thrust per unit of power, and modern systems can have values around 20 to 30 mN kW^{-1}. Photovoltaic cells, at least for the immediate term, are the only practical choice for the multi-kW needs of electric propulsion systems, and over long periods suffer from efficiency degradation at a rate of a few percent per year. Thus the peak thrust levels of electric propulsion systems generally varies with solar distance by somewhat more than an inverse square.

The demands of electric propulsion on a spacecraft's attitude control system are no more taxing than those arising from the use of impulsive manoeuvres, which would also require accurate attitude control for communication and, perhaps, solar power-raising needs. However, EM interference of communication frequencies by electric propulsion systems and the thermal regulation and accommodation of power-conditioning electronics are extant problems to be solved on a case-by-case basis (Jankovsky *et al.*, 2002).

2.2.4 Transfer orbits: chaotic transfers

A spacecraft's path under the influence of three bodies, such as the Earth, Moon and Sun, can show great sensitivity to small changes in its initial trajectory. In 1982 the spacecraft ISEE-3 was directed, using these non-linear dependencies, first towards the Sun–Earth L2 point, and then towards comets Giacobini–Zinner and Halley with the use of relatively small manoeuvres. Put simply, the Solar System is threaded by a complex web of trajectories based on the presence of Lagrange points associated with each pair of planetary bodies. Relatively small manoeuvres are needed to pass from orbits about one Lagrange point, in, say, the Earth–Moon system, to a Lagrange point in the Sun–Earth system; this non-intuitive process description is explained by Koon *et al.* (2000). This scheme was used by the Genesis mission, which orbited the Sun–Earth L1 point and returned to Earth via the Sun–Earth L2 point, a considerable orbital change brought about by little more than appropriate use of the weakly bound orbit about the Sun–Earth L1 point. However, these methods can result in long transport times, and such schemes are not robust to changes in launch date or manoeuvre underperformance.

2.3 Arrival strategies

A mission can require that a lander be delivered to a planet's surface directly from an interplanetary trajectory. The Luna sample return spacecraft and the Genesis, Stardust and Hayabusa missions, all used such a direct arrival scheme. Earth, the common target body for these craft, offers a relatively dense and deep atmosphere for aerodynamic braking to be an efficient way of slowing the returning craft. Arrival at an airless body requires one manoeuvre to slow the craft from its hyperbolic path so that it is captured by the target body. Further manoeuvres are used to lower the apoapsis of this new orbit, and to control other aspects of the orbit such as its inclination and periapsis position.

2.3.1 Aerocapture and aerobraking

For missions to atmosphere-bearing planets the presence of an atmosphere provides the mission designer with the opportunity to remove the hyperbolic excess of the inbound spacecraft. Aerocapture refers to the passage of a spacecraft with excess hyperbolic speed through a planet's atmosphere in order to reduce its speed, and to thus achieve an orbit. Aerobraking is the more general word used to describe the use of an atmosphere to reduce a craft's speed. These techniques, especially aerocapture, require that the target atmosphere's density profile is well known before the encounter. Long-term studies by spacecraft around Mars show that that planet's thermosphere has daily, seasonal and dust-storm-induced density variations that can be as large as 200% (Keating *et al.*, 1998) and which are not easily predicted. For aerocapture missions a spacecraft might require autonomous trajectory control via aerodynamic or reaction-control systems.

3

Entering atmospheres

The entry of a spacecraft into a planetary atmosphere has led to an iconic image of the space age; that of a capsule being roasted in a fireball streaking across the sky. The second familiar image is that of a pilot experiencing progressively heavier 'g' loads and these processes are common for all objects entering a planetary atmosphere. A planetary mission designer has to understand how these phenomena vary with characteristics of the entry craft and the target atmosphere. This section will illustrate these relationships and the engineering solutions that may be adopted for atmospheric entry.

3.1 Entry dynamics

A useful simplification is to disregard the spherical nature of the target planet; a reasonable premise because the atmosphere is often crossed in a short span of time and space; roughly a few minutes in length and a small fraction of the body's radius in extent. Similarly, the atmosphere can initially be treated as being non-rotating, isothermal, chemically homogeneous, and in hydrostatic equilibrium at some temperature T. In this case, the density at some height above a reference surface has the familiar exponential form:

$$\rho = \rho_0 \exp\left[-\frac{z}{H}\right] \tag{3.1}$$

Here H is the density scale height, and is defined as

$$H = \frac{kT}{mg} \tag{3.2}$$

where m is the mean molecular mass of the atmosphere in a uniform gravitational field, g, and k is Boltzmann's constant. For massive planet atmospheres this assumption of constant gravitational 'g' is a fair approximation. For example, during its entry into Jupiter the gravitational acceleration experienced by the Galileo probe varied by ~0.1%, although the probe experienced far higher aerodynamic decelerations. A counter-example is found in lighter planets with larger scale heights; in the region where the Huygens Titan probe encountered aerodynamic decelerations greater than $1\,\mathrm{m\,s}^{-2}$, its gravitational weight varied by over 40%, a consequence of the extensive atmosphere of Titan and the inverse-square law of gravity.

Current entry vehicles are both rotationally symmetric and passive, in the sense that they are not capable of making deliberate changes to their trajectory. We will also assume that drag is the only aerodynamic force applied to the craft, which acts parallel to its velocity vector at all times. An entry craft of mass m, and effective cross-sectional area S, passing through an atmosphere of density ρ, then experiences an acceleration that can be modelled by

$$\frac{\mathrm{d}V}{\mathrm{d}t} = \frac{-\rho S C_\mathrm{D} V^2}{2m} \tag{3.3}$$

where C_D is the drag coefficient. In the Newtonian or free-molecular flow regime at very high altitude, this is typically ~2.1 for all shapes, but falls to lower values that depend on the Mach and Reynolds numbers and the exact shape and angle of attack in the denser parts of the atmosphere. The instantaneous deceleration can be written as

$$\frac{\mathrm{d}V}{\mathrm{d}t} = -\frac{\eta V^2}{H} \tag{3.4}$$

where the new parameter η is

$$\eta = \frac{\rho(z) S H C_\mathrm{D}}{2m} \tag{3.5}$$

Note, the absence of gravity in this model leads to a straight trajectory, and the flight path angle γ is therefore constant, thus

$$\frac{\mathrm{d}h}{\mathrm{d}t} = -V \sin\gamma \tag{3.6}$$

which, after re-writing the dimensionless 'altitude' parameter η as

$$\eta = \frac{\rho_0 S H C_\mathrm{D}}{2m} \exp\left[-\frac{z}{H}\right] \tag{3.7}$$

allows us to re-write Equation 3.3, using the identity

$$\frac{\mathrm{d}V}{\mathrm{d}\eta} \equiv \frac{\mathrm{d}V}{\mathrm{d}h}\frac{\mathrm{d}h}{\mathrm{d}\eta}$$

as

$$\frac{dV}{d\eta} = -\frac{V}{\sin \gamma} \tag{3.8}$$

Integration of Equation 3.8 with respect to η yields

$$\frac{V}{V_0} = \exp\left[\frac{-\eta}{\sin \gamma}\right] \tag{3.9}$$

Finally, this may be substituted back into Equation 3.4 to give the deceleration

$$\frac{dV}{dt} = -\frac{\eta V_0^2}{H} \exp\left[\frac{-2\eta}{\sin \gamma}\right] \tag{3.10}$$

Clearly, the peak value for deceleration occurs when the exponential term is a maximum, which happens when 2η has a value of $\sin(\gamma)$. Substitution into Equation 3.8 shows that the peak deceleration of a spacecraft in this simple model depends only on the scale height of the atmosphere, not on the drag coefficient or mass of the spacecraft. In Table 3.1 the peak deceleration is calculated for vehicles entering atmosphere-bearing targets in the Solar System.

The foregoing discussion describes a naïve view of an entry craft's trajectory. In reality there are a number of important factors to be considered. Firstly, atmospheres are neither static nor isothermal. However, the exploratory nature of current spacecraft necessarily means that they will encounter atmospheres that are not well understood in terms of their spatial and temporal variability. It is then a challenge to size the decelerator system so that the craft experiences sufficient deceleration for the chosen type of mission. In spite of the increasing availability of computing power, many critical aspects of entry processes (such as transonic stability tests, the behaviour of catalytic surfaces in reactive gas flows, etc.) are best suited to experimental analysis. Thus, the bottle-neck in

Table 3.1. *The peak decelerations at various flight path angles for entry at escape speed and lift-to-drag of 0*

Body	Peak deceleration (m s^{-2})		
	$\gamma_0 = 2.5°$	$\gamma_0 = 5°$	$\gamma_0 = 15°$
Venus	54	100	321
Earth	123	247	735
Mars	18	36	107
Jupiter	982	1960	5828
Saturn	145	291	864
Titan	1	2	6
Uranus	129	257	763
Neptune	224	447	1330

development generally is one of performing an adequate number and range of tests that validate a given entry capsule design, often under conditions of high airflow speed and density that are reproducible only in shock-tube facilities that are expensive to operate and have long lead-times for tests. For this reason entry-capsule design tends to be conservative although aerothermal modelling is not the only area in which experimental testing is unavoidable. In lower-speed portions of the entry process, decelerators such as parachutes or airbags may have to be deployed, and used heat shields jettisoned. These events generally happen well after the peak heating and deceleration phases of entry and will not be discussed further.

3.2 Thermodynamics of entry

The kinetic energy of an entering spacecraft is mostly dissipated as heat. In Table 3.2 the specific energies associated with two types of entry path are calculated for atmosphere-bearing bodies of the Solar System. The first value in each row shows the energy associated with entry from a hypothetical circular parking orbit,[5] and the second shows the energy for arrival from outside that body's gravitational influence. In this last case the speed of entry is taken as the escape speed of that body.

The heat generated during entry is more than sufficient to destroy any object if all of the heat were to be absorbed by the spacecraft. To show this, Table 3.3 lists the heat needed to warm various compounds to their melting points, and their heat of vaporization; carbon and beryllium clearly are excellent theoretical candidates for a thermal protection system (TPS). Carbon's high melting point and low comparative toxicity and cost make it the practical choice of these two elements. The rôle of vaporization will be discussed later.

Table 3.2. *The energy per unit mass associated with arrival at a planet's surface*

Target	Arrival from orbit (MJ kg^{-1})	Interplanetary arrival (MJ kg^{-1})
Venus	17.9	54
Earth	20.8	63
Mars	4.2	13
Jupiter	590	1770
Saturn	209.7	630
Titan	1.2	3.5
Uranus	75.5	226
Neptune	92.0	276

[5] With an orbital radius 1.5 times that of the particular body's radius.

The production of heat from kinetic energy depends on the environment of the spacecraft. At the edge of an atmosphere the gas density around the entering craft is so low that the gas flow around the vehicle is ballistic at the molecular scale; this is often referred to as free-molecular flow. Parts of the craft are shadowed from the oncoming gas flow and forward-facing faces experience direct molecular collisions. In a denser gas, the molecular mean free path will be shorter than scales characteristic of the craft and a different flow type emerges. Here, the air ahead of the craft is slowed, compressed and heated. If the craft exceeds the local speed of sound then a shock field develops around the front of the vehicle. Air moving through the shock is rapidly heated and compressed. The strength of this shock field is dictated by, amongst other things, the geometry of the entry body. Narrow spear-like objects tend to have relatively weak and sharply pointed angular shock fields draped downstream from their noses. Blunt objects develop stronger and broader shocks that in turn influence larger masses of air by virtue of their large cross-sectional area. A frequently used parameter for describing the relative aerodynamic load experienced by an entry craft is the ballistic coefficient,[6] which will be used later in Chapter 4.

At a qualitative level, it can be seen that the heat load experienced by a hypersonic object can be lessened if its energy of entry is dissipated into a larger mass of air. Therefore, the large enthalpy change across a blunt object's shock reduces the energy that is absorbed by the object. Conversely, a slender object, shrouded in a weaker shock field, experiences an air flow that has been slowed comparatively little and so absorbs a larger fraction of the entry energy. However, the designer of an entry heat shield cannot pick a given geometry with impunity. A working spaceprobe must be accommodated within the envelope of the entry shell, stowed robustly in some manner. If a non-spherical entry shell is used, the centre-of-mass of this configuration must lie adequately below the aeroshell's centre-of-pressure, which in turn moves the spacecraft and its dense components (batteries, etc.) closer to the leading face of the entry shell. If the offset between the centres of mass

Table 3.3. *The heat needed to warm materials from $\sim300\,K$ to their melting point, and the enthalpy required to vaporize those substances*

Material	Melting point (K)	ΔH_{warm} (MJ kg^{-1})	$\Delta H_{vaporize}$ (MJ kg^{-1})
Beryllium	1550	3.4	32.5
Carbon	3775	6.7	29.7
Copper	1360	0.4	4.7
Tungsten	3685	0.6	4.5

[6] Defined as $M/(SC_D)$, a measure of the craft's areal density.

and aerodynamic pressure is made too small, then the craft may be unstable to disturbances and make large pitching movements, exposing non-shielded parts to the energetic airflow. As an example, a craft's transition from supersonic to subsonic speed causes changes in the wake flow which in turn can be coupled to the craft, destabilizing it. Some entry craft, such as that of the Genesis sample-return mission, are designed to deploy small drogue parachutes at supersonic speeds to provide extra stability through the transonic region (Desai and Lyons, 2005).

It is also worth mentioning here that atmospheric density profiles can be derived from entry accelerometry, using Equation 4.3, an assumption of hydrostatic equilibrium, and integrating the acceleration with appropriate boundary conditions to obtain velocity and altitude. Temperature and pressure profiles can also be derived using the ideal gas law and knowledge of the mean molecular mass of the gas. Accelerometry is usually included in entry vehicles anyway for engineering purposes, to provide a 'g-switch' to initiate the parachute descent sequence. The first use for atmospheric science was on Venera 8 (Cheremukhina *et al.*, 1974), but high sensitivity accelerometry was pioneered on Viking (Seiff and Kirk, 1977) and has also been implemented on Pioneer Venus (Seiff *et al.*, 1980), Galileo (Seiff *et al.*, 1998), Pathfinder (Seiff *et al.*, 1997) and Huygens (Colombatti *et al.*, 2006), among others. The detailed processing of the data must take into account a number of error sources and perturbations, as well as the three-dimensional nature of the problem (e.g. Withers *et al.*, 2003).

Simple models for the heating rates experienced by entry craft will necessarily neglect many important phenomena. In Table 3.4 only convective heating is considered, and topics such as the variation of the heating rate with the flow regime (turbulent or laminar) and real-gas properties of the atmosphere are not examined.

At subsonic and low-Mach numbers the gas ahead of the craft is primarily heated by being rapidly compressed by the craft. The vehicle is then immersed in a hot flow of gas and absorbs heat by convection. This form of heating is not applied uniformly to the craft, but is a function of the local geometry and the

Table 3.4. *Summarizing the principal features of this simple model for ballistic entry*

	Value	Speed at peak (m s^{-1})	Altitude at peak (m)
Peak heating	$\propto \sqrt{\dfrac{\rho}{R}} V^3$	$V_0 \exp\left[-\dfrac{1}{6}\right]$	$H \ln\left(\dfrac{3\rho_0 SHC_D}{m \sin \gamma_0}\right)$
Peak deceleration	$\dfrac{V_0^2}{2He} \sin \gamma_0$	$V_0 \exp\left[-\dfrac{1}{2}\right]$	$H \ln\left(\dfrac{\rho_0 SHC_D}{m \sin \gamma_0}\right)$

nature (laminar or turbulent) of the flow. In contrast, at hypersonic speeds the atmosphere interacts with, and is heated by, a shock field some distance ahead of the craft, rather than by the craft itself.

At high entry speeds the temperature rise in the shock wave around the craft may be sufficiently intense for radiant heating from the hot gas to be equivalent to the convective heating rate. For Earth entry this occurs at speeds above $10 \, \mathrm{km \, s^{-1}}$ for bluff objects, as is shown in Figure 3.1, adapted from Sherman (1971). Note that the same equivalence in the heating processes occurs at higher speeds for objects with smaller radii, but for such craft the temperatures in the shock would be far higher, potentially compromising the temperature limits of the TPS.

The brevity of the heating process can be seen in aerothermal models and in experimental data. The modelled stagnation heat flux for a Martian entry craft with a ballistic coefficient of $150 \, \mathrm{kg \, m^{-2}}$ is shown in Figure 3.2; note that the peak heating occurs somewhat before the instant of peak deceleration as predicted in Table 3.4.

3.2.1 Flow chemistry

Entry trajectories leading to air temperatures of up to 2000 K cause molecular excitation, dissociation, and partial ionization of the gas ahead of the vehicle.

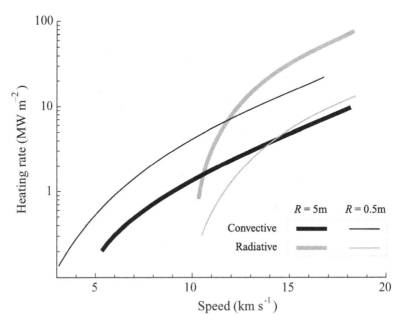

Figure 3.1. Here radiative and convective heating rates are compared for two spheres of different radii entering the Earth's atmosphere.

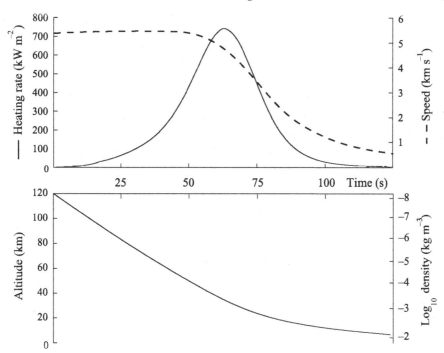

Figure 3.2. Predicted heating history for a cone–sphere Mars entry vehicle with 60° half-angle forebody.

Hypersonic entry craft, in general, produce flow fields that are described by a Damköhler number[7] close to unity, and so the chemistry in the air around the craft varies as a function of distance along the flow. From the TPS designer's viewpoint the chemical species in the flow are of no interest except for their potential to deliver heat to the TPS through recombination. Ionized gas species can recombine at cool (<1800 K) material surfaces and dump the enthalpy of molecular formation into the TPS. The degree of catalytic behaviour depends on the material temperature and gas species, but fully catalytic materials can experience heating rates several times higher than non-catalysing surfaces for the commonly encountered species of CO, C and O (Marraffa and Smith, 1998).

3.3 TPS technologies

The great value of a spacecraft lander and the highly energetic entry path that it takes often leads to a conservative TPS design that can accommodate uncertainties in atmospheric structure or mission performance. Prior to the flights of the

---

[7] A ratio of characteristic timescales for chemical reactions in the flow, and the duration of the flow itself in crossing the vehicle length.

American Space Shuttles, there was no need for reusability in spacecraft TPS and simplicity was the watchword. While purely heatsink-like shields have been proposed in pre-production studies, actual planetary missions have all used ablating materials. When heated sufficiently, such substances vaporize with the gas escaping from the underlying solid. As was seen in Table 3.3, carbon has many advantages as the main material component in such a system because of its high enthalpy of vaporization and high melting point.

Ablating TPS materials also lead to a reduction in the heat absorbed by the craft through convection because the ablation products are blown into the boundary layer around the craft, buffering the hotter incoming gas that has passed through the shock field. The Apollo and Luna programmes were the earliest missions to employ sample-return capsules and both systems used an ablating material that formed a crust of carbon-rich charred material, in itself a poor thermal conductor. Since those early missions, ablator-based protection systems have become widespread. Their efficiency and simplicity suggest that they will continue to be a preferred system for high-speed atmospheric entry. In Table 3.5 a number of entry-craft thermal-protection systems are described, with their composition detailed in the second column where possible.

3.4 Practicalities

Mechanical factors are also pertinent to the design of a TPS. Vibration during launch, exposure to hard vacuum during cruise, and aerodynamic loading during entry are just some of the hazards that a TPS must pass through without its design margins being compromised. To an extent, many of these points can be simulated in ground-based thermal vacuum chambers, but some processes are more difficult. For example, the ablating materials used for heat shields are composites of some sort, with phenolic resins or epoxies providing a matrix in which refractory particles or fibres can be embedded. Bulk thermophysical properties of the finished TPS can be tuned by controlling the recipe used in its manufacture, but the internal structure is also of concern. During ablation the gas released by pyrolysis has to escape from the TPS, otherwise the material could spall or delaminate, allowing hot gas to impinge on the entry craft's structure. A schematic of the Hayabusa return capsule is shown in Figure 3.3, adapted from Yamada *et al.* (2002), showing how each layer of resin-impregnated carbon fibre is slitted to give gas paths throughout the TPS.

Aerothermal research is still seeking to improve the efficiency of a TPS either through the use of novel materials, improved modelling, or different techniques.

Table 3.5. *Some TPS systems used in planetary probes. Where known, the manufacturer and marque of the TPS material is listed. Here, mass fraction is the ratio of TPS mass to that of the entry body. It has been noted that there is an approximately linear relation between TPS mass fraction and peak heat flux*

Project	Forward heat shield TPS materials	Peak heat flux (MW m^{-2})	Heat shield mass kg/kg fraction	Entry speed (km s^{-1})
Viking	Phenolic honeycomb filled with mixture of silica micro spheres, cork, and silica fibres [SLA-561V]: *Martin Marietta*	0.2	127/1185	4.6
Pioneer Venus	Carbon phenolic composite *General Electric Co.*	69 (large probe) 106 (small probes)	36/315 10/75	~11.5
Galileo	Carbon phenolic composite *Hughes Aircraft Company*	300	152/346	48
Venera	Asbestos composite over honeycomb *NPO Lavochkin*	unknown	~900/ ~1500	~11
VeGa	Asbestos composite over honeycomb *NPO Lavochkin*	unknown	unknown	~10.8
Mars Pathfinder	Phenolic honeycomb filled with mixture of silica micro spheres, cork, and silica fibres [SLA-561V]: *Lockheed Martin*	1	64/585	7.3
Beagle 2	Cork particles bonded in phenolic resin [NORCOAT Liège]: EADS	0.7	6/69	5.6
MER	Phenolic honeycomb filled with mixture of silica micro spheres, cork, and silica fibres [SLA-561V]: *Lockheed Martin*	0.5	78/820	~5.3
Huygens	Silica fibres in phenolic resin [AQ60]: EADS	~1	79/350	6
Hayabusa capsule	Segmented carbon phenolic composite ISAS	~17	7/19	12

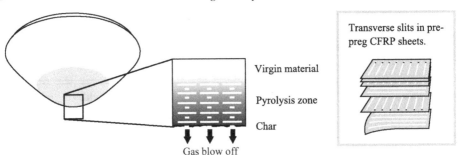

Virgin material

Pyrolysis zone

Char

Gas blow off

Transverse slits in pre-preg CFRP sheets.

Figure 3.3. The lay-up style of carbon fibre reinforced epoxy in the Hayabusa sample capsule.

Examples of studies in these fields are:

- Aerobraking in planetary atmospheres is an attractive alternative to propulsive orbital capture, and leads to the partial 're-use' of a TPS during each atmospheric pass. Materials such as C/SiC, which have high infrared emissivity and reject much of the radiative heat-load, can act as coatings or main TPS components but are challenging to produce in complex shapes.
- Modelling of high heat fluxes and the associated radiation fields is almost exclusively performed in one dimension. The absorption and emission balance in the multi-species and spatially varied flow around a vehicle is still a non-trivial computational process.

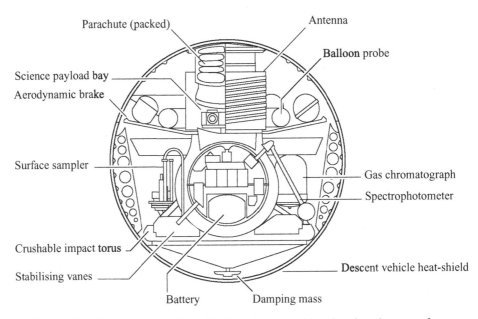

Figure 3.4. Cross-section of the VeGa entry assembly, showing the mass for oscillation damping mounted near the apex of the entry shell.

- Mechanically complex TPS materials may be required by missions that do not involve passively delivered surface-impacting landers. Such materials may arise in the form of hinged control surfaces capable of withstanding and interacting with hot, dense flows for autonomously targeted entry craft, or in extendable/inflatable structures that are too large to be launched as a solid entity.

The dynamics of atmospheric entry may in some cases (particularly non-spinning probes) stimulate oscillations of the probe attitude. While these are not always large enough to cause concern, several probes have carried internal damping masses held in a flexible mounting structure. Figure 3.4 shows a cross-section of the VeGa entry assembly, with the damping mass visible at the lowest point of the probe.

4

Descent through an atmosphere

4.1 Overview and fundamentals

The descent through the atmosphere is often the only part of a planetary probe mission, as for example the Pioneer Venus and Galileo probes; on other missions it is just the last stage of a long journey prior to surface operations. The key parameters are the altitude of deployment – usually the altitude at which the vehicle ends its entry phase, as defined by some Mach number threshold – and the required duration of descent.

The duration of descent for an atmospheric probe is often dictated by an external constraint on the mission duration, such as the visibility window of a flyby spacecraft that is to act as a communications relay. This imposes an upper limit on the descent duration – it may be that (as for the Huygens probe) some part of that mission window is desired to be spent on the surface.

The instantaneous rate of descent (and thus the total duration) is determined at steady state by the balance between weight and drag. The former is simply mass times gravity; the latter depends on ambient air density, the drag area of the vehicle and any drag-enhancement device such as a parachute or ballute. The drag area is usually expressed as a reference area and a drag coefficient. Often these parameters and the mass are lumped together into the so-called ballistic coefficient β.

Often the dynamic pressure of descent is used to force ambient air into sampling instruments such as gas chromatographs. In steady descent, the dynamic pressure can be equated to the ballistic coefficient times ambient gravity.

4.2 Extreme ballistic coefficients

There are situations where it may be desired to maximise β, and thus the descent rate. Example applications are balloon-dropped microprobes on Venus, payload delivery penetrators and probes for the deep atmospheres of the outer planets.

Considering the first, the desire is to reach the surface with a miniature probe before the balloon that released it has drifted out of sight. Because the probe is small, its ballistic coefficient is low (for a given mass density, mass increases with size more steeply than area, so small things have low mass/area ratios). Further, to protect the probe from the very hot lower Venusian atmosphere it may be coated in a layer of (low density) insulation that increases the cross-sectional area.

The natural tendency is to lengthen the probe such that its ballistic coefficient is increased. However, this increases the total surface area and thus increases the heat input into the probe. For a given descent duration, this increases the amount of thermal ballast needed, the insulation performance required, or the maximum temperature that can be tolerated by the equipment.

This underscores the issue, highlighted in a different context by the DS-2 Mars microprobes, that achieving high packaging densities is often important in small vehicles. Meeting a volume constraint with small vehicles is often more challenging than meeting the mass constraint.

Most of the descent takes place at terminal velocity, which can be considered steady-state (strictly, since the air density increases with decreasing altitude, the terminal velocity drops with time, although typically rather slowly). The terminal velocity V_t can be computed thus

$$\frac{1}{2} S C_d \rho V_t^2 = Mg \qquad (4.1)$$

g can often be considered constant (except on Titan, where the scale height is not negligible compared with the planetary radius). Often the drag coefficient (usually of order 1), the mass and the reference area S are lumped in a single parameter β, which equals $2M/(SC_d)$ (NB: sometimes the factor of 2 is not included in the definition – care!). This parameter has dimensions of mass per unit area, and values of 10–100 $\mathrm{kg\,m^{-2}}$ are typical.

Substituting this parameter, then we have

$$V_t^2 = \frac{\beta g}{\rho} \qquad (4.2)$$

Usually, one or more profiles of $\rho(z)$ will be specified to a project, a nominal profile and two extremes (as for the entry analysis). A very crude profile can be generated with the assumption of constant temperature and composition, such that in hydrostatic equilibrium the atmosphere follows an exponential law, with

$$\rho(z) = \rho(0) \exp\left(-\frac{z}{H}\right) \qquad (4.3)$$

Table 4.1 *Key parameters for parachute descent on various planetary bodies having atmospheres*

Body	g_{surf} (m s^{-2})	P_{surf} (bar)	T_{surf} (K)	M	ρ_{surf} (kg m^{-3})	H (km)	$V_{term100}$ (m s^{-1})	t_{scale} (s)
Venus	8.9	90	740	44	64.4		4	4225
Earth	9.81	1	288	29	1.21	8.4	28	296
Mars	3.7	0.007	200	44	0.02	10.2	141	72
Jupiter	24.9	2	180	2.3	0.31	26.1	90	290
Saturn	10.4	1	120	2.1	0.21	45.7	70	650
Titan	1.35	1.5	94	28	5.37	20.7	5	4125
Uranus	10.4	1	75	2.3	0.37	26.1	53	491
Neptune	13.8	1	70	2.8	0.48	15.1	54	281

where $\rho(0)$ is the surface density and H is the e-folding distance, or density scale height, given (for an ideal gas) by

$$H = \frac{RT}{gM} \tag{4.4}$$

where R is the universal gas constant, $8314\,\mathrm{J\,kg^{-1}\,K^{-1}}$, T the absolute temperature and M the relative molecular weight of the atmosphere. For Earth, the scale height is $8314 \times 288/(9.8 \times 28)$, equal to $\sim 8.7\,\mathrm{km}$.

The time to fall through the bottom scale height is $\sim H/V_t(0)$, and each additional scale height above takes a factor $\sim e^{-0.5} = 0.6$ times as long. Typically, planetary probes descend through about four scale heights, corresponding to a variation of ~ 100 in density.

If released from rest (as, for example, when a parachute line is cut) the vehicle will reach a new terminal velocity with a characteristic timescale of $\sim V/g$.

Atmospheric temperatures vary significantly with altitude. In thin atmospheres (essentially, stratospheres) where absorption of sunlight at high altitude is the controlling factor, temperatures may increase or stay roughly constant with height, and there is relatively little vertical motion. Below some altitude, however, temperatures are controlled by the vertical transport of heat from either the hot depths of the giant planets, or the surface where sunlight is absorbed. In this tropospheric regime, temperatures fall with increasing altitude, often at a roughly linear rate that is equal to or below the adiabatic lapse rate, $dT/dz = \Gamma = -g/c_p$, where c_p is the specific heat of the gas; for Earth in dry air, $\Gamma = -9.8/1000 \sim -10\,\mathrm{K\,km^{-1}}$.

Table 4.1 gives parameters for the planets, together with typical speeds and timescales.

Figure 4.1. Staging of the Pioneer Venus large-probe entry shell and parachute. Entry interface was defined at 200 km altitude, with the main deceleration lasting about 38 s. At about 70 km altitude, a mortar deployed a pilot chute, which removed the aft cover and deployed the main chute. The entry shield is then allowed to fall away, and after 16.5 minutes (at 48 km) the main parachute was jettisoned, allowing the probe to fall faster. Free descent took another 39 minutes. This descent sequence was also used (with some modifications in timing and the addition of a small stabilizing parachute in the last segment) on the Huygens probe.

The term $V_{\text{term}100}$ is the terminal velocity for surface atmospheric conditions (pressure, temperature, density and gravity P_{surf}, T_{surf}, ρ_{surf} and g_{surf}) for a spacecraft with a ballistic coefficient β of $100\,\text{kg}\,\text{m}^{-2}$. The term t_{scale} is the time taken to fall through one scale height H (the e-folding distance for pressure).

It can be seen from the table that Venus and Titan are rather similar in having exceptional descent times – both the Pioneer Venus large probe (Figure 4.1) and Huygens on Titan jettisoned their main parachutes to increase the ballistic coefficient and descend more rapidly.

4.3 Drag enhancement devices

In some applications, the ballistic coefficient employed at entry may be adequate for descent. In this case, the entry shield may remain attached, although care must be taken that the subsonic stability is adequate. The Pioneer Venus small probes and DS-2 Mars microprobes used this approach.

An aft skirt may be attached to the vehicle, to expand its cross section and thus reduce descent rate. This approach was applied on the Venera landers, which had a sharp-edged braking disc; the sharp edge providing a well-defined flow separation position which improves stability over rounded shapes. Veneras 13 and 14 also added a sawtooth edge to their landing rings to further control the flow and enhance stability. The aerodynamic properties were tested in a water tunnel and by free-flight drop tests.

An exotic possibility, not yet applied to planetary spacecraft but used in a variety of 'smart' munitions on Earth, is to use a small wing asymmetrically mounted such that autorotation occurs. This method of arresting descent, familiar in samaras (seed-wings, e.g. Lorenz, 2006) like maple and sycamore, is effective and lends itself to compact packaging. Provided that the rotation is tolerable (it may, in fact, be desirable to scan sensors) this may be a promising approach for small instrument packages ('dropzondes', 'microprobes').

4.4 Parachute types

The design of parachutes and related systems is a somewhat arcane science (e.g. Knacke, 1992; Murrow and McFall, 1968) of sufficient complexity that empirical testing remains the only trustworthy design tool.

Different parachute geometries are available with different inflation performance, drag coefficient, stability, manufacturing cost and so on. The lowest cost type of parachute is the cruciform – this is easily manufactured as two strips of fabric sewn at an orthogonal intersection. These are used widely in retarded bombs and submunitions, but are not usually used on planetary probes due to their generally poorer stability.

A key feature of a parachute is its porosity, at both the macroscopic (gaps in a ribbon parachute or ringsail, or in a disk-gap-band chute) and microscopic (porosity in the fabric) scales. The porosity allows some part of the flow to go through, rather than around, the parachute, and is crucial in controlling its inflation characteristics and its stability in operation – a chute without adequate porosity will exhibit undesirable oscillations. The microscopic porosity in particular is sensitive to the Reynolds number, so particular caution is required in applying test data to different flight conditions.

Circular (i.e. 'flat') canopies typically have wider oscillations than conical types and are therefore rarely used. Conical parachutes have triangular gores, and so form a conical shape (although in inflated operation, the cone tends to be rather rounded). Conical ribbon chutes have good supersonic characteristics and are strong since, generally, materials can be made stronger as ribbons than as broadloom fabric. Ribbon-type parachutes can have high porosities.

Disk-gap-band (DGB) chutes are a variant of circular canopies that have better stability characteristics: the gap allows a through-flow which stays better attached to the canopy, avoiding asymmetric flow separation which can cause oscillations. Each gore is approximately triangular, with a rectangular segment to form the band. Even though the DGB chute for the Mars Exploration Rovers was derived from the previous Viking and Pathfinder designs, its different size and operating conditions were such that testing found that the chute would not reliably inflate. A modification (e.g. Steltzner *et al.*, 2003) that enabled successful operation was to increase the size of the gap.

Ringsail parachutes are something of an intermediate between a ribbon chute and a hemispherical chute, with one side of the panels in each gore being free, allowing a flow through the canopy. These chutes offer good drag performance and were used on Beagle 2 (Northey, 2003) and on Apollo. However, there has been less testing and experience with this type of chute, which is not well-suited to supersonic inflation. Note that the drag coefficient (and the drag area) of a parachute is referred to its constructed size (i.e. the gores laid flat), since the inflated diameter is less well-known (but is usually a factor $\sim \pi/2$ smaller), and indeed can vary with time. Drag coefficients for most parachutes of the order of 0.5–0.6 are typical for subsonic conditions.

Most planetary probes have broadly similar parachute-inflation conditions (Mach number and dynamic pressure) which restrict the choice of design. Conical parachutes and disk-gap-band types are essentially the only types used on planetary spacecraft, in part due to the base of experience obtained with them and the significant costs of qualifying new designs in extreme environments.

Parawings are rectangular parachutes, with cells that inflate in the ram-air flow to create a lifting surface (these systems are therefore also termed ram-air parachutes). These are being considered for precision-landing applications on Mars, and for sample-return on Earth, one being used on the Genesis capsule, for example.

A descent system contemplated for the Gemini manned capsules was the Rogallo wing. This is essentially the original hang-glider, with a kite-like diamond flexible wing whose span and chord are maintained by a rigid frame comprising a keel and a spar. (Modern hang-gliders have more sophisticated aerofoils and shorter chords – achieving rather better glide performance at the cost of complexity and somewhat reduced stability.)

A variety of other parachute types are available, including variants with stiffening battens (guide-surface parachutes), ringsails, etc. Only a parachute expert would have particular reasons for using these systems, and given the large testing background and demonstrated reliability of conical ribbon and DGB parachutes, they are unlikely to be used.

4.4.1 Parachute components and manufacture

The fabric elements of a parachute canopy are usually referred to as gores (Figure 4.2). The lines between the gores also extend down as suspension lines to convey the forces to the payload. Generally, these lines meet at an apex and a single line carries the load. This single line, which is typically driven by the need to avoid payload wake effects on the parachute, is called a strop, and may also include some shock-absorbing elements to alleviate the peak loads during parachute inflation. (Sometimes the term 'riser' is applied to the strop, although this can also be applied to the suspension lines.) Finally, the riser usually splits into several lines for attachment to the payload to improve damping of attitude oscillations of the payload. This split line is termed a bridle. Note that the aerospace industry, generally driven by military requirements, often terms the payload a 'store'.

The original parachute material, silk, is still used in terrestrial applications. For planetary probes, temperature considerations and more importantly outgassing issues, force the use of synthetic materials like polyester.

Kevlar, having a very high strength-to-weight ratio is often used for risers but is an awkward material to sew and is therefore rarely used for the canopy itself. Polyethylene ('Spectra') is used in similar areas. Polyfluoroethylene (Teflon) has

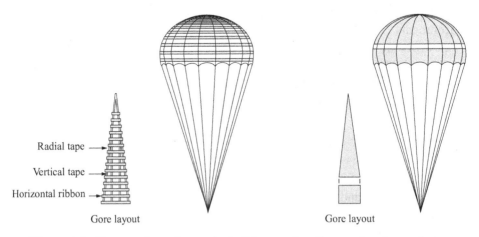

Figure 4.2. Construction of a conical ribbon and a disc-gap-band parachute. Gores are manufactured flat but form a curved surface when the chute is inflated. Note that the inflated diameter is a factor of $\sim\pi/2$ smaller than the constructed diameter.

inadequate strength for load-bearing applications, but is useful in ancillary components e.g. parachute bags, because of its low friction.

Polyesters are probably the most widely used materials for planetary parachutes. Dacron is a common polyester material. It has good strength properties, and can tolerate a wide temperature range. Polyester material was initially selected as the material for the Huygens probe parachutes, but appropriate lightweight fabric was not available and so nylon was used.

Nylon is another common terrestrial parachute material, but it has poorer outgassing properties than polyester. Furthermore, it is much less tolerant of temperature extremes. Note that in applications where planetary protection is a concern, the parachute canopy may represent the largest surface area of the probe system, and may require extensive cleaning treatments to bring the total probe bioload down to permitted levels (e.g. although nylon or other materials might work well at cold Martian temperatures, they could not survive the $>100\,°C$ sterilization procedures needed during the development programme).

Table 4.2 lists some key features of some planetary descent vehicle parachutes.

Table 4.2. *Features of some descent vehicle parachutes; NB: care must be taken to interpret parachute system masses; the canopy and lines themselves may weigh rather less than the container and deployment mortar*

Parachute	Quoted diameter (Do) or area	Mass	Comments : Mach number, dynamic pressure
Mercury drogue	6 ft	6.4 lb	30d conical ribbon
Mercury main	63 ft	56 lb	ringsail (+4.2 lb lines)
Apollo drogue	2×16.5 ft	50 lb	25d conical ribbon, 10–204 psf
Apollo main	3×83 ft	135 lb	ringsail, 30–90 psf
Venera 4 brake	$2.2\,m^2$	?	
Venera 4 main	$55\,m^2$	?	
Venera 5,6 brake	$1.9\,m^2$	?	
Venera 5,6 main	$12\,m^2$	?	
Venera 7 main	?	?	'glass nitron'; reefed by cord designed to melt at 200 °C to slow descent before landing
Venera 8 main	2.5 m	?	
Mars 2,3 auxiliary	2 m	?	Developed by N. A. Lobanov *et al.* at NII PDS
Mars 2,3 main	6.7 m	?	
Viking main	16.2 m	56 kg incl. mortar	Dacron DGB 100–500 Pa

Table 4.2. (Cont.)

Parachute	Quoted diameter (Do) or area	Mass	Comments : Mach number, Dynamic pressure
Pioneer Venus pilot	0.76 m	?	M~0.8 3300 Pa
Pioneer Venus main	4.94 m	?	20d conical ribbon polyester
VeGa 1,2 pilot	1.4 m^2 (D = 1.3 m)	?	
VeGa 1,2 brake	24 m^2 (D = 5.5 m)	?	
VeGa 1,2 main	3 canopies 60 m^2 each (D = 8.7 m)	?	
Galileo pilot	1.14 m	0.36 kg	20d conical ribbon dacron M~0.9–1, ~6000 Pa
Galileo main	3.8 m	3.7 kg	20d conical ribbon dacron
Pathfinder	12.7 m	17.5 kg	+7 kg bridle; large mass deployed at 700 Pa
Huygens pilot	2.59 m	0.7 kg	nylon DGB M ~ 1.5, q ~ 400 Pa
Huygens main	8.31 m	4.6 kg	nylon DGB (Total system incl. mortar, pilot, stabilizer etc. ~12.1 kg)
Huygens stabilizer	3.03 m	~0.76 kg	nylon DGB
MPL, Phoenix	8.4 m	?	polyester DGB
MER	14.1 m	26.4 kg total	nylon/polyester DGB M ~ 1.8 ~ 1.9, 730–750 Pa
Beagle 2 pilot	1.92 m	0.5 kg	nylon DGB
Beagle 2 main	10 m	2.764 kg	nylon 28 gore ringsail

4.5 Testing

Testing of descent control systems is notoriously difficult, since it is hard to reach the conditions of aerodynamic similarity (Mach and Reynolds number). Arguably more important even than these, since parachute inflation relies on interactions between the flowfield and the parachute itself, is dynamic pressure.

Validation of the aerodynamic stability of the DS-2 entry vehicles was performed largely by computational fluid dynamics studies, augmented by only a couple of wind-tunnel tests.

Note that good sets of aerodynamic coefficients, as a function of incidence angle and Mach number, are required not only for accurate trajectory predictions, but also for precision in the recovery of the atmospheric density profile from entry accelerometer measurements.

Full parachute systems tests therefore require lofting the test assembly to considerable altitude (40 km). Such range tests require extensive preparation of the test articles and range instrumentation, as well as good luck with the weather. If the test fails, isolation and elimination of the failure mechanism may take time and require one or more further tests, putting the project schedule in jeopardy.

4.6 Additional components of a descent control system

There is more to controlling the descent than the parachute itself.

Actuation usually means the ignition of a mortar charge which launches a folded pilot chute through a break-off patch in the aft cover, such that the chute can inflate well clear of the turbulent wake of the probe. The pilot chute serves to stabilize the probe at transonic speeds. It may also act to pull off the back cover and/or the main chute.

Actuation of the parachute may be triggered by the deceleration profile – since the overriding concern is to ensure deployment in a dynamic pressure regime that will ensure safe inflation and/or to avoid transonic Mach numbers at which the entry configuration may be unstable. Usually (following extensive modelling) the deployment is triggered at a time after some downgoing deceleration threshold has been crossed, measured by g-switches or accelerometers. In other circumstances, altitude as determined by a radar altimeter or barometer (or even time) may be an appropriate trigger, e.g. for staging a parachute.

The riser may incorporate a swivel, in order to decouple the spin of the parachute from the probe itself. This swivel (which may pose lubrication challenges) is usually necessary in order to permit a controlled rotation of the probe via spin vanes for scientific reasons (e.g. to pan instruments). Another feature may be 'reefing' the chute, whereby the opening of the canopy is initially restricted by additional lines to reduce the total drag of the parachute until some later time. This allows, for example, for a more rapid initial descent to minimize wind drift, or to reduce the initial deceleration loads. Reefing, or staging of one parachute to another, may require various pyrotechnic systems.

4.7 Mars – retro-rockets in atmosphere

Some additional remarks are appropriate about descent in the Martian atmosphere. Terminal velocity in the thin Martian atmosphere is typically too large to permit soft landing only using a parachute. Even the semi-hard landing of the Mars Pathfinder vehicle used a retro-rocket to null the descent velocity just prior to impact. A scientific concern is that the plume from the descent engines should not deposit fuel contaminants at the landing site, or significantly erode the surface material there. Nozzle design may need to take this into account (e.g. Mars Polar Lander had multiple-nozzle motors). Additionally, the retro system may need to be cut off some metres above the surface, to free-fall.

5

Descent to an airless body

There are two fundamental arrival strategies – from a closed orbit (circular or otherwise) around the target body, and from a hyperbolic or near-linear trajectory directly to the surface.

Landing places some significant requirements on the thrust capability of the landing propulsion. Obviously the thrust-to-weight ratio (in that gravity field) must exceed unity if the vehicle is to be slowed down. The ΔV requirements will depend significantly on the trajectory and thrust level chosen, and can in the case of a hover, be infinite; a lower bound is given by the impulsive approximation analogous to the Hohmann transfer between coplanar orbits – first an impulse is provided to put the vehicle on a trajectory that intersects the surface, on the opposite side in the case of a descent from orbit. A second impulse can then be applied to null the velocity at the impact site.

In practice the trajectory of the vehicle, the performance of the propulsion system and the topography of the target body are inadequately known for such a strategy to be performed open-loop, except in the case of landing on very small bodies where the orbital and impact velocities are low enough that the second, arrival ΔV can be safely provided by impact forces rather than propulsively. Thus some sort of closed-loop control is needed.

Compensation for varying propulsive performance (both due to engine performance variations, especially if feed pressure may vary in blowdown mode, and due to the progressively reducing mass of the vehicle) can be achieved by monitoring the spacecraft acceleration with onboard accelerometers. The NEAR spacecraft used this technique, with thruster cutoff after a fixed time as a backup against accelerometer failure. In fact because NEAR hit the ground while its expected descent profile indicated it should still have been falling, it tried to compensate for the upward force from the ground by firing thrusters downward, pushing the spacecraft into the regolith until the burn timed out.

For closed-loop control of the trajectory relative to the ground, some sort of navigation information is needed.

5.1 The gravity turn

One simple control strategy is to align the thrust vector with the velocity vector. If the initial state is at some altitude with a purely horizontal velocity (as from a circular orbit, for example), gravity will add a vertical downwards component. Continuous braking (or atmospheric drag for that matter) means the vehicle eventually loses its memory of the initial horizontal velocity, and the velocity progressively becomes dominated by the gravitational ΔV added. The trajectory therefore turns over and becomes vertical (it is readily visualised by throwing a table tennis ball). This type of guidance requires knowledge of the vehicle attitude and its velocity.

5.2 Efficient descent

In terms of propulsive efficiency, a slow descent like a gravity turn is poor, since the longer the descent takes, the greater a gravitational ΔV is added which the propulsion system must thrust against (hover is the limiting case). The most propulsively efficient descent from orbit would be an impulsive burn that completely nulls the horizontal component of velocity at the beginning, a ballistic free-fall to the surface, and an impulsive burn just above the surface to kill the vertical velocity.

The real world often does not allow such an efficient strategy, which would require perfect knowledge of the horizontal velocity, and an infinite thrust capability. It also relies on perfect timing of the final braking burn – too soon, and the vehicle comes to a halt at significant altitude and falls; too late and it crashes without slowing.

As an example of the altitude–efficiency relationship, the ΔV capability of Surveyor 5's vernier engines was compromised by a helium pressurant leak during coast to the Moon. The descent sequence was modified en route by ground controllers, to fire the solid retro motor later (and therefore more efficiently) leaving less ΔV to be met by the verniers. However, this strategy resulted in retro burnout at an altitude of 1.3 km, instead of the originally planned 11 km. There was clearly little margin for off-nominal retro performance or other errors.

5.3 Realistic trajectories

In practice, descent trajectories are intermediate between these extreme styles.

The usual approach is to null any significant horizontal and vertical motion above some nominal altitude. If a throttlable engine is used, then the vehicle is programmed to descend at a constant rate to some other nominal altitude, say 2 m above the surface at which the velocity is nulled, and then the engines are shut off and the vehicle free-falls a short distance to the surface.

Note that in general, even if a non-throttlable engine is used at a constant thrust, the thrust-to-weight ratio, and therefore the acceleration, will change as the fuel mass of the lander reduces.

5.4 Example – direct descent – Surveyor

The Surveyor landers performed soft-landings on the Moon, in preparation for Apollo (Figure 5.1). The lander was turned to a predetermined orientation using Sun and star attitude references. A pulse radar altimeter generated an altitude reference mark at 100 km altitude. After a predetermined delay (8 s) throttlable 'vernier' engines were ignited; their velocity increment at this point

Cruise attitude

Pre-retro manoeuvre 30 mins
before touchdown to align
main retro with flight path

Main retro start by altitude-marking radar
which ejects from nozzle. Craft stabilized
by vernier engines

60 miles, 6100 mph
(96 km, 2730 m/s)

Main retro burnout and ejection.
Vernier engines control descent

25 000 ft, 240 mph
(7.6 km, 107 m/s)

Vernier engines shut down

13 ft, 3.5 mph
(4 m, 1.5 m/s)

Figure 5.1. Surveyor descent sequence.

was modest – their function was to control attitude (using gyros for attitude determination) during the retro-motor firing. One second after vernier ignition, the large solid rocket motor was fired at 76 km altitude as the vehicle descended at 2.7 km s^{-1}, burning for about 40 s to leave the vehicle descending at just over 100 m s^{-1}. The motor, which comprised 655 kg of the 995 kg launch mass, was ejected 10 s later, at 40 km altitude. Subsequently the descent was controlled only by the vernier engines, using a four-beam Doppler radar altimeter and velocity sensor. (On Surveyor 1, one of the beams lost lock briefly, probably due to a spurious return from the retro motor casing as it fell away.)

The guidance law was essentially to perform a gravity turn, i.e. thrusting against the instantaneous velocity vector, with a constant deceleration of 0.9 lunar gravities. This law defines a parabola in range–velocity space, which was approximated in the lander guidance software as a set of straight-line segments.

Altitude marks were generated at 310 m and 4.5 m; the lander took 19 s to make that part of the descent. Thereafter, the vernier thrusters were turned off, and the spacecraft hit the ground 2 s later.

5.5 Examples: Luna 16 and Apollo

The soft-lander Luna 16, which returned samples of lunar soil to Earth, was placed first into a 110 km circular lunar orbit. It was then put into a 15 × 106 km orbit, with landing approach to be made from perilune. The vehicle simply killed its 1.7 km s^{-1} orbital speed and then fell vertically. The free-fall was monitored by a radar altimeter, and arrested at an altitude of 600 m ($\sim$200 m s^{-1}) by another burn. Again, note the low altitude of these manoeuvres to maximize propulsive efficiency.

The more massive Apollo vehicles had a lower thrust-to-weight ratio, and followed a more complex, shallow approach (which can be practised in the early 'Lunar Lander' arcade game).

5.6 Small bodies

In some respects the safe descent onto a small body like an asteroid is easier, since the ΔV requirement and the thrust requirements are low. However, the three-dimensional trajectory may be rather more complicated, since the descent from orbit may take a significant fraction of a rotation period (Figure 5.2), and thus the required thrust direction rotates a significant angle in inertial space.

A successful descent was accomplished by the NEAR spacecraft onto the asteroid Eros in 2001. A significant complication on small bodies is that their gravity fields are likely to be appreciably non-spherical. Light time and the

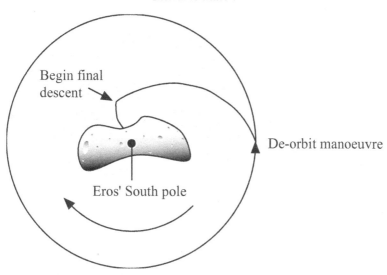

Figure 5.2. NEAR descent profile to scale from initial orbit (as viewed from the Sun), after Dunham *et al.* (2002).

limited autonomy and landing capability of NEAR (which was only designed for orbital operations) meant that the descent had to be performed open-loop, through a purely pre-programmed sequence. NEAR was in a near-circular 34×36 km retrograde orbit and performed a 2.57 m s^{-1} deorbit burn, changing inclination from 180° to 135°. Four separate braking manoeuvres were pre-programmed to execute at fixed times during the 4.5 h descent. Impact velocity was determined to be 1.5–1.8 m s^{-1} (vertical) and 0.1–0.3 m s^{-1} horizontal, around 500 m from the target point. The target was selected such that the descent trajectory (Figure 5.3) would maximize the number of low-altitude surface images; the longitude was selected such that the spacecraft could maintain continuous Earth pointing during descent while its body-fixed camera saw the surface of Eros throughout.

A soft-landing concept for Eros, employing electric and hydrazine thrusters to reduce the impact velocity to 0.7 m s^{-1}, was proposed as early as 1971 (Meissinger and Greenstadt, 1971). The vehicle would perform a closed-loop controlled descent monitored by a radar altimeter and three-beam Doppler radar.

5.7 Instrumentation

To land from orbit requires the controlled change (and therefore the measurement) of the spacecraft's dynamical state. During atmospheric entry, deceleration measurements (or during descent, pressure measurements) can provide

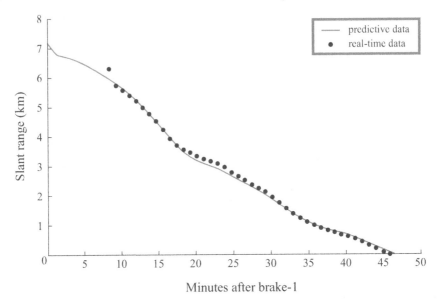

Figure 5.3. Slant range versus time during NEAR descent to Eros. The altitude at the start was about 5 km (from Dunham *et al.*, 2002).

convenient information on altitude, but on an airless body these cues will be absent.

Usually radar altimeters are used; these can also measure descent rate via the Doppler shift. With three beams making Doppler measurements, it is possible to deduce all three orthogonal velocity components. A usual strategy is first to null any sideways component of velocity, then to descend vertically.

If the entry conditions are such that the vehicle will be braked completely, it may be adequate to assume vertical descent (this may be the case on Mars). However, it may still be necessary to monitor and control the vehicle attitude during a later powered descent phase. Gyros may therefore be necessary to provide an attitude reference (usually descent is short enough that no attitude updates, as are needed on orbital platforms to correct gyro drift, are necessary). If suitable illumination exists on airless bodies (e.g. dayside landings on the Moon) it may be that optical horizon sensing can replace or augment gyroscopic determination.

The Viking landers included complete inertial guidance (i.e. a gyro-stabilized platform with accelerometers) in addition to a radar altimeter. Pathfinder was equipped with gyros: it used a radar altimeter to time the ignition of a braking rocket.

The Surveyor lander used a pulsed radar altimeter to generate an altitude reference at 100 km. At 80 km altitude, a separate RADVS (Radar Altimeter and

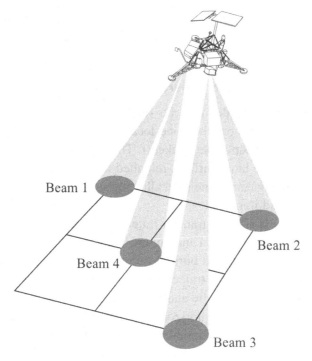

Figure 5.4. Surveyor RADVS (Parks, 1966).

Doppler Velocity Sensor) turned on, using a four-beam Frequency-Modulated Continuous Wave technique (Figure 5.4). A central beam at 12.9 GHz was used to measure the altitude, while the three outer beams at 13.3 GHz were used to determine the three components of the velocity. To accommodate the large variation in range (and therefore signal strength) amplifiers provided 40, 65 or 90 dB of gain. The lander also used gyroscopes as inertial references during the burn, and a star tracker for attitude determination prior to the burn.

Laser ranging is now also widely anticipated for Mars and small-body landers – most likely as a LIDAR (light detection and ranging), which is able to generate a range map of the region under the lander in order to assess its topography, and thus its suitability as a landing site, before committing to a possibly hazardous site. The asteroid mission Hayabusa employed a combination of the LIDAR instrument, a short-range LRF laser rangefinder, the ONC optical navigation camera and an FBS fan-beam sensor to inform the autonomous navigation system (Kubota *et al.*, 2003; Hashimoto *et al.*, 2003).

A scientific consideration for powered descent is the effect that impingement of the exhaust plume from the retro rockets might have on the surface material: the descent engines of the Viking spacecraft and the Mars Polar Lander were designed with flared or multiple nozzles to minimize these effects. Luna 16 similarly had two smaller engines for terminal descent. It is usual in any case for the last metre or two of descent to be made as a free-fall.

5.8 Powered re-ascent

Scientific return from a lander may be substantially enhanced if it can make surface measurements at more than one location. In principle, any vehicle that can make a soft landing can make a takeoff. This was performed on Surveyor 6, which after 177 h on the lunar surface, reignited its engines for 2.5 s, lifting it to 3 m high and translating 3 m to one side, allowing stereoscopic imaging and study of the original footpad imprints.

The most efficient horizontal transfer trajectory (in the absence of drag) is an impulsive–ballistic one, where a maximum thrust burn puts the vehicle on a ballistic trajectory, and a second burn close to impact brakes the motion. On a sufficiently large body (or short trajectory) where the 'flat earth' approximation applies, this trajectory is parabolic and the velocity impulse ΔV (applied at 45° to the horizontal) relates to the horizontal distance travelled D, as $\Delta V^2 = Dg$, taking a time $\sqrt{2(\Delta V/g)}$. The maximum altitude attained is just half the range. Clearly, an equivalent impulse must be applied on landing to bring the vehicle to rest.

Ascent may be required instead to reach orbit or a hyperbolic escape trajectory, to return samples to Earth for analysis. In these cases most of the same considerations apply as to powered descent, although issues such as the storability of propellant may come into play. Further, note that the algebra of rocket staging is such that the specific impulse performance of the last stage is strongly leveraged – an important factor in Mars sample return considerations. However, the ΔV to return from the Moon is essentially that of lunar escape velocity: the 2.7 km s^{-1} can be provided by a single stage using storable propellants, as in the 520 kg UDMH/nitric acid ascent stage used by Luna 16 to return a 39 kg entry capsule to Earth.

5.9 Hover

Although obviously not an efficient part of a trajectory, the ability to hover may be useful. It can be shown that a vehicle hovering with rocket propulsion that has a mass ratio of $1/e$ ($= 1/2.718$, ~0.367) can do so for a period in seconds equal to the specific impulse, which is also usually expressed in seconds. This neat result is a vivid demonstration of the otherwise unobvious dimensions of specific impulse. Stable hover requires throttlable propulsion. This can be achieved either with very careful nozzle and valve design (e.g. on the Apollo Lunar Module), or a more modern technique is to pulse-modulate thrusters having fixed thrust values (e.g. on the Phoenix lander).

5.10 Combined techniques – system engineering

At the price of system complexity, it may be that a combination of techniques offers the most mass-efficient, robust or cost-effective solution (rarely the latter, since complexity usually introduces cost). For example, rather than have a lander descend the last 100 m under retropropulsion, it may make more sense to free-fall from that altitude (requiring less fuel) and instead tolerate a higher touchdown velocity by using some sort of impact attenuator. Similarly, (in the case of landing in an atmosphere), a large parachute may permit a low vertical velocity and thus obviate the need for the terminal braking or airbags that might be needed by a system with a smaller parachute. On the other hand, a larger chute may lead to unacceptable landing dispersions due to greater wind drift during its slower descent.

6

Planetary balloons, aircraft, submarines and cryobots

6.1 Balloons

Traditionally, planetary exploration uses landers and rovers for *in situ* measurements and orbiters for remote sensing. Landers and the first generation rovers can conduct studies of very limited areas of the planet: square metres for landers and square kilometres for rovers. The main driver for selection of landing sites is safety and the safest sites are usually flat and not scientifically interesting. Besides, even the best imaging from the orbit cannot guarantee an obstacle-free site needed for safe landing.

Robotic balloons (aerobots) may significantly change the future of *in situ* planetary exploration. Aerobots can be used to study eight solar system bodies with atmospheres: Earth, Venus, Mars, Jupiter, Saturn, Uranus, Neptune and Saturn's moon Titan. Besides the Earth, Venus, Mars and Titan are the prime candidates.

Venus is the closest and the easiest planet for aerobots. The first planetary balloons ever flown were part of the highly successful Soviet-led VeGa (Venus–Halley) mission in 1985 (Sagdeev *et al.*, 1986; Kremnev *et al.*, 1986; Blamont *et al.*, 1993).

On Venus, aerobots may serve as the scientific platforms for *in situ* atmospheric measurement and for study of atmospheric circulation. They can be used to drop imaging and deep sounding probes at sites of interest and to acquire and relay high-rate imaging data. Balloon ascent from the surface is essential for a Venus surface-sample return mission.

On Mars, aerobots can fill the gap in resolution/coverage between orbiters and rovers. Powered aerobots (airships) can make controlled global flights for high-resolution radar, visible, infrared, thermal, magnetic and neutron mapping. They can be used for deployment of a network of surface stations. Tethered balloons could provide ultra high-resolution imaging of local areas for navigation of rovers and data relay to the main lander station. Solar-heated balloons could be used as atmospheric decelerators for low-speed landing and to conduct studies in summer polar areas. In the more distant future, airships could be used for human transportation.

On Titan, powered aerobots, and to a lesser extent free balloons, can perform long-duration low-altitude global flight for surface mapping, *in situ* atmospheric measurements, take surface samples and deploy landers and rovers for *in situ* surface studies.

One attractive feature of aerobots is their capability for deployment of large-size (but light-weight) structures that can be used to increase resolution and sensitivity of science instruments exploring the surface and sub-surface of the planet, and to increase communication data rate.

Aerobot technologies have advanced in recent years as a result of progress in envelope materials and design – technologies driven primarily by the needs of scientific balloons for the Earth's stratosphere. Technologies for deployment and inflation, navigation, control, communication and power are also developing rapidly in response to planetary applications.

6.1.1 Balloon basics and planetary environments

Any lighter-than-air (LTA) vehicle can be described by Archimedes' 2000-year-old principle of flotation:

$$B = \rho_a V g \tag{6.1}$$

where B is the buoyancy force, V the volume of gas inside the balloon, and ρ_a the atmospheric density. At equilibrium

$$\frac{B}{g} = M = M_b + M_g + M_p = \rho_a V \tag{6.2}$$

where M is the total mass of the aerobot, M_b the mass of the balloon envelope, M_g the mass of gas and M_p the mass of the payload. The denser the atmosphere, the smaller the volume of buoyant gas (and aerobot shell) needed to fly.

Using the ideal gas law, Equations 6.1 and 6.2 can be written as

$$M_b + M_p = (\rho_a - \rho_g)V = \frac{VP_a\mu_a}{RT_a}\left(1 - \frac{P_g\mu_g T_a}{P_a\mu_a T_g}\right) = \rho_a V\left(1 - \frac{P_g\mu_g T_a}{P_a\mu_a T_g}\right) \tag{6.3}$$

where ρ_g, P_g, μ_g and T_g are the density, pressure, molecular weight and temperature of the buoyant gas; P_a, μ_a and T_a are the pressure, molecular weight and temperature of the ambient atmosphere, and R is the gas constant. If the pressure inside the balloon exceeds the ambient pressure by ΔP (ΔP is called the super-pressure) then

$$M_b + M_p = \rho_a V \left(1 - \left(1 + \frac{\Delta P}{P_a} \right) \frac{\mu_g T_a}{\mu_a T_g} \right) \qquad (6.4)$$

This basic equation describes all types of balloons. Their classification is illustrated in Figure 6.1.

One more balloon type – Rozier – is a combination of a light-gas and a Montgolfière balloon. Cases with $\mu_g < \mu_a$ and $\Delta P = 0$ describe light-gas zero-pressure balloons:

$$M_b + M_p = \rho_a V \left(1 - \frac{\mu_g T_a}{\mu_a T_g} \right) \qquad (6.5)$$

The buoyancy of these balloons increases with the temperature of the buoyant gas. In a steady flight of a balloon made of transparent film without an additional heat source, $T_a = T_g$ and

$$M_b + M_p = \rho_a V \left(1 - \frac{\mu_g}{\mu_a} \right) \qquad (6.6)$$

For a fixed mass of gas, the inflated volume of zero-pressure balloons varies with ambient pressure, i.e. with altitude. If the zero-pressure balloon is displaced from its equilibrium altitude (where gas fills all balloon volume), e.g. by upward vertical convection currents or by heating, it has to vent gas to avoid stress in the envelope and maintain $\Delta P = 0$. When the disturbance action stops, ballast has to be dropped to bring the balloon to steady flight (now at higher altitude). Venting of gas and use of ballast significantly limits zero-pressure balloon lifetime and their use in planetary exploration.

The pressure of the buoyant gas of superpressure balloons in steady flight exceeds ambient pressure; the balloon envelope is filled completely and has a

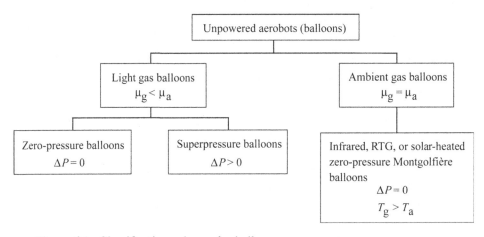

Figure 6.1. Classification scheme for balloons.

fixed volume. These balloons can remain afloat for long periods of time. On Earth, some superpressure balloons have stayed aloft for several years. Use of strong (and heavier) materials and more demanding balloon design is the price to be paid for long-duration flight. A number of materials (polyester, Kapton, nylon, PBO films and different composites) can be used for superpressure balloons. A 'pumpkin' shape design, where tendons take most of the superpressure load, significantly relieves requirements on the balloon material and allows the use of weaker films.

Superpressure balloons are described by Equation 6.4. Superpressure ΔP can be calculated as

$$\Delta P = \frac{M_g R T_g}{\mu_g V} - P_a \qquad (6.7)$$

ΔP increases with the temperature of the buoyant gas and balloon performance is driven by the radiative environment and properties of the balloon material. Even visually transparent films significantly absorb infrared radiation. To design superpressure balloons, radiation fluxes in the planetary atmospheres should be known or carefully evaluated.

The buoyancy of ambient gas balloons ($\mu_g = \mu_a$, $P_g = P_a$) is created by heating of the gas and depends on an overtemperature ΔT. The governing equation is

$$M_b + M_p = \rho_a V \left(1 - \frac{T_a}{T_g} \right) = \rho_a V \left(1 - \frac{T_a}{T_a + \Delta T} \right) \qquad (6.8)$$

On Earth, balloons can be heated by solar radiation during the day and by Earth/atmosphere infrared radiation at night. On Mars, buoyancy can be produced only by solar heating (see Chapter 8) and solar Montgolfière balloons can be used only for daytime missions. Concepts for Titan include using ambient air warmed by the waste heat from an RTG (see Section 9.3).

6.1.2 Planetary environments

Three primary candidate planets for aerobot missions (Venus, Mars and Titan) have very different environments (see Table 6.1).

The deep atmosphere of Venus exhibits extreme atmospheric parameters. The high temperature and pressure in the lower atmosphere strongly limit the lifetime of surface and near-surface vehicles: without nuclear power-driven refrigerators or high-temperature electronics, the lifetime would be ~2 to 3 h. High-temperature materials with good gas barrier and strength properties are needed for near-surface LTA vehicles. On the other hand, the environment of the higher troposphere is quite mild and comparable with the troposphere of the Earth. This region is the most favourable for aerobot missions (the VeGa balloons

Table 6.1. *Planetary atmospheric environment parameters for Venus, Mars, Titan and Earth*

Parameter	Venus	Mars	Titan	Earth
Acceleration due to gravity (Earth g)	0.9	0.37	0.14	1
Main atmospheric gas	CO_2	CO_2	N_2	N_2
Surface temperature (K)	735	230	94	290
Surface pressure (atm)	92	0.0067	1.5	1.0
Surface air density (kg m^{-3})	64	0.015	5.4	1.2
Solar flux at the upper atmosphere (W m^{-2})	3200	700	13	1300
Solar flux near the surface (W m^{-2})	5	700	$\sim$1	600
Altitude of tropopause (km)	$\sim$65	11	40	17
Pressure at tropopause (mbar)	97	2.7	200	90
Temperature at tropopause (K)	240	190	70	220
Diurnal temperature variations near the surface, $\Delta T/T$ (%)	<0.3	30–50	<1	<10
Winds at the tropopause (m s^{-1})	80–100	20–30	15	20–30
Winds in lower atmosphere (m s^{-1})	1–3	5–20	$\sim$1	5–20

flew at 54 km at 0.5 bar and $\sim$30 °C). The main challenge is the sulphuric acid clouds that cover 100% of Venus.

On Mars, the low density of the atmosphere in combination with large thermal variations requires light-weight and strong materials for long-duration aerobotic missions – a combination that is not easy to obtain. Although the proven balloon materials could be used for low-payload-mass aerobots, composite materials, new balloon designs ('pumpkin' shape), and advanced fabrication technology (so-called 3-DL or three-dimensional laminate technology, which is used for fabrication of sails for round-the-world yacht races) offer the most potential to improve the efficiency of aerobotic missions. The Martian troposphere is similar to the stratosphere of the Earth; this similarity provides the basis for terrestrial stratospheric flights to test Martian aerobot systems.

The combination of high density (4.4 times larger than on the Earth) with low gravity ($\frac{1}{7}$ of the Earth value) and low temperature contrasts makes Titan almost ideal for long-duration aerobot missions. The balloon materials become stronger at the extremely cold temperatures but adhesives that remain non-brittle at these temperatures are required.

Table 6.2 shows the typical parameters of aerobots to lift a payload of 10 kg on Venus, Mars, Titan and Earth. The values were calculated using the aerobot equations (including Equations 6.1 to 6.3). For the sake of comparison, the areal density of the balloon material is assumed to be $\sim$20 g m^{-2} for all planets, reflecting current technology (the VeGa balloon material was $\sim$340 g m^{-2}, which is comparatively heavy).

Table 6.2. *Typical parameters of planetary aerobots*

Parameter	Venus, 1 km	Venus, 60 km	Mars, 5 km	Titan, 1 km	Earth, 1 km	Earth, 4 km
Atmospheric density, (kg m^{-3})	61.56	0.489	0.010	4.80	1.13	0.010
Temperature of atmosphere (°C)	454	−10	−51	−181	−2	−33
Payload mass (kg)	10	10	10	10	10	10
Balloon diameter (m)	0.72	3.70	20.65	1.73	2.83	21.41
Balloon volume (m^3)	0.2	26.5	4610	2.7	11.9	5140
Balloon mass (kg)	0.84	1.79	31.6	1.02	1.37	33.9
Mass of buoyant gas (He) (kg)	1.16	1.25	4.46	1.97	1.94	7.44
Total floating mass (kg)	12.0	13.0	46.1	13.0	13.4	51.4
Payload mass as percentage of floating mass (%)	83.4	76.5	21.6	77.1	75.2	19.5
Mass of entry vehicle (kg)	36	39	138	39	N/A	N/A

Atmospheric density dominates the balloon size: a Mars aerobot requires a balloon over 150 times larger (in volume) than the Venus aerobot at 60 km, and over 1500 times larger than the Titan aerobot near the surface. A mass efficiency (ratio of payload mass to the total floating mass that includes mass of payload, balloon and buoyant gas) is 75–80% for the Venus and Titan aerobots (it was ~30% for the VeGa balloons) and only ~20% for the Mars aerobot. Use of hydrogen instead of helium for the buoyant gas will increase the efficiency of the Mars aerobot to 24%. The most radical way is to use lighter envelope materials: an areal density of 12 g m^{-2} will nearly double the mass efficiency.

Because of the dense atmospheres of Titan and Venus, payload mass is not as critical as on Mars. It is unlikely that in the immediate future Martian aerobots will be able to lift more than 20–30 kg of payload.

6.1.3 Deployment and inflation of planetary aerobots

Just as all lander and rover missions have many features in common, so it is with aerobots. The most common mission scenario would be: launch of an interplanetary bus with the aerobot system enclosed in an entry vehicle, cruise phase to the planet, targeting at a selected area on the planet, separation of the entry vehicle, entry and deceleration in the atmosphere, deployment and inflation of the aerobot, release of the heat shield and ascent (or descent) to the floating altitude where the active phase of the aerobot mission starts (Figure 6.2).

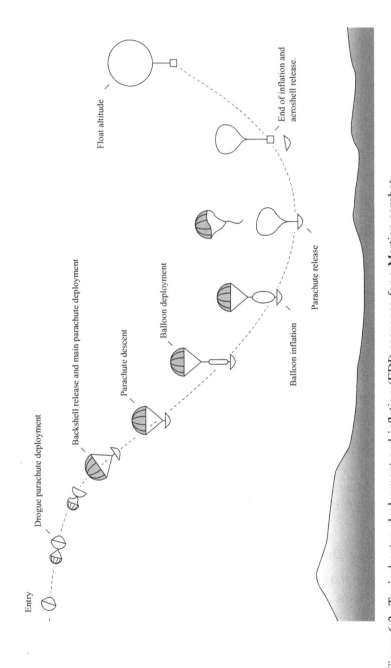

Figure 6.2. Typical entry, deployment and inflation (EDI) sequence for a Martian aerobot.

The launch of planetary balloons from the surface is impractical. Aerial deployment and inflation is more mass efficient and less risky since it does not require an additional (and costly) landing system, or the risky procedures of soft landing and the deployment and inflation of the balloon at the surface. However, aerial deployment and inflation is the most critical and least modelled part of the mission because of the complexity of the aerodynamic processes involved.

The feasibility of aerial deployment and inflation of balloons made of robust heavy material was demonstrated in the VeGa balloon mission. Future missions require much lighter and more efficient materials. Successful aerial deployment and inflation of a 3 m balloon (approximately VeGa balloon size) made of 17 times lighter material (12.5 μm Mylar film) was demonstrated in 1998 (Figure 6.3).

This test validated the concept of aerial deployment and inflation of the modern thin-film balloons, which are applicable to Venus and Titan missions.

The deployment and inflation of a Martian aerobot is even more challenging, because balloons are two orders of magnitude larger (in volume), descent velocities during deployment are 7–10 times faster, and balloon inflation should be completed very rapidly (usually in 150 to 250 s) to ensure that the balloon will start to rise before impact with the surface. Successive failures in flight tests of aerial deployment and inflation in the Russian–French Mars Aerostat project (1987–1995) show the complexity of the problem.

Only recently (in the summer of 2002) was the deployment and inflation of a Mars balloon prototype successfully demonstrated in the stratosphere (Figure 6.4).

6.2 Powered aerobots (airships)

A natural evolution of the balloon is the dirigible balloon, or airship. Rather than being at the whim of the wind, an airship offers the hope of traversing the surface in a desired direction. Planetary powered aerobots will provide the capability of global and targeted access to almost any location. For long-duration missions the power should be provided by non-expendable sources of energy – solar cells (Venus, Mars) or radioisotopes (Titan).

The volume (and thus the cube of the size) of an airship scales with its mass, while the drag area scales only as the square. Thus if propulsive power scales with mass, a larger airship is able to achieve higher forward speed. Airships are competitive with aircraft for large, slow payloads and missions. One advantage of an airship over an aeroplane is that the airship has a fail-safe condition, in that if its propulsion fails, it continues to fly. Further, for scientific investigations, an

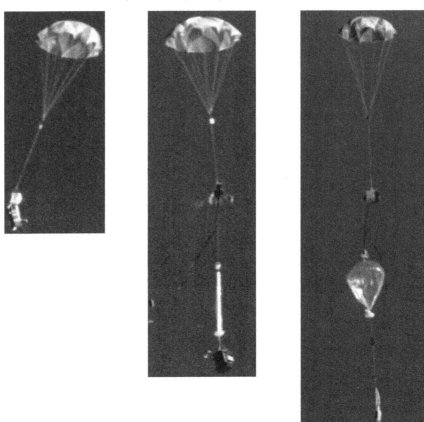

Figure 6.3. Tropospheric deployment and inflation test of 3 m Mylar balloon (August 21, 1998, El Mirage dry lake, California).

airship can remain motionless over a region of interest, unless local winds exceed its forward speed capability.

The available power will determine the possible speed of a powered aerobot. If T is the thrust of the airship propulsion (propeller), then in steady flight thrust is equal to aerodynamic drag and

$$T = \frac{1}{2}\rho_a C_d S U^2 \qquad (6.9)$$

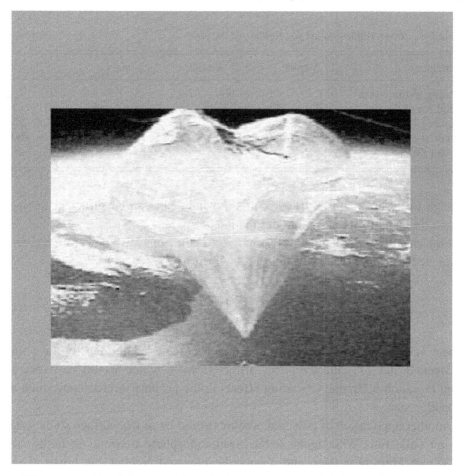

Figure 6.4. 10 m diameter Mars balloon prototype during inflation in the stratosphere at altitude 31 km (big island of Hawaii at left).

where C_d is the drag coefficient, S the reference area and U the airspeed of the airship. The effective mechanical power is

$$P = TU = \frac{1}{2}\rho_a C_d S U^3 \qquad (6.10)$$

For illustrative purposes, Table 6.3 shows power requirements for aerodynamic shaped aerobots with the same diameter as in Table 6.2, for air speeds $3\,\text{m s}^{-1}$ and $15\,\text{m s}^{-1}$. It was assumed that the drag coefficient is ~0.2 (a conservative value) and the efficiency of transformation from electrical to thrust power is $\sim50\%$.

The required power grows very rapidly: theoretically as the cube of the speed. For relatively small aerobots, the available power can be of the order of tens to hundreds of watts, and their air speed would likely be 1 to $7\,\text{m s}^{-1}$. It is not

Table 6.3. *Power requirements for planetary aerobots*

Parameter	Venus		Venus		Mars		Titan	
Floating altitude (km)	1	1	60	60	5	5	1	1
Speed (m s^{-1})	3	15	3	15	3	15	3	15
Required thrust (N)	22.5	560	4.7	118	3.0	75	10.1	254
Required electrical power (W)	135	16 000	28	3550	18	2260	61	7600

enough to fly upstream in 10 to 20 m s^{-1} winds, but can be sufficient to steer across the wind to the desirable destination.

Empirically for Earth airships (Lorenz, 2001) the propulsive power for a given-mass airship to move at a given speed is given by

$$P = 3M^{0.6}V^{1.85}\left(\frac{\rho}{\rho_0}\right)^{0.33-0.5n} \tag{6.11}$$

where n denotes how propulsive efficiency scales with density (a value between 0 and 1) – the exponent differs from 3, which would be expected from Equation 6.10, due to scaling effects in the propulsive efficiency, such as propeller size.

Another application of powered aerobots could be *in situ* surface studies and sample collection. When winds in the lower atmosphere are small (as in the case of Venus, Mars near noon and Titan), the powered aerobot can hover above the selected site; the surface instrument package can be winched down for the surface measurements or sample acquisition and winched up to the aerobot later. The hovering can be controlled by image processing in the horizontal plane and by pressure data or altimeters in the vertical direction. The aerobot would be used as a flying rover, covering a much greater area than the traditional surface rovers.

6.3 Aeroplanes and gliders

Aeroplanes or gliders for the Martian atmosphere have reached a significant level of technical maturity, in that proposals sufficiently well-developed to be seriously considered in competitive NASA mission selections have been made. The Martian atmosphere is very thin – comparable with the Earth's high stratosphere – which makes aviation difficult, but not impossible. Rather high flight speeds are required to achieve sufficient lift.

Key aerodynamic parameters are the wing loading (i.e. the vehicle weight per unit wing area) and the lift-to-drag ratio (L/D). The lift L on a wing is expressed as

$$L = \frac{1}{2}\rho S C_{\mathrm{L}} V^2 \tag{6.12}$$

where V is the flight speed, C_{L} is the lift coefficient, S the reference area (usually the wing area) and ρ the air density. In level flight, this expression must equal the vehicle weight W. High wing loading (W/S) therefore requires high dynamic pressure ($0.5\rho V^2$), thus a high flight speed, or a very high flight speed if the air is thin (low ρ). The drag is similarly written

$$D = \frac{1}{2}\rho S C_{\mathrm{D}} V^2 \tag{6.13}$$

In steady flight, this must be balanced by a forward component of weight (in gliding, the glide slope will equal L/D or $C_{\mathrm{L}}/C_{\mathrm{D}}$) or by thrust from some sort of propulsion. The drag power (and thus an absolute minimum propulsive power requirement – propulsive efficiencies of propellers, etc. may in fact be quite low, especially in thin atmospheres) equals the drag multiplied by the forward speed. In level flight, this is easily calculated knowing $W = L$ and $D = L/(L/D)$ – thus the lift-to-drag ratio is a key determinant of performance. The drag power is therefore $P = VW/(L/D)$ – of course more power is needed if the vehicle is to climb.

The considerations above show how it is energetically favourable to fly a given mass with a large wing, and thus to fly slowly. Extremely power-limited aircraft on Earth, notably human-powered aircraft, have low flight speeds and very large (and light and therefore flimsy) wings. There are structural limits to the size of a practicably rigid wing (particularly for vehicles delivered to other planets where the wing must be folded inside a launch vehicle and/or entry shield) and some minimum flight speed may be required for traversing a given range in a fixed time.

For terrestrial subsonic propeller-driven aircraft at least, an empirical relation of required flight power is

$$P = 11 m^{0.8} V^{0.9} \left(\frac{\rho}{\rho_0}\right)^{-0.5n} \tag{6.14}$$

Two important dimensionless numbers apply to flight – the Mach number and the Reynolds number. The achievable L/D scales strongly with Mach number and Reynolds number; it is impossible to develop high lift-to-drag ratios at high supersonic or hypersonic speeds (the Mach number being the ratio of the flight speed to the local sound speed). The Reynolds number is a measure of the relative importance of pressure (or inertial) forces in the fluid to the viscous forces (Re = $vl\rho/\mu$, where v and l are the characteristic velocity and dimension, μ is the dynamic viscosity). At low Reynolds numbers, viscous effects predominate. The important point for prospective aeronauts is that the lifting performance of a wing or a propeller declines significantly at low Reynolds number. This is a significant degradation for high-altitude flight on Earth, and particularly for flight on Mars.

The conventional aircraft that have been proposed for Mars include very light long-endurance planes (which would have to fly in the polar summer to continuously capture enough solar power for flight), gliders, and hydrazine-powered propeller planes – the latter of these would last only a few hours.

6.4 Other heavier-than-air concepts for aerial mobility

While the platform of choice for wide-area surveys on the Earth is usually the aeroplane, for delivery of *in situ* instrumentation onto the planet's surface, a helicopter offers important advantages in that it can be placed at zero speed at a given location, regardless of whether that location is flat for a distance long enough to permit an aeroplane to land.

A helicopter achieves lift by propelling air downwards via a whirling rotor, which may be considered as a set of wings. The formidable control aspects of rotary-wing vehicles will not be discussed here, but the fundamental propulsive aspect deserves some comment. In order to hover, the vehicle must produce a downward momentum flux in the air that balances the weight of the vehicle. This momentum flux (in kg m s^{-2}) may be generated with a large-area rotor pushing a lot of air down slowly (as in a helicopter), or a small amount of air down at high speed (as in a Harrier jump-jet). The momentum flux is $A\rho v^2$ where A is the area of the jet and v the induced velocity. The energy imparted to the air is $A\rho v^3/2$ – thus a large-area, low-velocity solution offers the lowest power consumption for a given thrust. Hence a large rotor is better, although (especially for a rotorcraft delivered in an entry shell) structural considerations are likely to be the limiting factor.

Note that a helicopter uses less power in modest forward flight than it does in hover. One expression for this power is given by actuator disc theory:

$$P = \frac{(mg)^{1.5}}{(2\rho A)^{0.5}}. \qquad (6.15)$$

Note the dependence on rotor area and density, and an even stronger dependence on required thrust. In practice, rotor drag, blockage and other effects mean power is rather higher than this and an expression of the form

$$P = 100m^{1.1} \left(\frac{g}{g_e}\right)^{1.5} \left(\frac{\rho_e}{\rho}\right)^{0.5} \qquad (6.16)$$

predicts the power that must be installed on the vehicle, where subscript e denotes the corresponding values for Earth-surface conditions.

Various imaginative concepts have been proposed for local exploration around a lander on Mars. Small vehicles in particular suffer from the aerodynamic inefficiencies posed by the thin atmosphere and the low Reynolds number, and thus some of the proposed solutions are inspired by insects, which confront similar Reynolds number regimes on Earth. One moving-wing concept is the entomopter (that is to say, looking like a dragonfly). Another approach is to use conventional rotorcraft designs, but with rotors adapted for the Reynolds number regime – a so-called 'mesicopter'.

6.5 Submarines, hydrobots and cryobots

A submarine is in essence a special case of airship, except in a rather more dense fluid. Submarine vehicles have been considered for Europa's sub-surface water ocean (where the term 'hydrobot', as an analogue of aerobot, has been used), although delivery to this ocean which lies under an ice crust that appears to be 10–20 km thick in most places presents significant challenges. The scientific goal would be to study the temperature structure and composition of the ocean, paying particular attention to searching for biota or biological molecules. The canonical illustration associated with this mission concept shows the vehicle inspecting the fauna around a seafloor hydrothermal vent, which might be expected to exist by analogy with Earth.

A submarine vehicle might also be contemplated to explore Titan's frigid seas of liquid hydrocarbons; these lakes and seas predominantly of ethane and methane at 94 K would have a bulk density of 450–650 kg m^{-3} and might be up to a few kilometres deep.

Thermal control and guidance in the presence of winds and currents are essentially equivalent for submersibles and for airships, although at significant depth the hydrostatic pressure may be formidable and structural design may need to take this into account. One bar of pressure corresponds to a depth increment of 10 m of water on Earth, or about 120 m of the hydrocarbon ocean on Titan, and a similar value for Europa (although the hydrostatic pressure beneath 10 km of ice already introduces 1 kbar of pressure).

For mobility, the empirical expressions given above for airship-power scale reasonably well for submarines. In fact, although the delivery of such a vehicle poses significant challenges, submersible explorers have been proposed for Europa's sub-surface ocean.

The presence of icy masses on certain bodies in the Solar System provides for an unusual class of spacecraft lander that has its operating location not at the surface of a body, but beneath it. While the engineering problems of having a whole spacecraft tunnel its way mechanically through rock and regolith are beyond current state of the art, it is possible to conceive of systems that can melt through an ice layer. The term 'cryobot' has been coined for such exploration craft.

Such thermal drilling is of particular interest in two locations – the Europan surface, and the Martian polar ice caps (e.g. Zimmerman *et al.*, 2001). The latter may yield records of climate change, impact ejecta and volcanic activity.

While there is considerable heritage in the use of electrically powered melting probes for the exploration of terrestrial polar ice masses (e.g. Philberth, 1962; Aamot, 1967), there are two principal difficulties. First, especially on Europa, the absence of an atmosphere causes melt or even just warm ice to evaporate rapidly. Thus the drill must supply latent heat of evaporation as well as for melting, leading to a substantial increase in required power. Second, if an ice layer is dirty, the dirt can accumulate to form an insulating layer which reduces the penetration rate. Removing this material is challenging.

Note that the energy required for thermal drilling is typically much higher than for mechanical drilling. However, thermal drills are much easier to implement, and the heat need not even be generated from electricity, but could be provided directly from a sufficiently large radioisotope heat source.

7

Arrival at a surface

This chapter covers the final moments of descent towards, and contact with, a solid (or, as in the case of Titan, possibly liquid) surface. We deal with the issues of surviving impact to deliver a working spacecraft to the surface. This usually requires some sort of prior deceleration achieved during descent. Active guidance, navigation and control can also be performed to avoid hazards and locate a safe landing site. Having arrived, the impact may be damped within the vehicle alone, or by also using the deformability of the surface.

7.1 Targeting and hazard avoidance

Thus far, planetary landers have been flown 'open loop' in terms of their horizontal targeting with respect to the surface. While feedback control is employed to regulate descent rate to achieve close-to-zero speed at zero altitude, only the horizontal speed tends to be controllable, not the location.

The Mars Exploration Rovers incorporated a camera system (DIMES – Descent Image Motion Estimation System) to sense sideways motion, and a set of rocket motors (TIRS – Transverse Impulse Rocket Subsystem) to null the motion to maximize the probability of successful airbag landing; Surveyor and other lunar landers similarly used multibeam Doppler radar and thrusters to null horizontal motion. However, the latitude and longitude co-ordinates of the landing site were simply those that happened to be under the spacecraft when its height became zero. These were within an expected delivery ellipse specified by entry conditions and uncertainties, etc., but were not controlled.

Such fatalism is unacceptable in situations where close proximity to small sites of scientific value is required, or where a heterogeneous target region may have some sites of acceptable topography mingled with dangerous hazards. So far, the prime example of precision landing on another planet is that of Apollo 12, landing within sight of Surveyor 3, permitting the retrieval of exposed equipment from the latter. This example, however, exploited the sensing and control abilities of a human crew, which are generally unavailable. Some sample-return architectures for Mars, for example, have envisaged a rover acquiring sub-surface samples and delivering them to a separate sample-return spacecraft; the latter must have the ability to land close enough to the rover to be reachable by it.

Beyond the descent control instrumentation described in Section 5.7, additional sensing is needed to control location – much of it derived from work on weapons delivery. Inertial guidance (accelerometers and gyroscopes) may, if the delivery (entry) state is sufficiently well known, be adequate, provided that the terrain has been mapped already such that the target site is known in inertial co-ordinates. For enhanced precision, or when terrain data is not registered adequately in a co-ordinate frame, terrain-matching cameras may be used, again borrowing from their application in cruise missiles. However, it is boulders on a scale ($\sim$0.5 m) much smaller than that needed for overall navigation that pose the greatest mechanical threat to a lander, and thus their distribution is either prohibitive to map directly, or must be determined indirectly, e.g. from radar backscatter. Scanning LIDARs are thus being contemplated for making topography maps underneath a lander to identify safe landing spots.

For landers on airless bodies, rocket propulsion is of course necessary to control sideways motion and location. For Mars, at least, consideration is being given to steerable (ram-air) parachute systems.

7.2 Landing gear

Spacecraft that are intended to operate on a solid surface after landing require some sort of landing gear to allow the craft to come to rest undamaged in a stable position, ready for operations. The design should be able to cope with:

- the expected mass of the lander and the resulting impact overloads and weight
- the lander's expected motion and orientation on contacting the surface
- the expected range of terrains (topography and surface materials) that might be encountered
- any forces or torques to be reacted against after landing (e.g. due to anchoring, drilling, robotic arm operations, launch of an ascent stage)
- doing so with some suitable margin of safety

The landing gear may also carry sensors to detect surface proximity or actual touchdown (e.g. to trigger shutdown of the retro-rockets), and may even be a convenient platform to mount experiments that need to be in contact with the surface or within the field of view of the lander's cameras. The main configuration types are legged landers (usually three or four legs) and pod landers (ranging from near-spherical to egg-shaped or prolate).

In most cases the vehicle's vertical speed on landing will already be below some nominal value as a result of the deceleration systems employed during entry and descent, for example entry shields, parachutes and retro-rockets. These systems may on the one hand have been able to reduce the descent speed to nearly zero before contact with the surface, in which case the landing gear will need to cope with some nominal residual landing speed of only a few m s^{-1} at most (e.g. free-fall from the height at which the retros are shut down). On the other hand, the landing gear may have to perform a greater share of the task of decelerating the vehicle. In both cases the remaining kinetic energy has to be dissipated over some finite distance while minimising loads and mass. Approaches include the following:

- Damped elastic structures (e.g. piston-like landing legs)
- Plastic deformation: crushable material (e.g. honeycomb material, foam, balsa wood) or structures (e.g. buckling of tubular struts, collapse of retro-rocket nozzles)
- Fluid damping: control of a fluid pressure or flow rate (e.g. airbags)

Note that landing loads can also be reduced by shedding mass (e.g. systems whose function has been completed, such as fuel tanks or spent rocket stages) prior to landing. Examples of this include all Luna and Surveyor landers.

Horizontal components of velocity also need to be addressed. These may be present due to the direction of the descent trajectory, atmospheric winds or the swinging of a vehicle under its parachute. While generally undesirable on impact[8] and required to be minimised, some degree of horizontal motion (downrange or cross-range) may be necessary during targeting or hazard avoidance. Some rotation may also be present, due to attitude-control motion or, as in the case of Huygens, a slow spin about the vertical axis to scan instruments around during descent.

The performance of crushable materials may be characterised by their energy absorption capability per unit mass, or 'specific energy', E_s, equal to the crushing strength σ divided by the bulk density ρ. Vergnolle (1995) reviews a number of soft landing impact attenuation technologies and quotes E_s values for aluminium and carbon of 16 and 100 kJ kg^{-1}, respectively. Such materials are usually in the form of a honeycomb structure or foam. The total kinetic energy absorbed by a crushable component is the integral of force with respect to distance, i.e. the work done. The maximum allowable load, together with the kinetic energy, thus defines the minimum distance (stroke length) over which deceleration must occur. The maximum allowable load, together with the dynamic crushing stress of a candidate material, determines the cross-sectional area that will be required. The stroke length, cross-sectional area and density of the candidate material thus determine the mass that will be required – a value to be minimised.

---

[8] An exception here might be aircraft, which require horizontal motion to gain lift to soften the landing.

The impact speed, deceleration and stroke length are related by

$$\frac{v^2}{2} = ad \tag{7.1}$$

where v is the impact speed, a the deceleration and d the length of the deceleration stroke. The force required is represented by the product $a \times m$, where m is the mass of the vehicle. Where the deceleration system is based on a crushable material with a crushing strength σ and has a cross-section area S, then the deceleration force is $S \times \sigma$.

Crushable components may be incorporated into the struts of landing legs, the landing gear footpads and/or the base of the lander (e.g. Surveyor 1–7). In the case of pod landers the whole vehicle may be encased in crushable material (e.g. the balsa wood spherical capsules of Ranger 3–5, or the crushable shells of Mars 2, 3, 6, 7), since it may impinge the surface several times in different orientations after the initial impact, before coming to a final halt.

One can of course consider the surface of the target body to be a plastically deformable material itself, helping to cushion the landing with no mass penalty to the lander. However, the uncertainties and spatial variations in surface mechanical properties (e.g. from bedrock to windblown sand) of solid bodies of the Solar System are such that it is prudent to assume a non-deformable surface, dissipating none of the kinetic energy of landing. The damping system built into the lander can thus cope with any eventuality, rather than relying on a 'soft' landing. (Nevertheless, payload delivery penetrators do use deformation of the surface for braking, as described in the next Section 7.3.)

In contrast, the deformability of the surface *is* important for sizing of the footpads (or equivalent structures). Too small a footpad and the lander may sink too deep into the surface, jeopardising the mission. This was of particular concern for the first lunar landings, since there was a risk that the surface may have turned out to have been of such a soft, deep regolith layer that the subsequent crewed landers would have disappeared into the surface. There is a similar element of risk for Philae, due to make the first soft landing on a comet nucleus in 2014. Footpads are thus sized to penetrate no deeper than a reasonable limit upon landing, calculated using soil mechanics models of the surface bearing strength. In the event, the Surveyor 1, 3, 5, 6, 7 landers penetrated only 2.1 to 10.5 cm into the lunar surface, for landing speeds ranging from 1.4 to 4.2 m s^{-1} (e.g. Jones, 1971).

In many cases the landing gear will need to be deployed prior to landing. Landing legs are usually stowed for launch and cruise for reasons of space (e.g. accommodation within the launch vehicle or entry shield) and resistance to vibration. They are then opened out (downwards and outwards, or upwards and outwards) to provide a wide base for landing. Various configurations have been adopted over the years, many of which involve attaching the footpads to an inverted tripod of piston-like struts incorporating crushable material. The struts are jointed at their ends such that each landing leg assembly unfolds upon operation of a deployment actuator.

The higher the ratio between the width of the landing gear and the height of the centre of mass, the more extreme the landing scenario required to make the lander topple over. Knowledge of the surface topography at the scale of the lander also influences both the base width (steeper slopes require this ratio to be higher) and possibly the clearance required between the footpads and the underside of the lander (the underside should contact the surface either not at all or in a controlled fashion, e.g. via crushable components). Several models of landing stability criteria exist, e.g. Buslaev (1987).

Airbags need to be inflated shortly before landing, e.g. by gas tanks or chemical generators. Unlike legs, however, deployment too early during descent can be problematic, since loss of pressure due to leakage or cooling can reduce their performance. Conversely, too high a pressure can lead to unacceptably high impact loads and even rupture of the bags. After the lander has come to a final halt (which may be after much bouncing and rolling, of order 1 km for Pathfinder and MER), the airbags are either deflated and retracted by motor-pulled tendons (as for Pathfinder and MER), or are released and allowed to spring apart by means of their own elasticity (as for the two-bag system of Luna 9, 13 and the Mars 96 Small Stations, and the three bags of Beagle 2). Particular attention must be paid to the durability of airbag materials against impingement onto rocks – the pressure bladder being protected within outer abrasion layers.

The landing sequence of pod landers involves an additional manoeuvre after the vehicle has come to a halt on the surface, to bring it into its proper orientation ready for operations. Mechanisms to achieve this have taken the form of a system of three or four opening 'petals' – hinged flaps covering the upper surface of the lander, any one of which is strong enough to bring the lander upright (e.g. the near-spherical Luna 9, 13, Mars 2, 3, 6, 7 and Mars 96 Small Stations, and the tetrahedral Mars Pathfinder and Mars Exploration Rover landers) – or like a pocket-watch, where a disc- or lens-shaped lander has a single hinged lid to perform the same function (e.g. Beagle 2). Landing edge-on can be mitigated by providing an inflatable 'tyre' around the edge that can topple the lander one way or the other. Another possibility is to build a lander that can operate in any orientation, or one where only internal parts are brought upright (e.g. the Ranger 3, 4, 5 landers).

An apparently elegant solution, to achieve both landing and re-orientation with airbags alone, is to position the lander (and thus the centre of mass) off-centre inside the airbags, such that the lander rolls to a halt in the correct orientation. On ejection of the airbags the lander falls the short distance to the surface, remaining upright. Another approach, so far not implemented for planetary missions but under consideration for ESA's ExoMars lander/rover, is to use so-called 'dead beat' airbags on the underside of the landing platform. These are vented such that they provide a near-critically damped soft landing, i.e. without bouncing. Accelerometry can be used to govern the timing and rate of venting of separate cells of the bags, in order to cope with non-vertical motion such as horizontal velocity components and rotation, and to keep the landing overloads within acceptable limits.

The toroidal landing gear of the late Venera/VeGa landers employed a cunning combination of techniques. While acting essentially like a single, annular crushable footpad, the torus was hollow, with vent holes around the top. Upon landing, additional damping was thus provided as the dense atmosphere (which had found its way into the torus during the lander's descent) was expelled through the vent holes, thus avoiding the need for on-board provision of a working fluid.

For missions involving surface mobility, the roles of landing gear and loco-motion may be combined. For instance, the Mars Science Laboratory, due for launch in 2011, is proposed to be lowered to the surface from a hovering 'sky crane' descent system, the rover's wheels also acting as landing gear for the initial touchdown. An earlier example was the proposal to equip Viking-derived Mars landers with caterpillar tracks instead of footpads.

Other systems sometimes used in landing-gear designs-include hold-down thrusters (used particularly in low surface gravity to prevent rebound or toppling over, e.g. the LK lunar lander, Phobos DAS and Philae) and harpoon anchors (to prevent rebound and/or later ejection by outgassing in the case of cometary nuclei, e.g. Phobos DAS, Philae (Thiel *et al.*, 2001)). Philae also incorporates a 'cardanic joint' damping mechanism in its landing gear, at the interface between the tripod legs and the main body of the lander.

Although it is useful in the first instance to consider the one-dimensional case of a vertical landing onto a flat, uniform surface, a real design has to take into account the full three-dimensional nature of the problem. For example, there may be some transverse motion, rotation and tilt, and on impact there may be sliding, rolling and bouncing to consider. Models have been published for a number of previous landers, representing a range of configurations.

The landing dynamics of the Surveyor, Apollo Lunar Module and Viking landers are particularly well covered (e.g. Sperling and Galba, 1967; Jones, 1971; Zupp and Doiron, 2001; Doiron and Zupp, 2000). The impact and subsequent motion (slipping, then slipping and rolling, then rolling) of the near-spherical Venera 7, 8 entry probes was discussed by Perminov (1990), while the particular case of the later Venera landers was discussed by Buslaev *et al.* (1983), Grigor'ev and Ermakov (1983) and Avduevskii *et al.* (1983). The airbag-assisted landing of Mars Pathfinder is described by Spencer *et al.* (1999) and Cadogan *et al* (2002).

The impact of the Huygens probe on Titan was discussed prior to launch by Lorenz (1994) and was measured by several accelerometers (Zarnecki *et al.*, 2005).

The set of forces to be taken into account in modelling landing dynamics depends on the specifics of each case and the constitutive model assumed for the mechanical behaviour of the surface. The failure stress of the surface material when a normal force is applied over the contact area is usually expressed as a bearing strength. A dynamic component may also be included, being dependent on the bulk density of the material and the landing speed. The resistance offered by the ground may also be considered to increase with penetration depth. Sliding friction will also need to be taken into account if significant horizontal components of velocity are expected.

Although computer models can be employed to test possible landing gear configurations under many different situations (e.g. Zupp and Doiron, 2001 for the Apollo LM; Hilchenbach *et al.*, 2000, 2004 for the comet lander Philae), full-scale or sub-scale drop tests may still be employed to validate a landing gear design (e.g. Bazhenov and Osin, 1978). For low-gravity environments such as asteroids and comets, tests can be performed horizontally with the lander suspended sideways on a long pendulum, encountering a vertical surface. Alternatively a system of counterweights can be employed (e.g. such a rig is described for the PrOP-F Phobos hopper by Kemurdzhian *et al.*, 1993).

Sub-scale drop tests involve constructing a scaled-down model of the lander, performing drop tests and scaling the results to determine the behaviour of the full-scale lander design. Such sub-scale drop tests were performed for several landing vehicles, and for Huygens (e.g. Seiff *et al.*, 2005). Table 7.1 gives the formulae for scaling key parameters for sub-scale tests, for a linear scale factor λ.

Table 7.1. *Scaling factors and scale model values for physical quantities*

Quantity	Full-scale value	Scale factor	Model value
Length	l	λ	λl
Area	A	λ^2	$\lambda^2 A$
Mass	m	λ^3	$\lambda^3 m$
Moment of inertia	I	λ^5	$\lambda^5 I$
Time	t	$\lambda^{0.5}$	$\lambda^{0.5} t$
Velocity	v	$\lambda^{0.5}$	$\lambda^{0.5} v$
Linear acceleration	a	1	a
Angular acceleration	α	λ^{-1}	$\lambda^{-1}\alpha$
Force	F	λ^3	$\lambda^3 F$
Pressure	p	λ	λp
Spring constant	k	λ^2	$\lambda^2 k$

7.3 Penetration dynamics

Vehicles that use their kinetic energy of arrival at a planetary surface to emplace a payload at depth are called penetrators (sometimes called kinetic-energy penetrators or KEPs). In contrast to other landers, the kinetic energy is intended to be dissipated mostly in the surface material rather than the structure of the landing vehicle. While planetary landers reach the surface at speeds of the order of 1–$10\,\mathrm{m\,s^{-1}}$, penetrators arrive at speeds ranging from 60–$300\,\mathrm{m\,s^{-1}}$, depending on factors such as the desired depth, the mass and geometry of the penetrator, the expected surface mechanical properties, the shock resistance of internal components, and constraints imposed by the entry and descent from orbit or interplanetary trajectory. A number of concepts for hypervelocity penetrators, arriving at speeds in excess of $1\,\mathrm{km\,s^{-1}}$, have been studied but have not yet flown.

At the time of writing, the only examples launched so far in planetary exploration are the Mars 96 penetrators and the Deep Space 2 Mars Microprobes. Details of these and other projects are given in Chapter 19. Here we discuss briefly the impact and penetration dynamics of such vehicles, which is a key aspect of their design. Accelerometry measurements of the penetration event can also be of scientific interest, in that they probe the mechanical properties of the sub-surface material.

The field of impact penetration testing and modelling was reviewed briefly by Lorenz and Ball (2001). The relevant literature is distributed across a wide range of fields including planetary science, soil mechanics, impact engineering, military and aerospace sources. In the development of planetary penetrators, both modelling and testing play essential roles. Modelling can be used at an early stage to evaluate the performance of candidate configurations, and to access regimes that may be prohibitively expensive to reproduce many times in the laboratory, if it can be done at all. Much time can be saved by modelling, however while it is easy to obtain output, much attention needs to be paid to how the target properties are modelled. Moreover, planetary surface/sub-surface mechanical properties are both subject to uncertainty as well as small-scale variations, so a wide range of cases has to be modelled. For example, the surface material may be cohesive (either hard or soft), granular (e.g. sand or reoglith), icy or even a mixture of these.

Models of penetration may be divided into several categories, and can be applied both to predictions of a given penetrator design's behaviour in a specified target, and to constrain target properties from actual penetration measurements. The first category of model is purely empirical, namely fits of penetration depth against projectile and target properties. The work of Young (1969, 1997) is widely cited, although it does not build up from the underlying physical processes and lacks dimensional consistency, and so is perhaps more distracting than useful for interpreting measurements. It can be usefully applied to initial calculations of penetration depth, however.

The second type of model is purely physical – by making an idealised model of the forces on a penetrator, the dynamic behaviour can be predicted. This type of model is the oldest for which good records exist; the Robins–Euler, Poncelet and Résal equations dating from the 18th and 19th centuries are examples of these, with deceleration being related to a constant term, or a linear combination of a constant term and velocity raised to the first or second power. More recent models often use a 'cavity expansion' technique to model the forces (e.g. Yew and Stirbis, 1978; Forrestal and Luk, 1992).

Beyond the simple algebraeic/analytic models of penetration, various levels of numerical sophistication can be applied to penetration models – at one end of the spectrum the SAMPLL code, developed by Young to apply his empirical penetration equations stepwise to layers, is a trivial example. At the other end of the complexity spectrum are full 2- and 3-dimensional finite-element models and smooth particle hydrocodes. Many examples exist (e.g. Autodyn) although are often restricted in access.

The penetration process can be divided conceptually into three phases: initial impact, 'free flight' and the terminal phase. The initial impact may be complicated by the partial immersion of the projectile tip in the target as well as ejection of material from the target to form a crater. The second phase (which may in fact be vanishingly short in duration) is a more or less steady state (near-constant deceleration), although when shaft friction is of interest then the phase may be subdivided according to whether or not the shaft has completely penetrated. Finally, in the third phase the projectile comes to a halt. In some cases this may be indistinguishable from the free-flight phase, but in others elastic phenomena in the target lead to a final peak in the deceleration history as the target 'grabs' the projectile.

Penetration dynamics can be further complicated if the penetrator is of the forebody/aftbody design (as opposed to bullet-shaped), in which case the interaction of both parts with the ground, and indeed with each other, must be studied. An important constraint on the design is to achieve clean and reliable separation of the two parts and ensure that they do not subsequently hit each other as the forebody comes to a halt. The aftbody must also be made to stay close to the surface, to ensure communications and proper deployment of surface components such as cameras, meteorological sensors and solar cells. To achieve clean separation it is necessary for the mass/area ratio of the forebody to be at least a factor of several ($\sim$4) higher than that of the aftbody.

There is no substitute for sooner or later embarking upon a series of impact penetration tests. These may be performed on a range of development models from simple 'boilerplate' structural models to instrumented prototypes and engineering models. Acceleration techniques include dropping under Earth gravity (whether by hand, drop rig, helicopter or aircraft), conventional 'powder' guns and airguns. Gas guns are usually too energetic for this field, although they are widely used to study hypervelocity impacts. A novel technique was used for acceleration of the Mars 96 penetrators, where a sideways velocity component due to strong Martian winds had to be considered. A penetrator was suspended and dropped from a parasail pulled by a car. The Mars 96 programme also used elastic 'bungee' cord acceleration to afford higher impact speeds than could be achieved from a simple drop tower. The impact testing phase is usually performed as a moving test article decelerates in a target; however, 'reverse' techniques are also used. For example, a target sample can be accelerated and impacted onto a stationary penetrator, or shock tests can be performed with an airgun/piston system, which applies a well-characterised acceleration pulse to an initially stationary test article.

One aspect of testing that is important for the design of a penetrator is the robustness to internal shear forces on components, and non-zero angle of impact and/or angle of attack. While many internal components may be highly resistant to shock, this is usually only for acceleration along a particular axis. Solutions to improve shock resistance, as part of careful structural design, include 'potting' materials that encase components in a block of material or glass microspheres. Susceptibility to non-zero angle of attack may be tested using spinning guns, such as that used for the Lunar-A project (Shiraishi *et al.*, 2000).

7.4 Splashdown dynamics: Titan landers, Earth-return capsules

Impacts into liquid surfaces are not often considered for spacecraft: the applications are the return to Earth of manned or unmanned capsules (e.g. McGehee *et. al.*, 1959; Vaughan, 1961; Stubbs, 1967; Hirano and Miura, 1970), the impact of the Challenger crew module after its disintegration after launch (Wierzbicki and Yue, 1986), and landing on liquid bodies on Titan. A review was published by Seddon and Moatamedi (2006).

Although impact with bodies of water is a process with which many of us are familiar in recreational settings, the methods for quantitative estimation of the mechanical loads generated upon impact of vehicles with free liquid surfaces are not obvious.

The first real progress in this field is usually attributed to Von Karman, in the context of estimating landing loads on seaplane floats (Von Karman, 1929). Essentially conservation of momentum is applied, but this momentum is shared between the impacting spacecraft and some 'added mass' of water, with the added mass prescribed as a function of the spacecraft's penetration distance. The usual approach is to set the added (or 'virtual') mass equal to that of a hemisphere of water with a diameter equal to that of the spacecraft at the undisturbed waterline. Assume a mass M_0 for the probe, at vertical impact velocity V_0. As it penetrates, it becomes loaded with a virtual mass M_v of liquid, with the probe/liquid ensemble moving at a velocity V.

Applying conservation of momentum and ignoring drag, weight and buoyancy

$$(M_0 + M_v)V = M_0V_0 \tag{7.2}$$

differentiating

$$(M_0 + M_v)\frac{dV}{dt} + V\frac{dM_v}{dt} = 0 \tag{7.3}$$

The virtual mass M_v is usually taken as a fraction k (~ 0.7) of the mass of a hemisphere of liquid (Figure 7.1) with a radius R equal to that of the (assumed axisymmetric) body at the plane of the undisturbed liquid surface. Thus for a liquid of density ρ, the virtual mass is

$$M_v = \frac{2k\pi\rho R^3}{3} \tag{7.4}$$

For a general axisymmetric shape $R = f(h)$, where h is the penetration distance, it is easy to show that

$$\frac{dM_v}{dh} = 2k\pi\rho R^2\frac{dR}{dh} \tag{7.5}$$

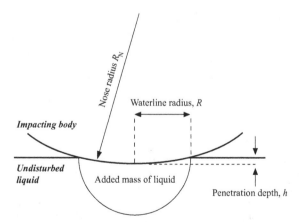

Figure 7.1. Geometry of splashdown dynamics showing the volume of liquid that forms the 'added mass' with which the impactor shares momentum.

noting that $dh/dt = V$ and $dV/dt = a$, hence

$$a = -\frac{V^2(2\pi k\rho R^2)\,dR}{(M_0 + M_v)\,dh}.$$ (7.6)

These equations are easy to solve numerically (indeed during the Mercury programme, the computation was performed by hand). Terms for drag, weight and buoyancy could be added, but do not significantly affect the peak loads.

For a spherically-bottomed vehicle with a radius of curvature R_N and a penetration distance h, this 'waterline' radius is given simply as

$$R = \sqrt{2R_N h - h^2}$$ (7.7)

and the equations can be solved analytically, to derive (for example) the peak loads.

The above method can be used to estimate the loads on a 75 kg human diving into a swimming pool. If the nose radius corresponds to the size of the head, the peak load is a little under 1 g: if, on the other hand, the nose radius is increased to,

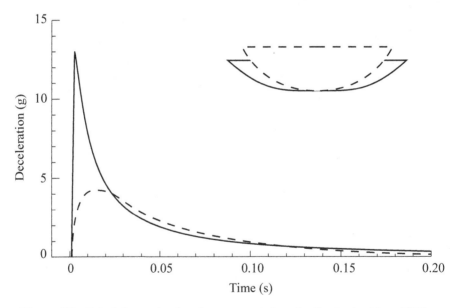

Figure 7.2. Splashdown deceleration computed as in the text with a 200 kg impactor at 5 m s^{-1} into an ethane ocean of density 600 kg m^{-3}. The solid curve shows the loads for a nose shape defined as $R = h^{0.3}$ (where $0 < h < 0.2$ is the 'depth' and dimensions are in metres) which is broadly representative of the shape of the Huygens probe descent module. The dashed line shows the corresponding calculation for a spherical nose of radius 0.65 m – the smaller nose radius leads to lower peak loads. The nose shapes are shown schematically in the inset.

say, 30 cm (i.e. a 'belly-flop') the loads increase to $\sim 6\,g$. This order-of-magnitude change in load is painfully apparent to those unfortunate enough to verify the nose-radius dependence experimentally.

This theory agrees remarkably well with measurements, as might be hoped for a model to estimate the loads to which heroic astronauts are to be subjected at the end of their flight. More sophisticated numerical methods can be used, but are probably unnecessary since the added mass approach is well-proven, reasonably accurate, and large margins are in any case prudently applied to landing loads. After this very short momentum-sharing stage of the impact, more conventional hydrodynamic drag and buoyancy come into play.

The case for impact on Titan is analytically identical, but with the density of liquid methane and ethane (600 kg m^{-3}) substituted for that of water. The loads calculated by this method (Figure 7.2) for the Huygens probe are a modest $10\,g$, in fact comparable with the loads during atmospheric entry.

It should be noted that post-splashdown dynamics, such as stability to capsize, may also be a consideration. A low centre of mass may therefore be important, or alternatively a shift of the centre of flotation (the point through which the buoyant force appears to act – the centroid of the displaced fluid) by the addition of inflatable flotation bags may be used.

The Mercury capsules in fact lowered the heat-shield surface beneath the capsule; this approach acting as an airbag to mitigate the loads applied to the crew. The bag would fill up after splashdown and act as an 'anchor', reducing the response to wave motions.

Broadly speaking, the practicalities of assembly of space probes are such that most are less dense than water. However, maintenance of long-term buoyancy requires that the gaps filled with air are not replaced by water – i.e. that the vehicle does not leak. The Mercury 4 capsule's hatch was released after splashdown (either by astronaut error or an uncommanded firing of the door pyro) and the vehicle sank.

8

Thermal control of landers and entry probes

For obvious reasons, most electronic and other equipment on Earth function optimally at around the same temperatures as humans. To achieve correct functioning in environments where the ambient temperatures are very different requires either development of equipment that can operate in these extreme conditions, or control of the equipment's temperature, usually by largely isolating it from the environment. This latter course of action is of course impossible for some elements, such as sensors designed to measure the environmental properties themselves.

On the surface of an atmosphereless body, the thermal environment is dictated, as in deep space, by radiative balance. The difference is that there are many radiating and shadowing surfaces around. Whereas in deep space only the Sun and a (spherical) Earth need be considered and calculations can be performed by hand, the evolving thermal environment on the surface of a rotating planet generally requires more elaborate computation using time-marching numerical methods.

If we consider our spacecraft as a sphere of radius r and heat capacity mc_p, then its rate of temperature change dT/dt will be given as

$$mc_p \frac{dT}{dt} = (1-a)\pi r^2 F + f_1 \pi r^2 \varepsilon \sigma T_p^4 + f_2(1-a)(1-A)\pi r^2 F - 4\pi r^2 \varepsilon \sigma T^4 + P_I$$

(8.1)

where F is the solar flux ($1340\,\mathrm{W\,m^{-2}}$ at Earth) and a the reflectivity of the spacecraft (see below). The term f_1 is a view factor describing how much of the sky is occupied by a nearby warm planet

$$\sim \left(\frac{R}{2(R+h)}\right)^2$$

(8.2)

where R is the radius of the planet and h the spacecraft altitude above it – this reduces to one half if the vehicle is on or near the surface, i.e. the planet occupies 2π

steradians of solid angle seen from the spacecraft. The term ε is the emissivity (see below) of the spacecraft surface, σ the Stefan–Boltzmann constant 5.67×10^{-8} W m^{-2} K^{-4} and f_2 is another view factor, describing the contribution of the reflected sunlight from the planet – if the spacecraft is over the equator at noon, or if the spacecraft is at very low altitude on the dayside, $f_2 = f_1$, but more generally $f_2 < f_1$, since the illuminated side of the planet may be obscured (i.e. the spacecraft sees a phase of the planet). The term A is the reflectivity or albedo of the planet and P_I denotes any internal dissipation in the spacecraft (e.g. due to the operation of its equipment). Setting $dT/dt = 0$ gives the asymptotic equilibrium temperature.

8.1 Surface coatings and radiation balance

In classical spacecraft design, the balance of radiation and internal dissipation is essentially the only issue, since there are no conductive or convective paths for heat transfer in free space. During a probe or lander's cruise through space to its target, this is also the case. Furthermore, in some environments such as the surface of the Moon or an asteroid (and even to a small extent, Mars), the radiative transfer pathways may be dominant and thus the surface radiative properties are crucial to maintaining acceptable temperatures.

The key properties for radiative balance are visible reflectance and thermal emissivity. For a given wavelength, the properties of emissivity and reflectivity are complementary, that is they sum to unity – any radiation that is not emitted or absorbed at the surface is reflected. However, usually the dominant illumination has a solar spectrum, peaking at around 0.5 μm, while the emission spectrum of a black body at typical spacecraft temperatures of 200 to 350 K has a peak at 10 μm. Thus, the 'thermal emissivity' and 'visible reflectivity' do not apply to the same wavelength and can therefore be considered approximately independent parameters.

These two parameters are usefully considered on a map (see Figure 8.1). Materials on the upper right part of the map (high emissivity, high reflectivity) will tend to be cold, since they efficiently reject heat in the infrared, while avoiding the absorption of sunlight. Similarly, materials near the origin will run hot.

In the thermal IR, the Earth's atmosphere has a brightness temperature of around 250 K, and this is the appropriate value for T_p, i.e. the temperature of the atmosphere at the altitude at which the thickening atmosphere becomes opaque, close to the tropopause (indeed the relatively uniform appearance of the planet at these wavelengths, independent of time of day, etc. is why Earth horizon sensors on satellites usually operate in the IR).

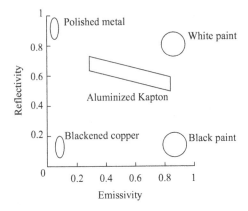

Figure 8.1. Map of radiative properties of surface coatings.

On or near the surface of a planet with a thick atmosphere, the incident flux F may have to be reduced due to absorption and scattering in the atmosphere. If the thermal optical depth of the atmosphere below is small, then T_p should be the surface temperature. If the atmosphere is thick enough for heat radiating down from the atmosphere above the probe to be significant (which might be considered another term similar to the f_1 term above) then it is likely that convective heat transfer will be even stronger.

Formally, the convective term has the form $+4\pi r^2 K(T_a - T)$ where K is a heat transfer coefficient between the atmosphere and the spacecraft, and T_a denotes the ambient temperature. The value of K will depend on the atmospheric density, and on windspeed and turbulence. This term is important on Mars, and can be difficult to predict (since the boundary layer on Mars is quite thick, so there is a strong windspeed and temperature gradient near the surface – thus the heat transfer will depend strongly on how much the vehicle projects above the surface). Coefficients of the order of $1\,\mathrm{W\,m^{-2}\,K^{-1}}$ are not untypical of afternoon conditions. However, in thick atmospheres this term is so large that the probe surface may be considered to be the same as the ambient atmosphere, and the thermal design must focus on isolating the interior of the probe from its surface.

8.2 Internal heat transfer

The temperature of the spacecraft surfaces may be driven by their radiative properties, and/or convection and conduction. To some extent the equipment inside the probe can be decoupled from these surfaces via insulators, subject to the structural needs to carry mechanical loads to the interior. Heat produced inside the spacecraft is due largely to dissipation in electrical systems. For cold environments, electric heaters may be added at key locations. Radioisotope

Heater Units (RHUs) are also used for providing heat, as was the case for Huygens; heat is supplied continuously, requiring no electrical power.

Russian designs tended to use pressurized equipment compartments (even in space) with fans to ensure adequate convective heat transfer.

Heat pipes are devices with variable heat conductance. In one direction thermal conduction through the walls and through the modest vapour in the pipe is quite small. However, if the 'wet' end of the pipe is the warmer one, the higher vapour pressure of the working fluid (typically ammonia or Freon) causes vapour, and thus latent heat, to flow to the cold end, resulting in much higher conductance.

In planetary surface applications, such as to maintain low temperatures around the supports of the Trans-Alaska oil pipeline (which are embedded in permafrost that could become unstable if allowed to thaw) the heat transfer liquid is trans-ferred to the 'hot end' by gravity. In microgravity, this will obviously not work, and a 'wick' is used to draw the liquid along the pipe by surface tension.

Deforming structures can also be used, either to modify effective thermal conductance by contacting surfaces, or by controlling surface exposure via louvres. Historically these were bimetallic strips – two metals with different coefficients of expansion. Today shape-memory alloys like Nitinol are sometimes used. Another approach (used on the Mars Exploration Rovers, e.g. Novak *et al.*, 2005) is a paraffin-wax actuator, used as a thermal switch to dump heat from the insulated interior when warm, and isolate it from the environment during the cold night.

8.3 Thermal environment during descent

The temperature of the air around a probe may change dramatically as it des-cends. For the deep atmospheres of the giant planets, and the lower atmospheres of Earth, Venus and Titan, the temperature rises almost linearly with depth.

In the deep atmospheres, the heat transfer by convection from the gas around the probe becomes prominent, and the infrared opacity of the gas prevents the probe from 'seeing' the cold of deep space. As a result, the probe will tend towards the ambient temperature.

In certain locations, on the mid-latitude surface of the Earth and on Venus at an altitude of around 50 km, this ambient temperature is close to the desired oper-ating temperature of electronic equipment. Elsewhere, however, the equipment must be held at a temperature offset from the ambient, and thus requires tem-perature gradients to be sustained within the probe itself.

For short-duration missions such as descent probes, the system may rely on the thermal transient. If it starts at a suitable temperature, and the exterior of the probe is suitably insulated, the probe will only warm or cool to unacceptable temperatures after the mission is over. This is the approach adopted in the short missions to date.

Passively sustaining a temperature gradient requires heat to flow across that gradient (like current through a resistor R with a potential held across it). In the case above, the temperature of the probe interior may be considered as the voltage on a capacitance, with the capacitance C being the analogue of heat capacity. The time constant with which this interior voltage approaches the externally applied voltage is simply RC. An acceptably long mission duration can sometimes be obtained by increasing either R or C or both – applying thicker or higher-performance insulation, or increasing the heat capacity of the interior, respectively.

As an example, the Pioneer Venus small probes had beryllium structural plates (e.g. Hennis and Varon, 1978; Lorenz *et al.*, 2005), to minimize mass and maximize heat capacity. Pressurizing the interior with xenon gas also reduced the gas conductivity. Note that the Pioneer Venus spacecraft were sealed entirely (which presented significant development challenges, particularly for the instrument seals) while the Galileo probe was unsealed, but individual equipment boxes inside were pressure-tight.

In practice there are limits on both of these design trends. While arbitrarily good insulation can be made at a cost, there are usually practical limitations to the utility of doing so – penetrations through the insulation are usually required to allow sensor access to the environment, for example. Increasing the mass of the interior is obviously ultimately limited by the mass capability of the launch and entry system; other limiting factors include the practicable density that can be achieved, such that a mass increase requires a volume increase too (and thus an increase in surface area and hence insulation mass).

Another approach that can delay the onset of unacceptable temperatures is the addition of thermal ballast exploiting phase changes (Russian papers usually refer to these as thermal 'accumulators'). The change of phase from solid to liquid or liquid to vapour is accompanied usually by the absorption of a large quantity of heat – ice being perhaps the most familiar example. A somewhat more convenient material than ice is lithium nitrate trihydrate. This material has a transition temperature of 303 K (29°C) and a latent heat of 296 kJ kg^{-1}. Soviet Venus landers (Figure 8.2) also incorporated 'sublimators' to reject heat. Similarly, the Apollo spacesuits rejected heat by evaporating water.

In the long term, the only way to prevent failure by overheating is to pump out the heat that leaks into the probe (or is generated within it – especially true of components with high local power dissipation such as transmitters). Since heat is being transported from cool to warm reservoirs, the expenditure of work is required. The generation of that work will itself require energy. On a small scale that might be achieved by stored energy inside the probe, but the tradeoff

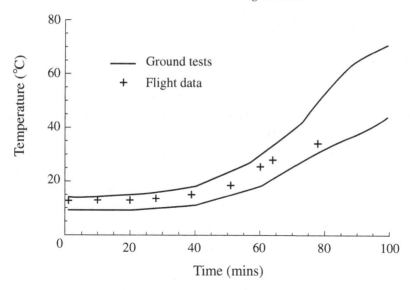

Figure 8.2. Temperature evolution of the gas inside the instrument container of the Venera 14 descent capsule. (Zelenov *et al.*, 1988b) Curves show the region determined during tests in ground-test chambers; crosses indicate the records from the descent on Venus.

between that approach and the application of phase-change ballast is unlikely to be favourable except in special instances, such as where only a small detector needs to be cooled.

To generate the work required for heat pumping would therefore require an external energy source. On the Venusian surface, with little available light (and ambient temperatures in any case too high for photovoltaics to function) this would require mechanical or thermoelectric conversion of heat flow from a heat source at a higher temperature than ambient. Thus a radioisotope power source outside the probe could be used to drive a heat pump to keep the interior cool. Ironically, the maintenance of long-term low temperatures requires parts of the system to be at exceptionally high temperatures.

Many of the principles discussed here are familiar to people who go on picnics or camping trips. Usually it is desired to keep food or beer at a suitably low temperature, and hence it is placed in a cooler. The thicker the insulation of the cooler, or the more cold beers placed with in it, the longer is the time before the beer gets too warm. A common enhancement is to deposit bags of ice in the cooler. Finally, for long durations, some kind of heat pump or refrigerator is required, driven by an external energy source.

One further approach worth mentioning parallels the design of low Earth orbit satellites which jump from high to low equilibrium temperatures as they go from day to night, but never encounter either extreme because of suitably long time constants.

Insulation requires some discussion. Much terrestrial insulation functions largely by suppressing convection. This effect is of little utility in very thin atmospheres (where the air cannot transport much heat anyway) or in very thick atmospheres, where the conductivity of the gas is large, allowing efficient heat transport even when convection does not occur. Venera landers used a porous silica material, machined into blocks around the outside and the inside of the pressure hull. Although the material itself was not airtight (to allow the material to breathe and prevent it from being crushed by the increasing pressure), the outside was coated to minimize forced convection. (Zelenov *et al.*, 1988a,b).

Solid plastics are an obvious approach, and were used on Pioneer Venus, although thermosoftening polymers like polyethylene and PTFE have very poor mechanical properties at temperatures above around 100 °C. Polymer foams may have rather better insulating properties, but these foams may have significant volatile contents, so outgassing may be an area of concern. The Huygens probe used a polyurethane foam Basotect (Figure 8.3).

Aerogel and fumed silica are very light foams with excellent insulation properties. These materials have so little solid material that they are often translucent, and to minimize the radiative transfer of heat through them may be doped with absorbing material (the Sojourner rover (Eisen *et al.*, 1998) used a doped aerogel as an insulator). Note that despite the name, aerogel is quite rigid and brittle.

Appropriate choice of structural materials can influence thermal performance significantly. The thermal conductivity of stainless steel is much lower than that of aluminium; titanium is similarly a poor conductor. Beryllium, while a very

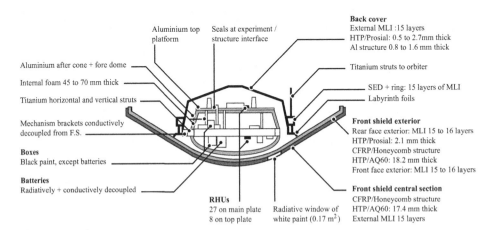

Figure 8.3. Thermal design features of the Huygens probe (from Jones and Giovagnoli, 1997).

difficult and expensive material with which to work (owing to the toxicity of beryllium dust), has excellent stiffness and heat capacity and was therefore used in the internal structure of the Pioneer Venus small probes.

One design for a Venus probe pressure vessel (38 cm diameter, able to hold a 15.6 kg payload) uses concentric spheres, the outer one of titanium alloy, the inner of stainless steel. The inner sphere distorts during entry deceleration, transferring loads via six short titanium struts that disengage (reducing heat leak) when the loads are removed. The space between the spheres is filled with a fibreglass felt and xenon gas. The effective thermal conductivity was $0.014 \, \mathrm{W \, m^{-1} \, K^{-1}}$ at $20\,°\mathrm{C}$ and $0.054 \, \mathrm{W \, m^{-1} \, K^{-1}}$ at $460\,°\mathrm{C}$, leading to a total heat leak to the payload of 84 W (Hall *et al.*, 1999).

8.4 Thermal testing

For testing at Mars ambient conditions, a chamber that can be pumped down to pressures of a few millibar (and that pressure controlled) is needed, probably also with a liquid-nitrogen-cooled jacket to represent the cold sky to which surfaces can radiate.

Special test chambers had to be constructed for the Pioneer Venus and Venera programmes – high pressure and temperature conditions are difficult and expensive to simulate, although the technology for static tests is quite straightforward, being used widely in the chemical industry. Rather more challenging is any simulation that must provide other environmental parameters, such as flow around the vehicle due to wind or descent, or illumination by the Sun.

8.5 Thermal modelling

Thermal mathematical models are a very important part of a mission. Such models are often used to develop and validate an initial design, and are refined as construction and testing proceeds. As new information becomes available, the effects on the spacecraft and its components can be evaluated with the model, and corrective design or procedure changes developed.

Temperature records are a major element of spacecraft housekeeping telemetry; electronic failure is often associated with overheating (as a cause or an effect) and the correct quantitative interpretation of the temperature evolution in terms of local power dissipation may require a thermal model (e.g. is battery 3 getting warm because it is on the Sun-facing side of the spacecraft, or has it shorted out as well?). Similarly, experiments to determine thermal properties of the planetary environment may require a detailed understanding of the heat flows inside the vehicle.

Commercial computer codes such as SINDA are generally used (since the overall problem, that of setting up nodes with specified heat transfer paths between them and appropriate boundary conditions, and propagating the corresponding differential equations forward in time, is constant; only the details of the nodes, links and conditions change from spacecraft to spacecraft). Each node is associated with a heat capacity, and the heat paths between them will depend on view factors and surface emissivities (for radiative transfer) and on the length, cross-section and conductivity (for conductive transfer).

Note that while spacecraft engineers are generally very familiar with the purely radiative and conductive heat transfer settings that occur in vacuum, the free (thermally driven) and forced (wind-driven) convective transfer that may occur between components or between components and the environment are less familiar and less easy to estimate, and can often only be determined with confidence by direct testing. Heat transfer coefficients of around $0.25\,\mathrm{W\,m^{-2}\,K^{-1}}$ were determined for the transfer between equipment boxes in the Huygens probe and its internal atmosphere at an altitude of 150 km (where the ambient pressure is around 3 mbar), while near the surface with a pressure of 1.5 bar, the heat transfer from equipment boxes is estimated at $3.5\,\mathrm{W\,m^{-2}\,K^{-1}}$ (Doenecke and Elsner, 1994). In contrast, in the thicker Venusian atmosphere, convective coupling were extremely strong, $150\text{--}1000\,\mathrm{W\,m^{-2}\,K^{-1}}$, such that the outer surfaces of the Pioneer Venus and Venera probes were essentially at ambient temperature. It is believed that the internal heat transfer of the Galileo probe was underestimated, leading to higher than planned temperatures and rates of change during descent; factors that required recalibration of the scientific instruments.

If we consider a spacecraft to be made of i plate-like elements of mass m_i, with a heat capacity c_i, then each block can radiate heat, have heat conducted into it, absorb heat, and generate heat internally. These heat transfer rates, P_R, P_C, P_A, P_I, alter the element's temperature history as;

$$m_i c_i \frac{\mathrm{d}T_i}{\mathrm{d}t} = P_R + P_C + P_A + P_I \qquad (8.3)$$

The amount of heat radiated from each element depends on its temperature T_i, cross-sectional area A_i, the Stefan–Boltzmann constant[9], σ, and the element's emissivity, ε_i as:

$$P_R = 2A_i \varepsilon \sigma T_i^4 \qquad (8.4)$$

Considerable effort goes into choosing materials and surface finishes to control the heat balance of a spacecraft by passively rejecting or absorbing heat where necessary. If the ith element is joined to another element, marked by index j, with

[9] $\sigma = 5.67 \times 10^{-8}\ \mathrm{Wm^{-2}\,k^{-4.}}$

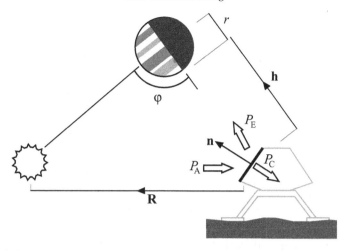

Figure 8.4. Geometry of the thermal model discussed above.

temperature T_j, and a thermal connection K, then the conduction term for the element i is

$$P_C = \sum_{j=i+1}^{j=n} K_{ij}(T_i - T_j). \tag{8.5}$$

The term describing power absorbed by the element is more complicated, as a spacecraft lander may be exposed to different radiant sources; the Sun, and hopefully at least one nearby planetary body. Other parts of the spacecraft which can be 'seen' by the element are neglected here for simplicity.

$$P_A = EA_i(1 - a_i)\frac{n \cdot R}{|R|} + A_i\frac{r^2}{(r + |h|)^2}\varepsilon_P\sigma T_P^4(1 - a_i)\frac{n \cdot h}{|h|}$$
$$+ A_i\frac{r^2}{2(r + |h|)^2}Efa_P(1 - a_i)\frac{n \cdot h}{|h|} \tag{8.6}$$

In Figure 8.4, $\mathbf{n}$ is the unit normal vector of the element, and $\mathbf{R}$ and $\mathbf{h}$ are the direction vectors to the Sun and a nearby planetary body, respectively. The first term in Equation 8.6 is the radiant heat absorbed from the Sun, giving an intensity, E, at the spacecraft's position. The second term shows the power absorbed from a nearby body of temperature T_P, with ε_P being the emissivity of that body and a_i the plate's albedo. The third term is the power absorbed via reflection from that object. Here, a_P is the albedo of the reflecting planet, and f is a view factor that accounts for the variable amount of lit planet that the spacecraft element can 'view'. Examples of how this view-factor varies with the spacecraft-to-planet separation can be found in Wertz and Larson (1999).

9

Power systems

A sobering thought experiment is to contemplate a world without electricity. Not only is electricity exploited as a convenient means of delivering mechanical or thermal energy to remote locations, but electricity is vital in information transmission and in sensing and control. Although the first planetary missions contemplated involved launching to the Moon a vehicle containing flash powder with which it would optically announce its arrival, and some early spacecraft used clockwork timers to sequence operations, every mission actually flown has been electrically powered.

In this chapter we first consider the overall requirements on the probe's power system, and how these requirements favour the various means adopted to meet them. The power supply and storage possibilities are then discussed, with particular reference to planetary probes. A general reference for power considerations is the book by Angrist (1982).

It is instructive to consider the electrical power requirements of various household devices to place spacecraft requirements in context. A modern PC may consume perhaps 200 W; a laptop perhaps an order of magnitude less. The Viking lander ran on 90 W. The Huygens probe's batteries supplied around 300 W for about 5 hours. The Sojourner rover had a solar array that delivered a mere 15 W.

9.1 System requirements

The total energy requirement of a mission (i.e. its integrated power requirement) is the most fundamental parameter for designing the power system. If that energy requirement is sufficiently low, the primary energy source can be practicably provided as chemically stored energy – it is practical to drive a wristwatch with a battery, but a washing machine or stove cannot be practically driven this way. The condition 'practicable' is determined usually by mass and volume constraints.

For higher energy requirements there are two solutions. One option is to extract energy from the environment – providing a power source rather than an energy source. In principle, the energy provision of such a power source is infinite (just run the source for longer to acquire more energy). The practical examples here are solar power (usually by photovoltaic arrays) and possibly wind power for Martian surface systems.

The other option is to use an energy source that has a higher energy density (watt-hours per kilogram) than chemical storage – radioisotope sources. Since reliability concerns or operations costs limit missions to a fraction of the half-life of the most common isotope, ^{238}Pu, radioisotope devices may be considered power sources rather than energy sources.

Most spacecraft in Earth orbit have higher energy requirements than can be practicably provided as stored chemical energy – usually only systems that perform their function within a few hours or days, such as launch vehicle stages and planetary entry probes, can be powered this way.

The choice of power source can be usefully demonstrated on a map (Figure 9.1).

Key design parameters for primary power are specific mass and (for solar power) specific area. Solar array performance for arrays of a few kW at 1 AU is typically $65–80\,\mathrm{W\,m^{-2}}$, with specific mass ranging from around $25\,\mathrm{kg\,kW^{-1}}$ (flexible) to $50\,\mathrm{kg\,kW^{-1}}$ (folded). To first order, performance is a factor of 3 poorer on the Martian surface. Higher power per unit area is possible using high-efficiency multi-junction solar cells, although at greater fiscal cost.

9.2 Power and energy budgets

Power and energy, like other resources in a design, are tracked and managed through budgets. In the simplest probes, this may simply be a case of adding up the power used by each subsystem and determining how long the probe can

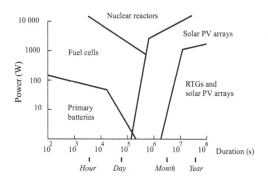

Figure 9.1. Regime diagram showing the typical applicability of various energy sources as a function of power and duration.

operate with a given energy supply. More generally, various margins and factors need to be tracked – degradation of batteries in flight, for example – which make resource tracking somewhat more involved.

Often the power requirements vary through a mission – for example if a relay satellite is used, energy-consuming downlinks may occur several times a day, while electrical power from solar arrays will only be available for some of these opportunities. To determine the required array and battery size needs a detailed study. Such operational considerations may also run into thermal constraints – e.g. under summer conditions the Mars Exploration Rovers had enough power to conduct additional operations during the day, but thermal dissipation would have caused overheating.

9.3 Radioisotope sources

Radioisotope thermoelectric generators (RTGs, Figure 9.2) are attractive power sources – reliable, compact and with only modest sensitivity to orientation and environmental conditions. They are, however, rather heavy and costly: much of the cost is due to the burdensome safety regulations associated with radioactive materials.

It is usually considered that a radioisotope power source quietly provides power at a well-determined rate that falls off exponentially with time due to the decay of radioactivity. In fact, as always, reality is much more complicated.

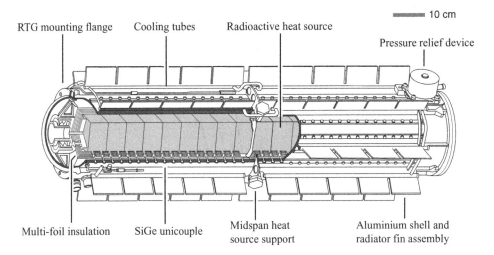

Figure 9.2. Radioisotope thermoelectric generator, as flown on Galileo, Ulysses, Cassini and New Horizons.

First, while the heat produced by the radioactive decay of a fixed amount of a pure element does indeed follow an exponential curve, it must be remembered that the daughter products of radioactive decay are themselves radioactive, and these daughter products have different activities and half-lives from the primary element (almost invariably ^{238}Pu in the form of PuO_2; this material has a half-life of 86 years and a specific power via alpha-decay of $410\,W\,kg^{-1}$).

Secondly, while the heat being produced continuously by the source follows a deterministic (close to but not quite a pure single exponential curve) decay with time, the conversion of that heat into electrical power is less deterministic. The conversion, usually by a set of thermoelectric converters (usually a 'thermopile' of semiconductor slabs) depends on both the intrinsic performance of the converters, and on the environment.

The intrinsic performance depends on the thermal design of the converter, i.e. how much heat flows through the thermopile, and how much just leaks conductively through the inert housing material. Notionally, one would want as little heat to be wasted as possible.

The performance also depends on the solid-state physics of the converter material. The latter's properties are uniquely related to the semiconductor

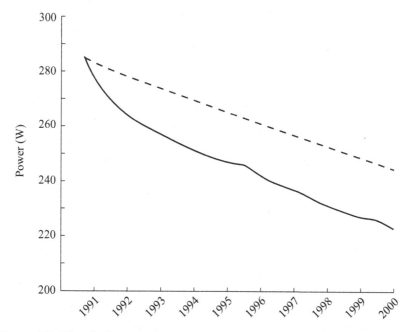

Figure 9.3. Electrical output power of the radioisotope thermoelectric generator on the Ulysses solar probe, versus time. The power produced (solid line) declines markedly with time, and more steeply than a simple exponential decay with a half-life of 86 years (dashed line) would suggest.

material, and may degrade with time through thermal or radiation effects (Figure 9.3). (As an aside, it may be noted that the thermoelectric coolers used on some modern microprocessors are essentially the same as the converters used on spacecraft – just used in reverse.)

The overriding concern with radioisotope power sources is safety, not just from the purely radiological standpoint, but also from the severe chemical toxicity of plutonium. This element appears to generate a level of concern far out of proportion to the probability of release, and the legal costs to a project of confronting objections by protesters can be formidable. Launch and disposal of RTGs in the USA requires a launch order signed by the President.

For outer Solar System exploration where solar fluxes are simply too low to be practicable, radioisotope power sources are mandated.

Although the specific thermal power of the source material is high, the electrical specific power is rather low – a little over $2 \, W \, kg^{-1}$ for the small 35 W RTG used on Viking. The larger designs with more efficient converters used on Galileo and Ulysses provided 285 W at about $5 \, W \, kg^{-1}$. This specific power is some 80 times lower than the fuel itself – the conversion efficiency is of the order of 5%, and only a fraction of the RTG is fuel – a large mass fraction must be devoted to the thermoelectric converters, the radiators for heat rejection, and especially to shielding to prevent release of material in the event of an accident. A new generation of RTG, the multimission RTG (MMRTG) is being developed (Schmidt *et al.*, 2005), nominally a 42 kg unit providing 126 W at beginning of mission under Mars surface conditions (128 W in deep space, where the radiators can operate more efficiently), i.e. $3 \, W \, kg^{-1}$.

Higher specific powers may be obtained through energy-conversion techniques more efficient than thermoelectric devices. Alkali metal thermionic emission technology (AMTEC) is one possible technique, although at a modest technology readiness level at present. Perhaps more promising in the long term is a Stirling engine, a reciprocating heat engine.

9.4 Solar power

Solar-array technology, driven largely by the demands of high-power direct broadcast communications satellites, has advanced considerably over recent decades. Previously, conversion efficiencies of only around 10% were typical for arrays made with crystalline silicon. Nowadays, efficiencies of upwards of 15% are more typical, with laboratory demonstrations of cells around 25%. These high efficiencies require multiple-layer cells with different conversion materials matched to different parts of the optical spectrum.

Among the environmental degradation mechanisms of solar arrays are radiation damage (principally to the charge-carrying characteristics of the semiconductors), UV damage (typically by opacification of coverglass adhesive), thermal effects (flexing leading to failure of cell interconnects) and mechanical damage by dust or debris impact. The latter aspect is probably of most relevance for planetary landers.

The design of solar power systems for landers has significantly different constraints from those for conventional satellites. First, environmental disturbances such as gravity mean the gossamer structures used on the large arrays (which often cannot bear their own weight), now common for satellites, cannot be used – there are therefore severe mass penalties for solar arrays with dimensions much larger than the body of the vehicle itself. Secondly, the orientation of the arrays with respect to the Sun is likely to be controlled by the

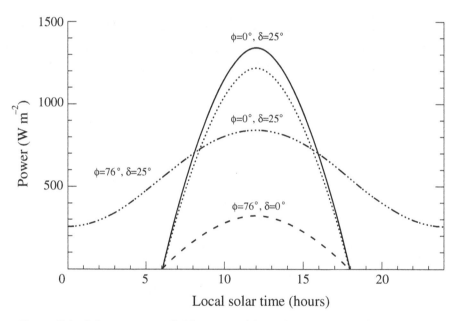

Figure 9.4. Solar power available on an airless planet at 1 AU from the Sun ($S = 1340 \, \mathrm{W \, m^{-2}}$) as a function of time, spacecraft latitude (ϕ) and solar declination (δ, i.e. latitude of the subsolar point). Solid line is for both the Sun and the lander at the equator giving a 12 hour day. The dotted line shows a lander on the equator during summer or winter, with the Sun 25° away from the equator. With a polar lander ($\phi = 76°$) the power can be zero all day during polar night ($\delta < -14°$); during midsummer (dash – triple-dot) there can be illumination all day, and indeed the daily averaged energy (the area under the curve) is highest for this situation. The curves represent the direct sunlight on a flat, horizontal plate, with no atmospheric absorption, scattering or surface reflection losses, all of which could be significant.

location of the vehicle on the rotating planetary body on which it sits (Figure 9.4). Depending on the planet's obliquity and the local season, the Sun may be perpetually above the horizon, never, or somewhere in between. In this latter usual case, the vehicle will undergo profound thermal cycling and require significant battery capacity to perform operations or even maintain keep-alive power and heating at night.

The performance of photovoltaic cells varies with the temperature of the cells, the level of illumination, incidence angle, shadowing and scattering, and radiation damage. Incidence angle effects include not only the obvious projected area effect, but also the increased Fresnel reflectivity of surfaces at low incidence.

Cells operate better at low temperatures, developing a slightly higher voltage. The voltage – current characteristic of a cell is shown in Figure 9.5. Usually a control circuit called a peak-power tracker is used to present the appropriate impedance to the solar array such that it operates at the knee of the curve.

Usual figures of merit for solar power performance are $W \, kg^{-1}$ and $W \, m^{-2}$. Note that a realistic design must use figures appropriate for an array – not just a

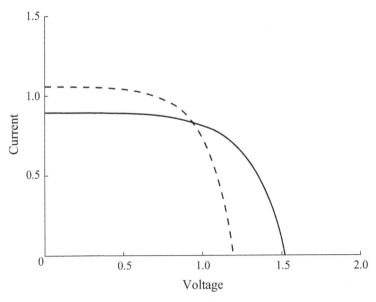

Figure 9.5. Output from a typical solar cell. Depending on the impedance presented by the spacecraft's power-conditioning circuits to the cell, it will operate somewhere along the solid curve for given conditions of illumination. The area of the rectangular box defined by the operating point and the origin determines the output power – this is maximized at the knee of the curve. The dashed curve shows the different operating characteristics of the cell at higher-than-normal temperatures.

cell. Part of a cell's area is obscured by electrodes, and some space between cells is needed for interconnects, diodes, hinges, etc. The active fraction of an array area might therefore be 90% or less. Care needs to be taken in designing a system for a planetary environment. For example, the spectrum of light reaching the cells through the dusty Martian atmosphere may be rather different from that of the unfiltered Sun, and thus the spectral responsivity of the cells can be important in predicting the degradation to dust loading (e.g. Landis *et al.*, 2004).

In specific cases, rather good performance can be obtained. For example, the Sojourner rover had a single array of area $0.22\,m^2$ providing $16.5\,W$ at Mars noon or $45\,W$ at 1 AU. The impressive performance of $204\,W\,m^{-2}$ is due in part to the use of very efficient GaAs on Ge cells, and in part due to the high active area fraction, since there were no hinges, etc.

9.5 Battery technology

9.5.1 Primary batteries

So-called primary batteries, where chemical energy is irreversibly converted into electrical power, typically have much higher energy densities than rechargeable or secondary batteries. For low-energy missions, such as planetary atmosphere probes, primary batteries are often the system of choice. They are convenient in that they impose few ancillary requirements such as attitude, and are robust.

The highest energy density cells commonly used at present are lithium thionyl chloride (Li $SOCl_2$) or lithium sulphur dioxide (Li SO_2). Lithium thionyl chloride cells were used on the DS-2 Mars Microprobes, and were qualified for $20\,000\,g$ impact; $LiSO_2$ cells were used on the Galileo and Huygens probes.

Performance of the cells depends on several factors. The most obvious is temperature, both in use (typically for only a few hours), and during the storage period before the mission itself (typically several years). Ideally the cells are stored at low temperatures, to minimize the self-discharge which can increase by as much as 20% for a $5\,K$ increase in temperature, while their capacity in operation is improved by moderate temperatures, perhaps by 10% for a $10\,K$ increase in temperature. Usually the only way to estimate the in-flight performance confidently is to store cells from the same batch as the flight cells at cruise temperatures and measure the capacity.

Primary cells often build up an oxide or similar coating on their electrodes. This coating usefully protects the electrodes during a long storage or cruise period, but retards ion migration and thus limits the in-use battery performance. It is usual, therefore, to burn off this coating, or 'depassivate' the batteries, by applying a brief short circuit to the battery, drawing a large current shortly before the prime mission itself.

9.5.2 Secondary batteries

Among secondary batteries, nickel–cadmium have been the most common for Earth satellites; more recently nickel–hydrogen cells have become popular owing to their rather higher specific energy. Silver–zinc batteries offer rather higher specific power than the other types, at the expense of a lower cycle life. Silver–zinc batteries were therefore used on Mars Pathfinder, since its technology demonstration objectives were expected to be fulfilled within a few weeks (as, indeed they were, although contact became unreliable after the battery failed).

Careful control of the charge state of batteries is crucial to prolonging their life, particularly in the case of NiCd cells. In general, life is maximized by shallow charge cycles (i.e. a low depth-of-discharge (DoD) – not using all the capacity of the cells) although this has the effect of slowly degrading the cell capacity (the 'memory effect'). Occasional battery reconditioning – a very deep discharge, which carries some risk of cell damage – can recover most of this lost capacity. Battery specific energy can vary by a factor of 2, so the values in Table 9.1 should be considered representative.

These figures refer essentially to the cells themselves. Batteries will have casings (usually rather substantial ones, since they are massive components which therefore need secure attachment), cabling, diodes, etc. As an example, the $LiSO_2$ batteries used on Huygens have a rated capacity of 2400 W h, and a mass of 24 kg – a specific energy half that quoted above; similarly the Sojourner battery had a capacity of 150 W h and a mass of 1.24 kg.

Table 9.1. *Battery parameters*

Battery type	Chemical formula	Specific energy W hr kg^{-1}	Cycle life 75%–25% DoD
Silver–zinc	AgZn	100	75–2000
Nickel–cadmium	NiCd	30	800–30 000
Nickel–hydrogen	NiH	60	4000–30 000
Lithium–sulphur dioxide	$LiSO_2$	200	
Lithium–thionyl chloride	$LiSOCl_2$	200	

9.5.3 Fuel cells

Fuel cells can be considered a subset of primary batteries. The distinction is that the electrolytes or reactants are stored separately: for large energy requirements the packaging of reactants separately from the reaction vessel is more efficient and therefore results in higher energy densities. The most usual types use hydrogen and oxygen, although methanol/air technologies are under development for terrestrial applications such as mobile communications. The fuel cells used on the Space Shuttle generate 12 kW with a very respectable specific power of 275 W kg^{-1}.

The modest power requirements of planetary probes, coupled with the relatively high cost and complexity (particularly with regard to hydrogen storage) means they are rarely used on planetary probes – the only relevant case being the Apollo lander.

As an aside, the electrochemical conversion technology of fuel cells can be applied differently, using electricity (from solar panels for example) to convert CO_2 into oxygen. Similarly, a zirconia cell, like that used in fuel cells, was used on the MPL TEGA instrument to sense the presence of oxygen in gases evolved from a soil sample.

9.6 Other power sources

As on Earth, it may be possible to derive power from ambient sources other than sunlight. Wind power has been proposed for the Martian surface environment. Development models of a wind-powered rover for Venus were also built and tested in the Soviet Union during the 1980s (the KhM-VD and KhM-VD2, from VNIITransMash). The available power of the windstream relates to the air density and the cube of the windspeed (thus 8 times more power is available if the windspeed doubles). Exploitation of the mechanical power of wind for locomotion (via balloon, tumbleweed rover, etc.) appears more likely in the near term than for electrical-power generation. It may also be possible to exploit temperature changes (either diurnal changes on a lander, or the temperature change with depth in a deep atmosphere for a descent probe or balloon) to derive usable energy.

9.7 Power and thermal control

One of the largest consumers of electrical power, both on spacecraft and in households, is heating. While electrical heating is convenient and controllable, electrical energy is a more expensive asset than thermal energy, thus wherever a purely thermal-energy storage or source system can be used (either radioisotope heater units, or phase-change materials) instead of electrical energy for thermal control, it may be advantageous to do so. Similarly, if thermal control is a significant driver, e.g. during Martian night, it may make sense to oversize solar arrays and dissipate the 'extra' power as heat in the lander.

9.8 Nuts and bolts

We have discussed in this section the system-level issues associated with providing energy to a planetary probe. There are of course many subsystems required to make everything actually work; subsystems that may require significant design and development effort. Details of these, which would include power switches, current limiters, peak-power trackers for solar arrays, battery charge regulators and so on, are beyond the scope of this book, however. Requirements are often driven by non-nominal cases – for example a regulator that works perfectly well for years of a main mission may be inadequate to provide enough current for a fraction of a second to fire pyrotechnic devices at the start of the mission (pyrotechnic devices are often therefore wired directly to the unregulated battery terminals).

The mass of a spacecraft harness is often forgotten, but can be quite significant – e.g. the harness on the 200 kg Huygens probe itself weighed some 12 kg – a not insubstantial amount.

10

Communication and tracking of entry probes

Telecommunication is one of the most important functions of entry probes: it transmits to Earth all the science and engineering data that are the main goal of the mission. Tracking of the probes is another function that can help to analyse the probe dynamics during the entry and descent, providing independent science data on the atmosphere of a planet.

During entry, if communications are to be attempted at all, only status tones or very low data rates are possible. This is because the attitude during entry and descent may be very dynamic, preventing pointing of high-gain antennas. Depending on the wavelength of the communication link and the aero-thermochemistry of the plasma sheath, transmissions may be completely blocked for a short period (the entry 'blackout').

During the highly dynamic entry phase data rates in direct-to-Earth (DTE) links are usually very small due to the great distance to the Earth and the use of low-gain antennas on probes. A relay link (Figure 10.1) uses a much shorter distance to the relay orbiter to boost the received signal strength though using a less efficient receiving antenna than on Earth. The probe data received on the orbiter is re-transmitted to the Earth using the high-gain antenna of the orbiter.

Motion of the probe affects the frequency, amplitude and phase of the signal at the receiving station. The entry process includes phases that are significantly different from the point of view of the communications link. Pre-entry is essentially the last phase of the cruise; the probe, though usually spinning, moves under gravitational forces and its trajectory is highly predictable.

As soon as the probe enters the atmosphere, aerodynamic forces affect its motion, resulting in large deceleration and a corresponding change of the Doppler frequency. The trajectory uncertainty greatly increases due to insufficient knowledge of the planetary atmosphere, errors in entry-point location,

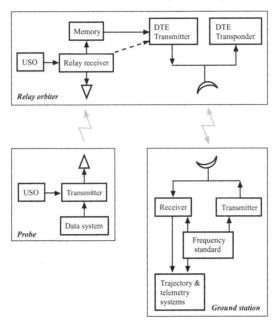

Figure 10.1. Relay communication system.

and aerodynamic motion of the probe. At some point plasma forming around the probe may block the signal and the reception might be interrupted.

After re-emerging from plasma blockage the probe will continue to decelerate, experiencing often rapid and not very predictable trajectory variations resulting from wobbling, wind and turbulence conditions, active propulsion and, finally, landing on the surface. At this, the most critical phase that will often determine the mission success, the reception of the probe signal is highly important for data analysis. After the landing-probe motion ceases or becomes very slow, the signal behaviour generally becomes smooth and predictable again. In some cases, however, interference of the direct signal to the Earth or relay spacecraft with a reflected ray (i.e. multipath) can cause sharp nulls in the pattern of radiation detected in the far field. As planetary rotation, or the motion of a relay spacecraft, causes the receiver to fly through this pattern, sharp drops in signal strength can occur; such drops were observed for example on the Huygens probe.

Antenna pointing for the DTE link becomes the main issue especially for rovers. For the lighter-than-air atmospheric probes (balloons and airships) the motion after deployment is usually smooth and does not restrict link performance. By contrast, communication with fast-manoeuvring airplanes may encounter significant challenges.

Table 10.1. *Definitions of radio-frequency bands*

Frequency Band	(GHz)	Notes
P	0.25 to 0.5	Previous, since early World War 2 British radar used this band but later switched to higher frequencies. These frequencies, and sometimes up to 3 GHz are commonly termed UHF (Ultra High Frequency) – e.g. Viking Lander (381 MHz) and MER relay
L	0.5 to 2	Long wave. Used in spaceborne radar
S	2 to 4	Short wave. Common for interplanetary missions, especially low-gain e.g. Pioneer Venus (2.3 GHz), Huygens, Galileo probes
C	4 to 8	Compromise between S and X. Used mainly for terrestrial satellite communications and radars – not common on planetary missions
X	8 to 12	Used in World War 2 for fire control, X for cross (as in crosshair). Common as the main up/downlink for interplanetary missions e.g. Voyager (8.4 GHz). Now becoming progressively supplanted by Ka band
Ku	12 to 18	Kurz-under (i.e. lower in frequency than the main water absorption in the Earth's atmosphere). Used for e.g. Huygens radar altimeter, Cassini Radar.
Ka	26 to 40	Kurz-above (higher in frequency than the main water absorption). Becoming more widely used due to the high gains achievable with small aperture. Cassini downlink

10.1 Entry probes: communication basics

The frequencies used for space communication lie in bands coordinated by the ITU (the International Telecommunication Union, a branch of the United Nations). The designation of the wavebands generally derives from radar development in World War 2 in the UK and Germany (Table 10.1).

Frequencies below about 100 MHz cannot be used for space communication as the Earth's ionosphere absorbs or reflects the radiation. Caution must similarly be exercised in choosing frequencies for planetary missions. For example, although Mars' ionosphere has a lower density and thus allows lower frequencies through, frequencies of 50 MHz and lower may be unusable depending on solar activity and time of day. At Jupiter, the synchrotron radiation from its Van Allen belts provides a significant background radiation, while ammonia contributes absorption in the atmosphere at specific wavelengths.

The efficiency of a data transmission link is defined by the RF energy required to transmit one bit of information at a certain probability of error. On a probe, it is controlled by the RF power of the transmitter, antenna gain in the direction of the

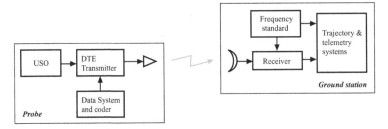

Figure 10.2. One-way DTE communication system.

receiving station (either on Earth or on the orbiter relay) and the telemetry coding scheme.

The accuracy of trajectory measurements, primarily Doppler velocity and VLBI (Very-Long Baseline Interferometry) position measurements, is controlled by the frequency stability of the signal radiated from the probe. To the first order the Doppler shift measurements f_D determine the line-of-site velocity V_D as

$$V_D = \frac{c f_D}{f_0} = \frac{c(f_r - f_0)}{f_0} \qquad (10.1)$$

where c is the velocity of light, f_0 the frequency of the signal radiated from the probe and f_r the frequency of the signal received on the other end of the link (Earth or relay orbiter). Unknown bias, drift or short-term instability of the probe's signal frequency δf will result in a corresponding error in the velocity $\delta V_D = c \delta f / f_0$. For example, a relative frequency instability of 10^{-9} will produce a velocity error $0.3\,\mathrm{m\,s^{-1}}$.

Any probe communication system consists of a set of basic elements: a transmitter that includes an exciter and power amplifier, an oscillator, a receiver, a coder, an antenna. The one-way DTE link consists of an oscillator, a transmitter and an antenna (Figure 10.2). An ultra-stable oscillator (USO) generates the reference signal for the transmitter, data from the probe transmits directly to Earth. The accuracy of Doppler velocity measurements is governed by the bias and stability of the USO (Table 10.2).

Accurate velocity measurements require the use of USOs operating during high deceleration loads (up to 500 g for Venus entry), and rapid change of the temperature inside the probes. Earlier Venera probes had oscillators with temperature instability up to 10^{-6}, short-term instability $\sim 10^{-10}$, and required extensive test calibrations to extract the Doppler data, while the more advanced rubidium USO on Huygens had a stability of $\sim 10^{-12}$ (averaged over 100 s) (Bird *et al.*, 1997, 2002).

The main parameters of a transmitter are its output RF power and its efficiency, defined as the ratio of the RF power to the DC power consumption of the transmitter.

Table 10.2. *Radial velocity error due to oscillator-frequency instability*

Relative frequency instability $\delta f/f_0$	10^{-6}	10^{-7}	10^{-8}	10^{-9}	10^{-10}	10^{-11}
Radial velocity error, m s^{-1}	300	30	3	0.3	0.03	0.003

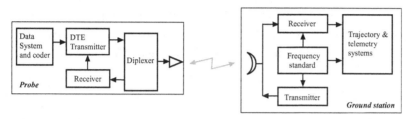

Figure 10.3. Two-way DTE communication system.

Addition of a command receiver will build the simplest two-way link that will allow control of the probe by commands from Earth and the upload of data for software modification. Improvements in trajectory measurements would require the use of a more complicated and expensive transponder (Figure 10.3). A transponder is the receiver–exciter that locks its frequency to the frequency of the received signal and multiplies it to form the transmitter frequency. Noise temperature and capabilities of signal tracking (bandwidth of acquisition and lock of signal, tracked rate of change of signal frequency) are the main parameters of the transponder.

The frequency stability in the two-way link is determined mostly by the stability of the reference oscillator on the ground station and propagation effects. A ground-reference oscillator can be much more stable than the probe oscillator. The transponder may also be used for range measurements that can improve knowledge of a planet's ephemeredes and rotation state (i.e. using the lander as a tracking beacon on the planet), as well as the location of the probe.

The antenna is the element that defines the spatial distribution of transmitted or received power; it can be the bulkiest element of the probe telecom system. Very often the antenna defines the capability of the link.

The beamwidth of the antenna should enclose the solid angle that encompasses the direction to the receiving station in the probe-fixed co-ordinate system. During entry and descent the attitude of the probe with respect to the receiving station is subject to rapid and large variations in all directions. Omnidirectional or low-gain antennas are used commonly for communication at these phases.

After landing, the direction to the receiving station in the probe-fixed co-ordinate system is still *a priori* unknown but it changes slowly (at least in the case of landers or LTA probes). With appropriate pointing it enables the use of high-gain antennas that provide significantly higher transmission data rates.

Antenna size controls other parameters. The beamwidth of the directed antenna depends on the antenna size and wavelength. For high-gain parabolic or flat array antennas the beamwidth measured at the −3 dB level can be estimated as

$$\Theta(\text{deg.}) \approx \frac{70\lambda}{D} \qquad (10.2)$$

where λ is the wavelength and D the diameter of the antenna.

Antenna gain is the ratio of power transmitted by the antenna in a given direction to the power transmitted by an ideal isotropic antenna. The maximum gain G is

$$G = \frac{4\pi S}{\lambda^2} \qquad (10.3)$$

where $S = kA$, the effective area of the antenna, A is the geometric area and k is a geometric efficiency factor, ~0.5–0.6. The expression for beamwidth becomes

$$\Theta(\text{deg.}) \approx 70\pi\sqrt{\frac{k}{G}} \approx \frac{170}{\sqrt{G}} \qquad (10.4)$$

NB: G is a dimensionless number, not in dB. Examples for high-gain antennas are given in Table 10.3.

Low-gain antennas and antennas for longer wavelengths are more often characterized by their gain. Antenna patterns for many omnidirectional antennas are well known. For low-gain antennas radiating primarily along their axis the previous formula can be used to estimate beamwidth. For short wavelengths and high directivity, parabolic ('dish') antennas are typically used on spacecraft, with the dish usually fed with a horn antenna at a Cassegrain focus (much like an optical telescope), or sometimes for structural reasons in an offset position. The Viking lander had an S-band dish. In general, landers and especially probes and rovers, because of their variable attitude, cannot point high-gain narrow-beam antennas, and so low- or medium-gain antennas are used, and packaging constraints make dishes unattractive even when high gain is required. Flat plate

Table 10.3. *Beamwidth and gain for high-gain antennas in the X-band ($\lambda = 3.6$ cm) and in the Ka-band ($\lambda = 0.9$ cm)*

Antenna diameter (m)	0.2	0.4	0.7	1.0	2.0	0.4	1.0
Wavelength (cm)	3.6	3.6	3.6	3.6	3.6	0.9	0.9
Beamwidth (°)	13	6.5	3.6	2.6	1.3	1.6	0.63
Gain (dB)	22	28	33	36	42	40	48

phased arrays are common – for example the X-band flat-plate array on Pathfinder had an on-axis gain of some 24 dBi.

Low-gain antennas include canted turnstile (crossed dipole), helical, patch and dipole antennas, with various gain, polarization and structural characteristics. Helical antennas are suited for circularly polarized signals (the Huygens probe used quadrifilar helix antennas for its two telemetry links, separated both in frequency – 2040 and 2097 MHz – and in polarization).

The antennas used on the Pioneer Venus probes operating at 2.3 GHz used a crossed dipole fed with a quadrature hybrid. The antenna elements were made of steel, and were covered in a PTFE radome (electrically transparent) to prevent heat and corrosion damage during entry and descent. The antenna gain was about 2 dB out to 60° off-axis, falling to 0 dB at the horizontal (Hanson, 1978). All Venera/VeGa landers and VeGa balloons used helical antennas.

The effective area of an antenna expressed via gain is

$$S = \frac{G\lambda^2}{4\pi} \tag{10.5}$$

It is important to note that with a given gain the effective area is proportional to the square of the wavelength. The lower frequency (and longer wavelength) low-gain antennas have greater effective area than high-frequency (and shorter wavelength) antennas with the same gain and beamwidth. The effective area of a low-gain antenna for several values of gain is shown in Figure 10.4.

The gain of the transmitting antenna is the primary parameter for transmission of signals. For reception, effective area is paramount. Since gain is inversely

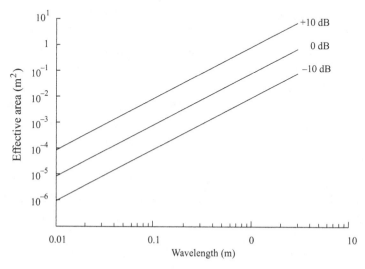

Figure 10.4. Effective area of low-gain antennas with gains −10, 0 and 10 dB.

proportional to the square of the wavelength for a given antenna size, a shorter wavelength would yield greater gain and thus a higher data rate for the same size of receiving antenna.

10.2 Main telecom equation

The power of the signal from the probe at the output of the receiving antenna P_r is determined by

$$P_r = \frac{P_t G_t S_r}{4\pi R^2} \eta_1 \tag{10.6}$$

where P_t is the RF power of the probe transmitter, G_t the gain of the transmitting antenna, S_r the effective area of the receiving antenna, R the distance from probe to receiving station and η_1 the losses. Two other forms of this equation are

$$P_r = \frac{P_t G_t G_r \lambda^2}{(4\pi R)^2} \eta_1 = \frac{P_t S_t S_r}{\lambda^2 R^2} \eta_1 \tag{10.7}$$

Although these equations are just different forms they show some features important for the probe link design.

First, given the transmitting-antenna gain and the area of the receiving antenna, the signal power does not depend on wavelength, i.e. on the frequency band. That is why the signal power in the DTE or relay link with omnidirectional or low-gain antennas on the probe and a high-gain antenna on the Earth station or orbiter does not depend on the frequency band.

Second, given the gains of the transmitting and receiving antennas, the received signal power is proportional to the square of the wavelength. That is why in the relay link with omnidirectional or low-gain antennas on both ends (probe and orbiter), UHF and even VHF bands have been used on all entry probes.

Third, given the areas of the transmitting and receiving antennas, the received signal power is inversely proportional to the square of the wavelength. That is why in the DTE link with high-gain antennas on both ends (probe and Earth station) shorter wavelengths from L-band to X-band have been used on all entry probes.

Practical application of these statements depends also on other factors: availability and efficiency of the on-board transmitters, the frequency bands of the ground stations, the wavelength dependence of absorption in the planetary atmosphere, etc.

A number of factors contribute to the signal losses η. We will list some of them; detailed description would be beyond the scope of this book.

Atmospheric losses include atmospheric absorption and atmospheric refraction losses in the atmospheres of Earth and of the destination planet. Refraction losses do not depend on wavelength and usually are a fraction of a dB for local elevation angles of the line-of-sight (both for the Earth station and for the probe on the planet) greater than 10–20°. Absorption losses could be wavelength dependent and usually increase with the link frequency. On Earth the most important source of absorption is precipitation. In the deep atmospheres of Venus and Jupiter pressure-induced absorption of carbon dioxide (Venus) and absorption of ammonia (Jupiter) are the greatest contributors at short wavelengths. For example, absorption of the X-band signal from a probe on the Venus surface is ~8 dB.

Antenna pointing losses, being a fraction of a dB for the Earth-based antennas, could be significant for the probe and relay orbiter antennas. Use of omnidirectional antennas is a cure that sacrifices the overall link budget.

Polarization losses are of order 0.2–0.3 dB for antennas with matched polarizations (linear or circular with appropriate orientation and rotation). They might increase to 3 dB if the received signal has linear polarization while the receiving antenna has circular, or vice versa (it usually happens at the edges of the antenna pattern of low-gain antennas). In unmatched circular polarization (one left-handed – another right-handed) the polarization losses may exceed 10 dB.

Hardware losses on the probe, in cables, diplexer, filters, etc., are usually of the order of fractions of a dB. The receiving system contributes to other losses that will be described later.

Actually the main parameter of the communication link is the ratio of the received signal power to the spectral density of noise at the output of the linear part of the receiver. Usually the output noise has a flat spectrum in signal bandwidth ('white' noise). The spectral density of the noise P_{N0} can be calculated as

$$P_{N0} = kT_{eff} = 290\,k\,(N - 1) \tag{10.8}$$

where k is the Boltzmann constant (1.38×10^{-23} W K^{-1} Hz^{-1}), T_{eff} the effective temperature of the system and N the noise-factor. The noise power is

$$P_N = P_{N0}\Delta F = kT_{eff}\Delta F \tag{10.9}$$

where ΔF is the appropriate bandwidth.

The effective temperature includes several components: noise radiation received from the ionospheres, tropospheres and surfaces of the Earth and the planet; the radiation of the galaxy; the Sun, and the noise of the system components – receiver (mostly noise from the input low-noise amplifier), waveguides, cables, etc.

Radiation received from atmospheres, the galaxy and the surface of the Earth depends strongly on the wavelength, the direction of the antenna pointing (both elevation and azimuth), local weather and other factors. Galactic noise increases with wavelength while noise from the troposphere decreases. This combined noise has a minimum in the S-band – one of the reasons why the S-band was selected for deep-space communications in the early stages.

A planet's radiation does not contribute significantly to the system noise if the receiving antenna beamwidth is much larger than the angular size of the planet. In the opposite case, if the receiving antenna beamwidth is less than the angular size of the planet (which could be the case for orbiter relay antennas or Earth antennas in the Ka-band) the radiation of the planet may become a major contributor to the system temperature, which may reach 400–600 K for Venus in 1–8 GHz and 10 000 K for Jupiter in the UHF. In general, if the solid angle of a planet as seen from the receiving station is θ_p and the solid angle of the antenna beam is θ_a, the contribution of the planet's radiation to the system noise temperature is

$$\delta T_{\text{eff p}} = \frac{\theta_p}{\theta_a} T_p \qquad (10.10)$$

where T_p is the equivalent temperature of the planet's radiation.

The ratio of signal power to nose spectral density has dimensions of energy. Finally,

$$\frac{P_S}{P_{N0}} = \frac{P_t G_t S_r}{4\pi R^2 k T_{\text{eff}}} \eta_1 \eta_2 \qquad (10.11)$$

or

$$\frac{P_S}{P_{N0}} = \frac{P_t G_t G_r}{(4\pi\lambda R)^2 T_{\text{eff}}} \eta_1 \eta_2 \qquad (10.12)$$

where $\eta_2 =$ additional losses in the receiving system which include losses in high-frequency components (waveguides, cables, filters, diplexers) and signal processing losses (carrier, subcarrier and symbol synchronization losses, etc.). In modern ground stations these losses are usually of order 1–3 dB.

10.3 Frequency measurements

Of the three signal parameters amplitude, frequency and phase, frequency measurements are most widely used for the entry probes, since they provide direct data on probe velocity. Typically, in the receiver, the signal bandwidth is shifted down to an intermediate frequency (IF) with a heterodyne signal formed by mixing the received signal with a reference oscillator.

In the closed-loop system the signal is then acquired and filtered with a phase-locked loop (PLL), and the frequency of the voltage-controlled oscillator (VCO) of the PLL is the resultant parameter that yields the probe Doppler velocity.

As a result of the highly dynamic and often poorly predictable behaviour of a probe, and consequently the signal, during the entry phase, the real-time PLL might not acquire the signal or may lose tracking. An open-loop system will improve reliability and flexibility in signal detection, filtering and frequency measurements, as well as data acquisition. In the open-loop system the IF signal is usually digitized with the sampling frequency exceeding $2\Delta F$ in the one-channel scheme, or exceeding ΔF in the two-channel conversion. The digitized signal is recorded for further digital filtering and processing resulting in frequency measurements. Multiple runs with different moving filters allow frequency measurements to be made in extreme situations. Critical data from two Soviet probes – Venera 7 after landing (Avduevsky *et al.*, 1971) and Mars 6 during the entry phase (Kerzhanovich, 1977), as well as radio science data on many US probes, have been retrieved with an open-loop system.

The root mean square (RMS) error of frequency measurements in a one-way system caused by system noise σ_f can be estimated as

$$\sigma_f = \frac{1}{\pi T}\sqrt{\frac{\rho_L}{2}} = \frac{1}{\pi T}\sqrt{\frac{N_0}{2P_C B_L}} \qquad (10.13)$$

where $\rho_L = P_c/P_{N0}$, B_L is the equivalent bandwidth of the PLL or digital filter, T the integration time, and index C means carrier. This equation can also be used to estimate the frequency error in a two-way link provided that the SNR in the up-link is significantly greater than the SNR in the down-link. The corresponding velocity error can be estimated using Equation 10.1. The same equations are applicable to relay links.

10.4 Data transmission

A generic diagram of probe data transmission is shown in Figure 10.5. Data collected on the probe can be coded to improve link performance. Coded data either modulate a subcarrier, which in turn modulates the carrier frequency, or modulate the carrier directly. On the down-end of the link, the acquired signal is demodulated and decoded. In cases where an orbiter or flyby spacecraft relays the data, it is received and decoded onboard, and then retransmitted to ground stations on Earth.

Of the different types of modulation (frequency modulation, amplitude modulation, etc.), BPSK (binary phase-shift keying) is the most widely used for entry probes. This modulation can be implemented with or without a subcarrier, and with carrier or without carrier. The highly dynamic behaviour of the signal

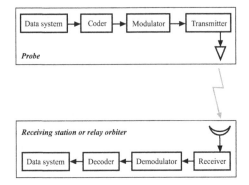

Figure 10.5. Diagram of data transmission.

frequency during entry and descent, and a small link margin, especially in DTE links, often makes a subcarrier with residual carrier modulation preferable; it is much easier, for example, to perform Doppler tracking on the unsuppressed carrier signal than on the modulated subcarrier with suppressed carrier.

The bit error rate (BER) is the main parameter characterizing the performance of a communication link. The ratio of the energy per bit of raw data to the noise spectral density E_b/N_0 at a given BER characterizes the efficiency of coding and modulation. Essentially what coding does (as with error detection and correction codes for memories exposed to radiation) is to add redundancy to the data by adding symbols. Although this increases the total number of bits to be transmitted, the error-correction ability more than compensates, such that the probability of uncorrectable errors is reduced overall (or the transmission rate for a given power can be increased; sometimes the effect is expressed as a 'coding gain' – the boost in power that would give the same improvement in data rate for a given BER). The penalty is in the additional hardware and/or software required for both the transmitter and receiver – dedicated hardware has historically been used, although software-based coders and decoders are becoming more common as the algorithms become more complex.

Uncoded PCM (pulse code modulation) has a threshold $BER = 10^{-5}$ at $E_b/N_0 = 9.6$. Various codes can improve this. The convolutional code with code length 7 and bit rate of half the symbol rate ($(7,\frac{1}{2})$ code) is one of the standards – in other words, the system transmits two coded bits for each data bit, with the bits determined by an algorithm with a 'memory' of 7 bits. At a $BER = 10^{-5}$ it improves link performance by 5.1 dB, i.e. equivalent to a three times increase of the transmitter power or antenna area, although 3 dB of this are spent in the increased bandwidth needed to transmit the two symbols per bit. One common algorithm used to recover data from convolutionally coded data is the Viterbi decoder – its advantage is that it has a fixed decoding time, lending itself to hardware implementation.

Reed–Solomon codes are also common and are particularly robust with respect to short-burst errors – a common code is (255,223) where 32 bytes of parity symbols are added to 223 bytes of data. This allows the correction of up to 16 error bytes. A common combination is first to apply a Reed–Solomon code and then a convolutional coding.

'Turbo codes' (another combination) can give an additional gain of 4 dB but are computationally demanding and require large frames of data (2000–8000 bits) for coding and decoding, which can be risky for the entry probes since possible losses of signal can affect a significant amount of data.

10.5 Link budget

The link-budget calculation is based on Equation 10.2 in logarithmic form, where all values are expressed in appropriate decibel units: e.g. value 23 dB (over 1 K) corresponds to a system temperature of 200 K. Examples of link budget calculations for DTE and relay links for a Venus entry probe are shown in Table 10.4.

Assumptions for the budget calculations are: space losses are defined as $(\lambda/4\pi D)^2$; no pointing losses for omnidirectional antenna of the probe; Venus' brightness temperature of 600 K in S-band and 500 K in the UHF; residual carrier transmission with 60° phase modulation, and greater processing losses on the relay orbiter. The required data margins are increased to 5 dB instead of the usual 3 dB to account for signal variations during entry and descent phases. In this example the relay link provides a gain of 150 times with respect to the DTE link.

Table 10.5 provides an overview of the probe links used on some planetary landers and entry probes. It is worth remarking that probes having a mission duration that is not large in relation to the two-way light time all have one-way communications; the mission profile is preprogrammed since there would be no time for intervention from the ground. All Venera, VeGa, early Soviet Mars landers, VeGa balloons and the Galileo and Huygens probes were of this type. While the Pioneer Venus Small Probes were also one-way, the Large Probe had a receiver for two-way Doppler ranging only, not for telecommand.

10.6 Tracking

Somewhat distinct from the function of receiving data from a planetary probe is determining its location. Although the same radio signal is usually used, the hardware is often different.

Location may be determined by a variety of means; it is important to understand what is meant by location – all locations are relative. On Mars, where comprehensive high-resolution mapping datasets exist, correlation of local images with orbital data is often used. On the other hand, precise Doppler and/or range measurements give a better location with respect to the Earth.

Table 10.4. *Examples of probe link budget calculation*

	Parameter	Units	DTE S-band link	UHF relay link
Link	Frequency	MHz	2300	400
	Wavelength	m	0.1304	0.75
	Range	m	1.05×10^{11}	2.5×10^{6}
	Space losses	dB	−260.1	−172.4
	Venusian atmospheric absorption loss at 50 km	dB	−0.1	0.0
	Venusian atmospheric refraction loss	dB	−0.1	−0.1
Probe	TX power	W	5.0	5.0
	Cable losses	dB	−0.3	−0.3
	TX antenna gain	dB	0.0	0.0
	Pointing loss	dB	0.0	0.0
	TX antenna ellipticity	dB	2.0	2.0
	EIRP	dBW	6.7	6.7
RX station	Receiving antenna diameter	M	64.0	0.975[1]
	Antenna gain	dB	61.2	10.0
	Beamwidth	deg.	0.14	53.84
	Pointing losses	dB	−0.1	−0.3
	Polarization losses	dB	−0.2	−0.2
	System temperature	K	25.0	100.0
	Venus temperature contribution	K	1.3	500.0
	Total system temperature	K	26.3	600.0
	Received power	dBW	−192.7	−156.3
	Noise spectral density	dBW Hz^{-1}	−214.4	−200.8
	P_s/Noise	dBHz	21.7	44.5
	Modulation index	degrees	60.0	60.0
	Carrier losses	dB	−6.0	−6.0
	Data losses	dB	−1.2	−1.2
	Carrier power/no	dBHz	15.7	38.5
	Loop bandwidth	Hz	1.0	200.0
	SNR in bandwidth	dB	15.7	15.5
	Processing losses	dB	−1.0	−2.0
	Data power/noise	dBHz	19.5	41.3
	Bit rate	bps	10	1500
	E_b/noise	dB	9.5	9.5
	Threshold $(7, \frac{1}{2})$ code	dB	4.5	4.5
	Data margin	dB	5.0	5.0

[1] Equivalent diameter.

Doppler measurements require a stable or known transmitter frequency (using an ultrastable oscillator or transponder) on the probe, and a receiver with a much more stable reference on Earth. This allows the line-of-sight velocity to be determined. Doppler navigation is also possible using the frequency history of the signal received from an orbiting transmitter, which depends on the altitude and

Table 10.5. *Main characteristics of probe links used on various planetary landers and entry probes, including transmitter and receiver frequencies and corresponding bit rates*

Craft	Link type	Frequency (MHz)	Tx power (W)	Antenna type	Modulation	Bit rate (bps)
Venera 4–8 probes	1-way DTE	922	20	Hemispherical spiral; bifilar conical on Venera 8	FSK	1
Pioneer Venus Small Probes	1-way DTE	2294	10	Crossed-dipole	?	16 to 64
Pioneer Venus Large Probe	1-way DTE (data)	2294	40	Crossed-dipole	?	128 to 256
VeGa AZ	1-way DTE	1667.8	4.5	Helicone	?	1 to 4
Galileo probe	Relay	1387.0 and 1387.1	2×23	Crossed-dipole	?	2×128
Huygens	Relay	2040 and 2097	2×10	Helicones	BPSK	2×8096[1]
Ranger 3, 4, 5 landers	DTE	960	0.05	Crossed-dipole	PM	analogue
Luna 9, 13	2-way DTE	183.538	?	4 blades/ petals	PM	?
Mars 6, 7 Landers	1-way Relay	122.8; 138.6	2×30?	4 blades/ petals	?	256?
Mars 96 small stations	Relay	437.100 and 401.5275	≥1	Quadrupolar	Manchester	8
Mars Pathfinder	2-way DTE	8420	12	Dipole array HGA (30cm dia.)	BPSK	4740
Beagle 2	2-way relay	401.6 and 437.1	5	patch	BPSK	2000–128 000 and 2000–8000
MER	2-way UHF relay	401.585625 and 437.1	15	UHF Monopole	PCM / Bi-Phase-L / PM	8000–256000 and 8000
	2-way DTE	8439.444445 and 7183.118056 (Spirit), 8435.370371 and 7179.650464 (Opportunity)	15	Dipole array HGA	PCM / Bi-Phase-L / PM	8000–256000 and 8000
Surveyor 1–7	2-way DTE	2295 and 2113	10 or 0.1	Planar array HGA or two conical omni-directional antennas	?	?
Lunokhod 1,2	2-way DTE	922.76	?	?	?	?

Table 10.5. (Cont.)

Craft	Link type	Frequency (MHz)	Tx power (W)	Antenna type	Modulation	Bit rate (bps)
Luna 16,20,24	2-way DTE	922.764	?	?	?	?
Venera 9–14 landers	1-way relay	122.8; 138.6	2×30	Quadrifilar helical	PSK	256 to 3072
VeGa lander	1-way Relay	122.8; 138.6	2×30	Quadrifilar helical	PSK	3072
Viking lander	2-way DTE	2294	20	76 cm diameter HGA	?	500
	2-way UHF relay	381	30	8-element crossed dipole	?	4000 and 16 000
MPL/Phoenix	2-way DTE	X-band	?	Parabolic dish	?	2100–12 600
	2-way UHF relay	401.585625 and 437.1	15	UHF Monopole and Quad Helix	PCM / Bi-Phase-L / PM	8000–256000 and 8000
Mars 96 penetrators	Relay	UHF	5	Spiral UHF antenna	?	8000
DS-2 Mars microprobes	1-way relay	UHF	?	Whip antenna with 'whiskers'	?	7000
Lunar-A penetrators	2-way relay	400	?	?	?	256–1024
Phobos DAS	2-way DTE	1672	?	Spiral	?	4–20
Phobos PROP-F	Relay	?	?	?	?	?
Philae	2-way relay	2208; 2033	1	Patch	?	16,000
MINERVA	2-way relay	?	?	?	?	9600

[1] In the event, data from only one of the two channels was received correctly on board Cassini.

horizontal miss distance. Simultaneous measurements in DTE and orbiter relay link would provide two components of the probe velocity. A third component (vertical) can be measured independently using either altimetry or pressure/temperature measurements. One-way Doppler measurements were the main source of data on planetary winds in all Venera, Mars 6 and Huygens probes.

An entirely different approach relies on the combination of data from several telescopes. The different phase of the signals as received on Earth allows the direction to the transmitter to be very precisely established. This VLBI technique was applied to the VeGa balloons, and also the Huygens probe.

11

Radiation environment

'Radiation' in the spacecraft environment context generally refers to subatomic particles in space. Of course, the Sun and other astrophysical sources yield electromagnetic radiation (hard UV, X-rays and gamma rays) that are somewhat damaging to materials and living things, but these effects are generally small. In this chapter we discuss briefly the sources of energetic particles and their effects on spacecraft systems (Trainor, 1994); effects on living things are discussed in Section 14.3

Note that because the missions of entry probes and landers tend to be short, and the radiation environment at or near a planetary surface is more benign than in orbit, the radiation hazard is generally not as significant a concern as it is for orbiters. Landers on airless bodies (the Moon, Mercury, and especially Europa) may be exceptions, due to secondary radiation from the surface. However, all landers will need a radiation tolerance in that they spend time, perhaps many years, in the space environment.

There are four principal sources of radiation that must be considered. First is any radiation source carried by the spacecraft, such as a radioisotope thermoelectric generator (RTG), radioisotope heaters or sources associated with instruments such as X-ray fluorescence spectrometers. A characteristic of RTGs is their neutron flux.

A second source is galactic cosmic rays (GCRs). These are high-energy particles, usually nuclei of high atomic number ('heavy-Z' or 'high-Z' particles) from astrophysical sources. These are damaging, both directly, and indirectly, in that they may produce a shower of secondary particles and quanta by a number of methods. Heavy particles (protons, nuclei), shatter nuclei into lighter particles that in turn generate cascades of short-lived radiations by collisions and pair production from Bremsstrahlung X-rays. Energetic electrons generate X-rays when striking shielding, via the Bremsstrahlung process, and generate no particulate radiations.

Particles from the Sun form another population. These are usually less energetic, of lower atomic mass, but far higher in number. The flux of solar particles can be strongly enhanced during high solar activity (flares, coronal mass ejections etc.) and deleterious effects on Mars-orbiting spacecraft have been noted.

Usually the strongest sources of concern are the particles trapped in a planet's magnetic field. This is particularly the case for the planet Jupiter (and a Jupiter flyby may be the dominant radiation dose for an outer-solar-system mission beyond Jupiter itself). In general the field concentrates the particles in toroidal 'radiation belts', and thus the orbital design of a mission around a magnetic planet must be done carefully to minimize the dose. Around Jupiter, the moons Io and Europa are immersed in these belts, and thus radiation hardness is essential: on Europa's surface it may be advantageous to bury a lander to gain some shielding effect from the ground.

On Earth-orbiting satellites, trapped particles in the Earth's magnetic field are responsible for the bulk of radiation problems. They tend to occur predominantly in the auroral ovals (i.e. latitude belts approx $60°-70°$ from the equator, where magnetic field lines funnel in towards the Earth's poles) and in the South Atlantic Anomaly (SAA). This region, roughly over Brazil, is one where the Earth's net field is rather weak, and trapped particles penetrate to lower altitudes leading to increased interaction with low-orbiting satellites.

Modelling the radiation effects on components is a challenging task. Not only are various components susceptible in varying degrees to the different sources, but the effects will depend in a complex manner on the mass distribution and thus the shielding effects around the relevant component. Optimum shielding materials depend on the expected radiation source: for example, tantalum is particularly effective at shielding against stray neutrons from RTGs. And more shielding (usually expressed as an equivalent thickness of aluminium) is not necessarily better, in that GCRs often produce even more damaging Bremsstrahlung upon striking the shield. While shielding may reduce the total dose in a radiation belt, during a long cruise in deep space a modest amount of shielding may in fact increase the radiation damage. Various simulation codes are available to model these effects.

The radiation hazard is in general worse in orbit than on a planetary surface, where the atmosphere can shield a large fraction of the incoming particles. Titan and Venus are particularly benign in this regard; Mars less so. Asteroids and comets may endure a comparable radiation flux to that received *en route* in heliocentric orbit.

Radiation doses are usually expressed in units of rads: (this unit prevails in parallel with the corresponding SI unit, the gray: 1 rad = 0.01 gray; 1 gray corresponds to 1 joule absorbed per kg. For comparison, a prompt dose of a few hundred rad is typically fatal to humans). The Galileo spacecraft was designed to endure a dose of around 150 kilorads. Around Europa, in, but not in the worst part of, Jupiter's radiation belts, a spacecraft would endure 4 megarads in one month. Note that the energy of an individual particle is usually expressed in electron volts (1 eV $\sim 1.6 \times 10^{-19}$ J).

Radiation damage usually manifests itself in effects on semiconductor devices, although very high doses can render optical components opaque or degrade the strength of materials. The main 'total dose effect' in electronic components is a steady increase in the gate voltage or leakage current. Ultimately, these parameters may exceed the levels at which the circuit will function as intended. A similar effect is seen in certain detectors like CCDs, whose dark current may increase ('hot pixels'). Some total dose effects can be at least partly cured by 'annealing', running at a high temperature for a short time.

In addition to steady total dose effects, there are 'sudden death' radiation damage mechanisms. One of these is the 'single event upset' (SEU), wherein the passage of a particle through a digital component alters the state of that component. A bit, most typically in a computer memory, is 'flipped', from '0' to '1' or vice versa. Where that bit is simply some data, such as a single pixel in an image, such a change is not usually catastrophic. However, if the bit is in a computer instruction, the effects may be profound and impossible to predict.

A principal protection against SEUs is to have rad-hard memory and processors. These critical functions are made less vulnerable to SEUs by, for example, the use of alternative substrates (e.g. silicon-on-sapphire) and by the use of larger gates. The energy required to flip a bit will depend on the operating voltage and the capacitance of the memory cell – more modern, high-density memories use lower voltages and smaller cells and can thus, in fact, be more vulnerable. A second approach is to use coded memory, whereby two distinct words differ by more than one bit-change. Thus, an inconsistent single bit can indicate that a memory cell has been flipped, and the incorrect bit identified and corrected by a software process or dedicated circuitry.

A final damage mechanism is not reversible, but is preventable. This is 'latch-up'. In this mechanism, the passage of a charged particle through a semiconductor creates a parasitic transistor. A large current can flow if the device is powered up, and the heating produced by the current will destroy the device. Latch-up protection consists of fast current-sensing logic that determines whether a latch-up has occurred, and if so, shuts the circuit down before heat has built up to damaging levels.

12

Surface activities: arms, drills, moles and mobility

While much can be achieved by purely passive observations and measurements of a planetary lander's immediate environment, some key science requires the landed system to interact with the surface mechanically. This may involve the acquisition of samples of material, either to be returned to Earth or delivered to instrumentation internal to the lander. Other instruments, while external, require intimate contact with target rocks – these include alpha-X-ray, X-ray fluorescence or Mössbauer spectrometers, and microscopes. Other interactions may include mechanical-properties investigations using a penetrometer, or current measurements of wheel-drive motors.

Thus a variety of mechanisms have been operated on planetary surfaces, including deployment devices and sampling arms of various types, together with drills, abrasion tools and instrumentation. Soviet/Russian landers have tended to feature simple but robust actuators, usually simple hinged arms, and often actuated by pyro or spring. These include the penetrometers on the Luna and Venera missions. Lunokhods 1 and 2 carried a cone-vane shear penetrometer that was lowered into the lunar regolith and rotated by a motor, to measure bearing strength and shear strength. The rovers made 500 and 740 such measurements, respectively, during their traverses across the lunar surface.

A more sophisticated arm was flown on the Surveyor 3, 4 and 7 lunar landers (Figure 12.1). The Surveyor soil mechanics surface sampler (SMSS) was a tubular aluminium pantograph, five segments long, with a total reach of 1.5 m. As its name suggests, it was primarily a soil-mechanics experiment (indeed, in many ways the whole mission was primarily a soil-mechanics experiment). The strength properties of the soil, deduced also from landing dynamics, were inferred by measuring the motor current required to dig a trench in the ground. On Surveyor 7 the SMSS was mounted differently to enable it to pick up and reposition the alpha-scattering experiment on the lunar surface.

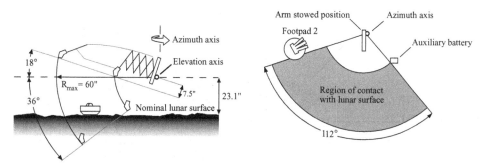

Figure 12.1. Space envelope of operation of the Surveyor 3 soil mechanics surface sampler (SMSS).

As an aside, lunar regolith is in fact a particularly nasty material to work with, having a wide range of particle sizes, and with grains being very angular (see e.g. Heiken *et al.*, 1991). It is thus able to penetrate many mechanisms, and can be highly abrasive once it does so. (It is believed that difficulties in moving Surveyor 3's camera mirror may have been due to dust ingestion – its thrusters apparently kicked up clouds of dust at landing.)

The Viking lander surface-sampler arm was successful at delivering samples to the biological analysis instruments. It was, however, rather fragile. It used a shoulder joint to point an extensible boom (a coiled prestressed tubular tape, much like those used for booms and antennas on spacecraft). The multi-purpose scoop on the end of the arm is shown in Figure 12.2. Although it was used for some trenching and 'bulldozing' experiments (Moore *et al.*, 1977), there were fears that it would be damaged in such operations. Viking Lander 1's robot arm was initially stuck until a pin was unjammed by repeated extensions (Spitzer, 1976).

The robot arm on Mars Polar Lander (Bonitz *et al.*, 2001) and Phoenix, has (as for Viking) the principal function of delivering soil to experiments on the lander deck. The 4-degree-of-freedom 'backhoe' design has a 2.2 m reach. The 5 kg arm is capable of exerting considerable force on the ground (some 80 N, enough, in principle, to drag the lander!) in order to cut a trench in a possibly ice-rich soil. The sampling scoop is fitted with 'ripper tines' to tear through such potentially hard material. The arm also carries a camera for close examination of the soil.

An additional function on MPL was to emplace a temperature-sensing spike (mounted on the 'wrist') into the ground, and to vary the height of an air-temperature sensor to measure the boundary layer temperature profile. The Phoenix arm carries instead a thermal and electrical conductivity probe.

The Sojourner rover performed some soil mechanics experiments with its wheels (rotation and motor currents being recorded, as well as the resultant trenches being imaged by the rover cameras and/or the lander camera) (Moore *et al.*,

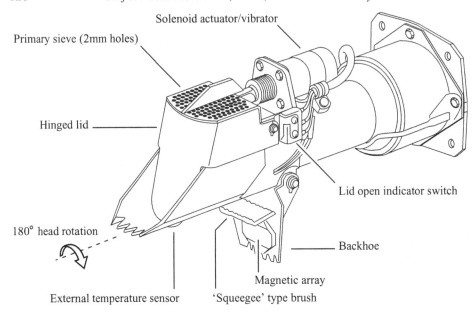

Figure 12.2. Viking lander scoop.

1999). In addition, it had an alpha-proton-X-ray spectrometer (APXS), which required direct contact with the rocks; emplacing this instrument was the rover's main scientific function (it was very much an engineering experiment overall).

The Mars Exploration Rovers had a similar overall goal, although with much more capability and instrumentation – an alpha-X-ray spectrometer being supplemented by a Mössbauer spectrometer and a microscope camera. Furthermore, rather than only attack exposed, and therefore generally dust-covered surfaces with these instruments, MER carried a rock abrasion tool (RAT – Gorevan *et al.*, 2003), which could brush dust off, and grind a large shallow hole to allow depth-profiling. The RAT and other instruments were emplaced by a small arm, the instrument deployment device (IDD, Tunstel *et al.*, 2005).

The Beagle 2 lander carried a well-instrumented robotic arm. At the end, a PAW (position adjustable workbench) was equipped with a stereo camera (with illumination), microscope, X-ray fluorescence spectrometer, Mössbauer spectrometer, rock corer–grinder, wide-angle mirror, wind-sensor and sampling device (Sims *et al.*, 2003).

The amount of energy required (which can be inferred from motor currents) to drill a given volume of material is a quantity that can be considered a measure of the strength of the material, and is thus a diagnostic of the rock type. For example, a rather weak rock strength of 10 MPa corresponds to $10^7 \, \mathrm{J \, m^{-3}}$ – thus to drill a 1 cm diameter hole to a depth of 10 cm requires roughly 80 J of energy – while a harder rock may require 20 times more energy.

Sampling (coring) drills were flown on the Luna Ye-8-5 and Ye-8-5M sample-return missions to acquire lunar soil samples for return to Earth. The first successful mission was Luna 16, which returned 101 g of material (Grafov *et al.*, 1971), while Luna 20 returned 30 g. The drill was a thin-walled tube carrying helical threads on its outside surface and a crown on sharp teeth at its cutting end; it was insulated and sealed prior to its 500 rpm operation, to permit the use of conventional lubricants. The Luna 16 and 20 drills reached 25–35 cm depth. A more advanced drill, used successfully on Luna 24, reached some 2 m into the ground and collected 170.1 g of material. The stratigraphy of the regolith column was preserved by stowing the acquired sample in a coiled plastic tube.

Drills for Venus have been flown with some success by the Soviet Union (Venera 11 to VeGa 2, e.g. Barmin and Shevchenko, 1983) but are at an earlier stage of development elsewhere. The challenge is to acquire a sample at the high ambient temperature and pressure and transfer it to the interior of the lander. A recent innovation is ultrasonic (vibratory, rather than rotary) drilling.

A novel sample acquisition technology was flown on the Beagle 2 PAW. The PLUTO 'mole' is a self-hammering percussive drill (Richter, 2001) that winds a spring to push a free hammer. The mole derives originally from Russian technology (e.g. Brodsky *et al.*, 1995; Gromov *et al.*, 1997).

Mobility is often an important, desirable and enabling aspect to a planetary mission, allowing scientific targets beyond the craft's immediate environment to be reached. Mobility may be required in atmospheric, surface and sub-surface environments. Aerial mobility (e.g. balloons) and ice-penetrating 'cryobots' were addressed in Chapter 6. Mobility across a surface takes us into the field of robotics (e.g. Ellery, 2000) and planetary rovers (e.g. Kemurdzhian *et al.*, 1993; CNES, 1993), which is deserving of a whole book in its own right; the details are beyond the scope of this work.

At the simplest end of the spectrum of complexity are relatively 'dumb' instrument-deployment devices, whose function is to transport sensors from a lander across the planetary surface beyond the radius accessible from the lander itself (e.g. by robotic arm). Such devices are usually tethered to the lander to provide power and data connections, which limits mobility but does minimize the need for power and communications equipment and autonomous control on the rover. The first such device flown was the PROP-M tethered walking rover flown on the Mars 2, 3, 6 and 7 landers in 1971 and 1973. All four missions were lost before PROP-M was to operate, however. Deployed by an arm from the lander, PROP-M was to perform penetrometry and densitometry measurements on the Martian surface material. It had the capability to sense (by means of 'whiskers' at the front) the presence of an obstacle, step backwards and turn to move around it.

The first planetary rovers were the two Lunokhod vehicles deployed by Luna 17 and Luna 21 in 1970 and 1973, following a launch failure in February 1969. These were teleoperated from the Earth (the relatively short two-way light time between the Earth and the Moon making this possible) and between them travelled a combined total of more than 47 km across the lunar surface. They returned many images and performed measurements of the lunar soil and surface environment, as well as carrying laser retroreflectors.

Two types of wheeled vehicle were used by the Apollo landings: the hand-drawn MET (modular equipment transporter, carried on Apollo 14 only) and the LRV (lunar roving vehicle, carried on Apollo 15, 16 and 17, see Cowart, 1973). An extensive literature exists concerning wheel–ground interaction ('trafficability') for lunar and planetary rovers (e.g. Bekker, 1962; Carrier *et al.*, 1991) and rover dynamics more generally (e.g. Avotin *et al.*, 1979; Kemurdzhian, 1986).

The first successful Martian rover was of course Sojourner (Mishkin, 2004), followed by the much larger Mars Exploration Rovers, Spirit and Opportunity (Squyres, 2005). In 2011 an even larger Martian rover, the Mars Science Laboratory, is planned to be launched, while ESA is planning to launch its own Martian rover on the ExoMars mission, due no earlier than 2018. Further lunar robotic rovers from the US, Japan and China are also in the early stages of planning as part of lunar landing missions.

For sample-return missions, additional challenges are introduced. For example, the constraints of the delivery and stability of the lander/ascent assembly are such that the ascent stage may be less slender than is optimal in the case of Mars; the ascent stage may incur a higher aerodynamic loss than would otherwise be the case. For Venus, aerodynamic losses are so large that any contemplated sample return mission would first use a balloon to climb above the densest part of the atmosphere before using a rocket stage. Unless the mission duration is very short, storable (i.e. non-cryogenic) rocket propellants need to be used. On the other hand, it has been proposed to perform *in situ* propellant production on Mars – to derive oxygen from the CO_2 atmosphere and thus only require the delivery of the chemical processor, a power source to drive it and a fuel.

To date, however, ascents from other planetary bodies have been relatively limited in sophistication. Until recently, they were confined to the Moon – the Luna 16, 20 and 24 sample return missions (and eight other, unsuccessful missions of the Ye-8-5 and Ye-8-5M craft), and the Apollo landers. In all cases the landers were squat vehicles and served as launch pads for the ascent stages. While the Apollo vehicles had sophisticated (astronaut) guidance, the Luna missions were confined to a longitude of 60°E where a vertical ascent assured a direct return to Earth without further course adjustment. Both the Lunas and Apollo used storable bipropellant engines (nitric acid/UDMH and dinitrogen tetroxide/ UDMH, respectively, UDMH being unsymmetrical dimethyl hydrazine).

Surveyor 6 took off briefly from the surface of the Moon, to land about 3 m away thus enabling it to image the footpad impressions from its original landing. This is a reminder that, in general, any system capable of soft landing may also have the ability to lift off again (subject to the ignition characteristics of its engines).

For Mars sample return, studies to date require an Apollo-like orbital rendezvous. This avoids the need to carry the interplanetary return propulsion down to the Martian surface, and to inject the Martian ascent stage back to Earth. Isolating the sample in a small capsule also has planetary protection advantages.

Ascents from small bodies are easy, possibly too easy (the Philae lander includes a hold-down thruster to prevent the lander drifting away in the low gravity as it is anchored onto the surface by harpoon). Indeed, jumping is a convenient, albeit risky, way to achieve surface mobility in low gravity (e.g. Kemurdzhian *et al.*, 1988; Richter, 1998; Yoshimitsu *et al.*, 2003; Scheeres, 2004). To date, the only example of an ascent has been the recent Hayabusa mission, which appears to have landed and taken off twice from Itokawa, if not perhaps with the full participation of ground controllers.

13

Structures

Planetary probes present a very diverse range of structural problems and solutions. This is in contrast to free-flying spacecraft (i.e. satellites and deep-space probes) which generally have a simple box or drum structure because there is only a single dominant loading (launch). On the other hand, landers and probes can range from resembling spiders to cannonballs, with the range generally being driven by thermal as well as structural requirements. Landers may be spidery open frames with equipment boxes bolted on, like the Surveyor landers; in contrast, entry probes for hot, deep atmospheres are constructed as pressure vessels and have thus been spherical in shape.

On most satellites the largest accelerations and thus structural loads are encountered during launch (typically 5–10 g): however, entry probes to Venus or Jupiter may encounter decelerations of 100–500 g. In such situations, load paths must be kept as short as possible to minimize the structural mass. The Pioneer Venus and Galileo probes (which had thermal constraints) used thick-walled pressure vessels supporting solid deck plates to which equipment was bolted. Spherical geometries are also appropriate where landing attitude is not initially controlled (e.g. Luna 9, 13; though note that the interiors of these spacecraft were pressurized, which also tended to favour a spherical design).

The Huygens probe did not need to exclude the atmosphere and therefore had an unsealed, thin-walled shell to preserve an aerodynamic shape and support light foam insulation. Huygens had three main sets of design loads (NB no impact or surface loads were considered). First are the launch loads, which are orthogonal to the probe axis since the probe is cantilevered sideways from the Cassini orbiter on its Titan launcher. These are transmitted through a support ring around the equator of the probe to a honeycomb platform onto which the units are attached. The same load path transmits the loads from the heat shield during entry (although these loads are in a different direction from the launch loads). Finally, parachute inflation loads – under Titan gravity, the probe weight and thus the parachute suspension load are quite modest – must be conveyed from the upper surface of the probe. This upper surface is also a honeycomb platform, and the loads are transmitted to the experiment platform via three stiff rods, as well as, in part, by the thin-walled (but stiffened) alloy shell.

The Soviet Luna 16, 20, etc. soft landers used an interesting structural design, with the large spherical propellant tanks towards the periphery of the vehicle, but presumably providing much of the required stiffness simply from the tank walls.

Structural and thermal design are intimately connected. The structure provides thermal leak paths from the outside to the equipment, and the designer might choose a more-or-less thermally conductive material to meet thermal needs, even when this might offer poorer strength to weight performance. The Pioneer Venus small probes used beryllium shelves for thermal reasons.

An interesting metric for an aerospace vehicle is its mass density. This parameter is directly relevant for capsules that may splash down on Earth or Titan, in that it determines whether they will float. Generally, probes tend to have densities of the order of $200–400\,\mathrm{kg\,m^{-3}}$; it is in fact difficult to attain much higher densities without explicitly adding ballast, largely due to the low volume-packing fraction associated with practical assemblies (adequate clearance must be maintained for access to connectors, for example). For a given shape of vehicle, there is also a direct relationship between the density, the size of the vehicle, and the ballistic coefficient.

The landed parts of the DS-2 microprobes were rather dense – indeed, being milled out of solid alloy, with a dense tungsten nose. The structure had to be stiff to withstand the very high impact loads; the tungsten nose was in fact not chosen for structural reasons, but to push the centre of mass as far forward as possible for aerodynamic stability. It should be noted that the difficulty, if subsystems grow or new equipment is added in an evolving design, is often a lack of volume in which to accommodate the growth, not a lack of mass.

It is often assumed that Venus and giant planet probes must have pressure vessels. In fact, deep sea instrumentation has been constructed and operated without using pressure vessels; simple plastic tubes containing the electronics are filled with oil and closed with bungs. The pressure is resisted by the incompressible oil which keeps the seawater out, but is transmitted to the electronic components. With the exception of a battery, which needed an additional vent hole, the components tolerated the pressure. While it is important to exclude hot, corrosive atmospheres, this exclusion requirement should not necessarily be interpreted as a requirement for a pressure vessel.

14

Contamination of spacecraft and planets

The transfer of material that is not native to a planet has been happening over the history of the Solar System, with meteorite delivery being a common example of this interchange. With the development of rocket launchers capable of injecting objects into interplanetary trajectories, mankind joined Nature in being able to alter another planet's composition. Generally spacecraft and their associated hardware are designed and assembled so as to minimize the amount of debris that they carry. This chapter examines the problems associated with the unintentional delivery of living or dead organic matter to celestial bodies; so-called 'forward contamination'. The topic is often referred to by the phrase planetary protection, and its scope includes not only the possible contamination of planetary bodies, but also the potential introduction to the Earth of materal from a non-terrestrial biosphere. Furthermore, the threat that planetary protection seeks to minimize is not restricted to the introduction of non-native organisms to another planetary body. Non-living material, such as DNA fragments and other complex bio-relevant molecules might trigger false-positives from equipment designed to detect extant or extinct life.

A practical definition of a living entity might be that the agent processes matter and energy in such a way that it can reproduce, and in doing so prosper in the face of environmental stresses. If the environment of the organism changes too radically then the organism may be killed or rendered dormant. Techniques that are intended to kill microbes may, if applied without sufficient vigour, only make the organism dormant. Many bacteria take on a spore-form in such stressful times, they become water-deficient and develop protective coats which make them able to withstand a wider range of stresses than when in their active form (Nicholson et al., 2000). Spores can be revived and grown to form colonies that may be counted visually, allowing the lethality of the killing process to be measured. In such a way the effectiveness of a sterilizing technique can be quantified by noting the conditions needed to kill off a given fraction of the original microbial population. A process, such as heating the sample to a given temperature, causes the death of all but 10% of an original microbial population after a period of time termed the 'D10' (decadal) value. If nine-tenths of an initial number of organisms are killed and if the remaining biota are unexceptional, a further 90% of that surviving population will die if the process is applied a second time. In Figure 14.1(a) three idealized plots are shown for the fractions of three hypothetical microbial groups that survive being exposed to different temperatures.

The survival curves suggest that a constant fraction of a given population is killed per unit of time. In Figure 14.1(b) the D10 durations of each curve are plotted (each curve in Figure 14.1(a) is associated with a point in Figure 14.1(b), with the wider error bar associated with T_a reflecting the wider variation in the durations needed to cause a 10-fold reduction in that population.

Plots of D10 with respect to some measure of a sterilizing process' vigour (such as temperature, T) tend to follow an Arrhenius-like rate relationship such as

$$\frac{1}{D10} \propto \exp\left(-\frac{E}{kT}\right) \tag{14.1}$$

Here E represents a deactivation energy, and values for spores are generally around 10^5 J mol^{-1}. Although the word 'sterile' is usually assumed to imply the total absence of any viable biota, the preceding suggests that no object can be proved to host no viable biota, only that none were detected. Determining the

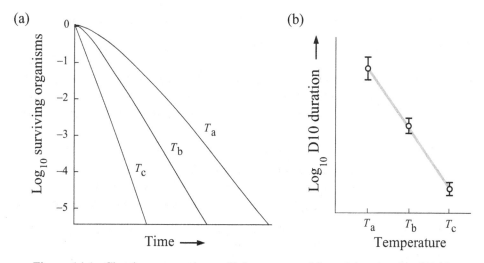

Figure 14.1. Charting a generic sterilizing process (a), and its associated D10 distribution (b).

actual bioload of a component frequently means reviving and culturing spores that have passed through a sterilization process. More sophisticated techniques that rely on reactions between chemical marker molecules and the organism can be used to detect individual living biota, but it cannot be assumed that the detection of contamination on an article of hardware can be performed with complete accuracy. Many bacteria are difficult to cultivate and spores may need differing revival conditions.

14.1 Sources of contamination

Spacecraft generally acquire their largest load of contaminating organisms from the personnel involved with the assembly and testing of the spacecraft. The organisms transferred through such forms of contact therefore belong predominantly to the species found on and within people. Human beings are host to around 200 species of microbial organisms, with bacteria the most common skin-borne organisms, followed by fungi and their spores. The most prevalent bacteria types are those associated with the human gut, skin, hair, mouth and nose. Using wash-and-strip techniques it has been shown that operators display a near constant load of organisms which is weakly affected by climate or season. Some individuals have an intrinsically higher loading of viable organisms in their skin effluvia, and in measurements involving full-environment chambers, people (wearing a sterile scrub suit, socks and cap) have been shown to shed cultivatable skin flakes at a rate of several thousand particles per minute in still air (Riemensnider, 1968). This flux of organic debris from the assembly personnel can be minimized with particular hygiene protocols, but most of the contamination control comes from the use of air filtration systems, careful planning of assembly areas, appropriate garb, and clean handling and working procedures. The main interfaces between people and spacecraft hardware are the hands, and gloves of either latex or polythene are commonly used along with standard cleanroom donning and doffing procedures. Table 14.1 shows the degree of contamination incurred during assembly of spacecraft-representative fasteners.

Clearly, the assembly of hardware in an absolutely sterile state cannot be easily achieved. If the presence of biological contamination is a given, then methods of removing or killing the organisms may have to be employed. As there is no unique sequence of steps by which a spacecraft can be produced from a collection of parts, there are many alternative schedules for the cleaning procedures necessary. These can rely on building with non-sterile parts and sterilizing the final assembly, or using sterilized components from the beginning and ensuring rigorous bio-load monitoring and process control.

Table 14.1. *The degree of contamination incurred during assembly of spacecraft-representative fasteners (adapted from Vesley et al., 1966)*

Process	Minimum, mean and maximum microbial colony count per average assembly process		
No hand care	4	122.6	380.8
2 min ordinary soap wash	2	13.3	56
2 min hexachlorophene wash	0	1.3	7.4
2 min hexachlorophene wash, and gloves	0	0.2	2

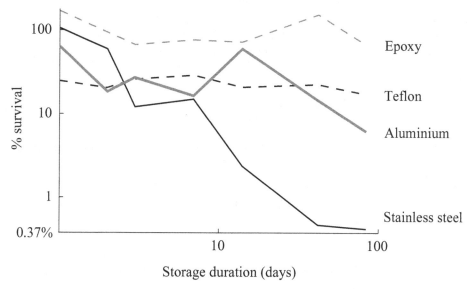

Figure 14.2. The die-off rates at 22 °C for skin-carried organisms on two metallic and non-metallic spacecraft construction materials.

Given that humans are the prime source for the biota deposited on spacecraft materials, the simplest way to place such organisms under stress is to subject the contaminated hardware to conditions drastically different from those found on and in the human body. The moist ecosphere of the human body means that the easiest way to stress a human-borne microbe is to put it in a dessicating environment. The data in Figure 14.2, taken from Vesley *et al.* (1966), shows the survival rates for spores on plates of aerospace materials that had been handled by a group of laboratory workers.

Having outlined the problem of spacecraft hardware contamination, it is useful to describe the present regulations, and then the techniques that can be employed to achieve these requirements.

14.2 Current regulations for spacecraft-borne bioload

As all space-faring nations are also signatories to the Outer Space Treaty, drawn up by the United Nations and released in 1967, they shoulder a legal obligation to

'... conduct exploration of them so as to avoid their harmful contamination and also adverse changes in the environment of the Earth resulting from the introduction of extraterrestrial matter'.

Building on this broad obligation to 'avoid' contamination of celestial bodies, the Committee on Space Research (COSPAR) has taken on the role of co-ordinating the regulations that are applied to space missions. Table 14.2 summarizes the position of the COSPAR planetary protection group as of October 2002, with the core recommendations being made at the 25th COSPAR meeting (De Vincenzi and Stabekis, 1984) with sub-categories for Mars missions being developed by the Space Science Board (SSB) of the US National Research Council.

14.3 Techniques for cleaning and sterilizing

Many techniques are available to the spacecraft engineer to ensure that a spacecraft has its bioload reduced and its biorelevant contamination minimized to acceptable levels (Ulrich, 1966). It is rare to find a situation that merits the application of only one method, and in general a suite of methods is chosen with particular processes being applied to specific subsystems according to their compatibility; see Debus *et al.* (2002), for a flight mission example.

14.3.1 Filtration and intrinsically clean assembly

Clean assembly techniques require that the bioload of components is monitored and tracked throughout the build process, with items being stored in sterile containers along with witness plates. Rigorous traceability of processes, such as soldering and fastener attachment, are needed with bioload reduction and monitoring being applied to tools and build environments where necessary, to ensure that the bioload of the finished spacecraft is understood with confidence during its assembly.

14.3.2 Thermal stress

Most known microbes cannot endure temperatures much in excess of 110 °C when alive, and few can survive in habitats with wide temperature ranges. However, in the dormant spore phase, both fungi and bacteria can endure wider temperature extremes, with extreme cold being far less of a threat than extreme heat. Heating both desiccates the already water-depleted spore, and damages the

Table 14.2. *An abbreviated description of current COSPAR planetary protection regulations*

Category	Mission target	Comments
I	The Sun, The Moon, Venus, certain classes of asteroids	Essentially no steps have to be taken to ensure compliance
	Bodies with no direct relevance to the study of life or chemical evolution	Terrestrial biota are killed by the destination's environments, and no *in situ* biota are expected
II	The gas giant planets, comets, TNOs, carbonaceous asteroids. Mission targets relevant to the origin of life or chemical evolution. If contamination is taken to such bodies, future missions should not have their science compromised	No specific changes are needed for missions hardware or design Documents should be prepared that detail the post-mission and failure contingencies for the mission
III	Mars, Europa. Mission targets are significant to the study of life's origin and chemical evolution	Detailed spacecraft construction documentation needed, may include an inventory of organic matter onboard. Cleanroom assembly, and implementation of bioburden reduction procedures beyond clean working. Orbit biasing to lower collision risks or whole craft bioload to be $<5 \times 10^5$ spores
IV	Landers to Mars and Europa. Spacecraft to these destinations could jeopardize the scientific return of future missions	Requires more detailed documentation than Category III, assays of bioburden, a probability of contamination analysis, and an organic matter inventory. Extra requirements may include trajectory biasing, assembly in cleanrooms, bioload reduction, and partial sterilization of landed hardware. The requirements are akin to those of the Viking landers, with the exception of whole spacecraft sterilization
IVa	Mars landers without payload to study extant life	Bioload to be no greater than Viking lander pre-sterilization levels, compliance to Category IV in general. Bioburden on exposed surfaces to be an average of <300 spores m^{-2}, and the total vehicle surface burden $<3 \times 10^5$ spores

Table 14.2. (Cont.)

Category	Mission target	Comments
IVb	Mars landers with payload for the study of extant life	Either the whole landed spacecraft is to be as sterile as the Viking landers, or to limits dictated by the payload's detection limit. Or, the sub-systems in the sample acquisition chain should be sterilized and prevented from being contaminated by other hardware. Bioload on exposed surfaces to be similar to Viking lander levels (by inference, a total of 30 spores)
IVc	Mars landers with payload for the study of extant life which visit regions of special scientific interest; such regions are places where terrestrial microbes may thrive or where native life may prosper	If the craft lands in a special region[1] then the entire landed craft must be sterile to Viking lander levels ($>112\,°C$ for 30 hours). If the craft lands outside this area, parts that can contact the region (wheels, arms, sensor covers, etc.) must be sterilized to Viking lander levels. The whole landed system may need to be sterilized if non-nominal arrival could cause contamination
V	All sample return missions to the Earth or the Moon. A subcategory 'unrestricted Earth return' applies for material from bodies thought to have no native biota	For restricted Earth return, any sample should be contained using a verifiable and fail-safe method after sample acquisition. No uncontained material from the mission's target shall be returned to Earth – the so-called 'breaking of the chain' of contact. Example missions would be those that deliver material from Mars or Europa to Earth

[1] Such as an ice-rich region of Mars, or an area showing extant hydrothermal activity.

DNA of the microbe. Thus heating is frequently used as a robust method of irrevocably disabling a microbe since the volume, rather than just the surface, of an object can be sterilized. Figure 14.3 shows the response of two bacilli to prolonged heating, the spores of Bacillus subtilis var. niger are often used as candidates for establishing thermal-kill procedures for hardware because of their high thermal resistance.

In Table 14.3 the survival rates are shown for Bacillus subtilis var. spores in different settings. It is notable that embedded spores tend to survive heating better than exposed organisms.

Table 14.3. *The resistance of spores to heating at 120°C in different settings (after Bruch, 1964)*

Spore and environment	D10 (hours)
Bacillus subtilis var. niger in asbestos patching cement	2.1
Bacillus subtilis var. niger in solid rocket propellant	2.5
Bacillus subtilis var. niger on paper: vacuum/air at 1 bar	0.3/0.91

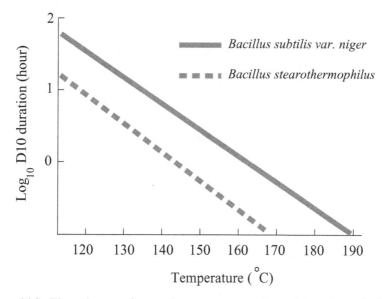

Figure 14.3. The resistance of spores from two common bacterial species to dry heat.

Heating brings with it the possibility of damage to components from mechanical tolerances being exceeded by expansion, and by degradation of material properties. For landers that are subject to whole-craft heating there are obvious benefits in establishing a low bioload prior to the final heat-treatment; a lower temperature/ duration process can then be used to meet the specific COSPAR regulation.

14.3.3 Radiation exposure

Both corpuscular and electromagnetic radiation can harm living and dormant cells. Damage to cellular molecules can occur either through the reaction of radiation-formed ions and radicals, or by direct absorption of the radiation. In each case the depth to which the radiation penetrates depends on the energy of the radiation and its ability to lose energy to the surroundings by scattering or by ion-formation. Of the many forms of radiation, hadronic (protons, neutrons, atomic nuclei) particles are able to generate ion-tracks most readily, with charged particles

Table 14.4. *The features of the types of radiation pertinent to this chapter*

Radiation	Effect on biological matter	Comment
Germicidal ultra-violet (UV) light with wavelengths between 100 nm and 300 nm	UV is absorbed by the DNA bases cytosine and thymine which can then link to each other rather than to their complementary adenines on the opposite side of the DNA	Produced by various discharge lamps (mercury, xenon, deuterium, hydrogen). Can damage elastomers and plastics
Beta particles	Ion-pairs are generated	Limited penetration depths
Protons and alpha particles	Intense ion-trails produced, leading to lethal chemistries upon recombination	Modest penetration depths for plausible energy particles
Gamma rays	DNA cleavage	High doses can alter glasses and damage semiconductor junctions

being the most efficient at ion-production. Such radiation has a high linear energy transfer (LET). To a lesser extent, electromagnetic radiation can also generate tracks of ions or radicals, but their weaker interaction with matter results in deeper penetration distances for the same particle energy, and so the induced ion-pairs are more sparsely scattered. The DNA repair mechanisms in cells are best able to mend single breaks in the molecule, and so radiations that generate dense localized ion-tracks are more likely to kill a cell or render a dormant spore unrevivable. Table 14.4 shows some features of the types of radiation pertinent to this chapter.

For biological material the most important species formed by radiation are the OH and O radicals, and an organism is harmed at the molecular level either by direct damage to its DNA or through the effect of radicals formed from water during the ionization process.

There are two main drawbacks to the use of radiation as a sterilizing agent; ionizing radiation can alter the electronic properties of semiconductors in an irrevocable manner, leading to memory cells that are unwriteable (frozen bits) or otherwise damaged junctions. In practice, semiconductors used for planetary spacecraft are often encapsulated in 'radiation-hard' packages so as to operate at higher background radiation levels. However, the trend in using more complex and modern integrated circuits leads to smaller junctions which in turn are more susceptible to radiation damage. A modern EEPROM wafer would have at least one junction irrevocably destroyed after exposure to 200 grays of β radiation (Shaneyfelt *et al.* 1994). This same dose would kill only nine-tenths of an E. coli colony, and a smaller reduction in more resistant species or spores (Figure 14.4).

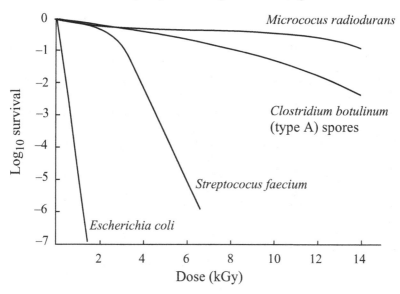

Figure 14.4. The effect of β radiation on the survival of common bacteria (data from Goldblinth, 1971).

14.3.4 Sterilizing chemicals

Certain gases, such as ethylene oxide, chlorine dioxide and paraformaldehyde are toxic to living and spore-form bacteria, either by alkylation or denaturing of vital cell architecture such as nucleic acids. The gases can be delivered to the spacecraft hardware at room temperature and pressure in some form of flow-controlled enclosure. Depending on the agent used, the enclosure around the hardware can be simple, and need only be relatively gas-tight if access to the sterilizing area is appropriately controlled. The properties of the above common sterilizing agents are listed in Table 14.5. The oxidizing nature of all of the compounds is reflected by their general incompatability with organic compounds and their toxicity.

14.3.5 Other gaseous sterilizing methods

Vigorous sterilizing agents can be generated through the electrical excitation of gases. These techniques are considered separately as they require more sophisticated equipment, such as a vacuum chamber, and some means of forming the sterilizing agent, which can be a partial plasma or a gas in an excited but neutral state. Low-pressure neutral vapours of hydrogen peroxide have already been shown to have a useful sterilizing effect (Rohatgi *et al.*, 2001) and at low pressure can be readily ionized by an electrical discharge. The plasma's ions, such as OH^- have a high lethality for bacteria in both spore and live phase. The process occurs at total pressures of less than 10 torr and no significant heat is generated upon exposure of

Table 14.5. *Chemicals employed in gas-phase sterilization*

Compound	Nature at room temperature	Comments
Ethylene oxide (epoxy ethane) CH_2OCH_2	Colourless gas, m.p. $-112\,°C$, b.p. $10.8\,°C$. Density 1.5 that of air. Flash point $-20\,°C$, auto-ignition temperature $429\,°C$. Flammable in air at concentrations above 2.6% volume at room temperature	Ether-like odour, corrosive to eyes, skin, mucosal surfaces. Acute exposure can cause headaches, nausea, convulsions. Carcinogenic in humans at high concentrations. Degrades some fluoro-elastomers, Viton® and Buna® variants
Chlorinedioxide ClO_2	Yellowish green gas, m.p. $-59\,°C$, b.p. $11\,°C$. Vapour density 2.4 times that of air. May decompose violently at concentrations $>10\%$ by contact with sunlight, mercury, platinum-group metals, carbon monoxide, short aliphatic molecules	Pungent odour, corrosive to eyes, skin, mucosal surfaces. Acute exposure can cause headaches, and nausea. A strong oxidizer; can degrade some elastomers and plastics
Paraformaldehyde $(CH_2O)_n$	White crystalline powder, m.p. $120\,°C$. Flash point: $71\,°C$, auto-ignition temperature: $300\,°C$. Flammable in air at concentrations between 7.0–73.0% at room temperature	Formaldehyde-like odour, corrosive to eyes, skin, mucosal surfaces. Suspected to be carcinogenic in humans at high concentrations

Can degrade EPDM (ethylene propylene diene, modified) polymers and Buna® variants |
| Hydrogen peroxide | Clear liquid, b.p. $150\,°C$. Vapour density 1.17 times that of air. High-concentration solutions decompose in the presence of heat and platinum-group metals. | Modest concentration solutions used ($\sim$30%) to form vapours at low pressures. Relatively low toxicity, wide material compatibility |

objects to the gas, with the whole operation occurring at room temperatures. Other plasmas, of helium (Fraser *et al.*, 1975), air (Lei *et al.*, 2004), oxygen (Mogul *et al.*, 2003) and nitrogen (Yoshida *et al.*, 2003), have also shown sterilizing capability.

14.4 Problems specific to spacecraft

The space environment is intrinsically hostile to organisms that respire or need an abundance of liquid water to survive. Exactly how hazardous this environment is depends on the organism being considered; a human will be killed in a few tens of seconds, but bacteria in the pores of the skin or in the gut will survive for much longer. This shielding principle also occurs in spacecraft, which tend to warm their electronic components to biologically benign temperatures. Several landers and probes have also had pressurized compartments and could have provided hospitable accommodation for microbial stowaways by slowing their desiccation.[10]

Once launched, a planetary spacecraft encounters environments that are generally inimical to terrestrial life. The spacecraft's bioburden will experience sterilization to some degree, with the aforementioned stresses of desiccation and radiation exposure being the most critical. In general, the bulk of a spacecraft's material does not experience prolonged temperature extremes, as spacecraft frequently need to keep electronic and mechanical systems at modest temperatures. Landers present special problems as it is possible that the craft may enter regions that raise the revival likelihood for spores. In the case of Martian landers, this could be a landing site at which water ice may be contacted (either directly by drills for buried ice, or traversing exposed ice deposits). Missions to distant ice-rich satellites of the outer planets face similar problems – potentially, a mission could require a remote device to be entirely immersed in liquid water. Clearly in such cases extreme measures to kill all microbes and then remove traces of their soluble biorelevant compounds should be considered.

14.4.1 The space environment – vacuum exposure

Exposing organisms to hard vacua leads to their desiccation through water evaporation or ice sublimation, and the irreversible polymerization of carbohydrates, lipids and nucleic proteins in spores and living bacteria. These Maillard reactions occur relatively slowly, with spores of Bacillus subtilis needing a D10 duration of the order of 10^4 days when exposed to low pressures in the range of 10^{-5} Pa (Horneck, 1993). Such durations are comparable to recent planetary missions and so exposure to vacuum does not in itself cause a significant reduction in a

---

[10] Early Soviet craft used the active circulation of partial atmospheres to even out thermal extremes in hermetic electronic subsystems.

spacecraft's bioload. As might be expected, the exposure of dead organisms and any biorelevant material such as digestion by-products to vacuum results in no substantial loss of material.

14.4.2 The space environment – ultra-violet radiation

The Sun's unfiltered spectrum contains shorter and more intense ultra-violet (UV) radiation than is seen at the Earth's surface. In Figure 14.5 the irradiance spectrum of solar light in space at 1 AU is shown along with that experienced at the surfaces of Mars and the Earth. The terrestrial ozone layer attenuates light below 300 nm, whereas the CO_2 in Mars' atmosphere blocks light with wavelengths less than 200 nm. The grey line is a relative absorption curve for DNA; when compared to terrestrial illumination the Martian lighting environment deposits much more energy into this important molecule. Other organic molecules, such as amino acids associated with biological systems are also rapidly degraded under unfiltered solar light.

14.4.3 The space environment – penetrating radiation

Outside the protective barrier of the Earth's atmosphere and magnetic field, spacecraft hardware is exposed to energetic electromagnetic and corpuscular

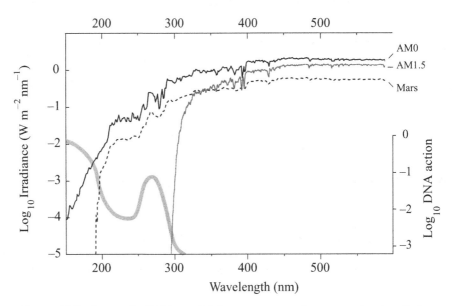

Figure 14.5. A generic DNA sensitivity spectrum and comparative irradiation spectra for Earth orbital space (AM0), Mars' surface (Mars), and the Earth's surface (AM1.5).

radiations. Solar X-rays are the most commonly encountered example of the former, and of the latter, cosmic rays (CR) and solar protons form the greatest hazard. In low Earth orbits, trapped radiation is a common threat to spacecraft, but as interplanetary probes generally spend relatively little time in near-Earth space, damage from trapped radiation in the Van Allen belts and the South Atlantic Anomaly should be smaller than the dose gained en route to the target body.

The Sun's 11-year activity cycle is associated with increased flare activity and raised fluxes and energies of solar particles. During these events the dose rate of protons can rise by five orders of magnitude, and be sustained for tens of hours,[11] delivering up to 2 grays per day at 1 AU. The intensity of such events falls off with distance from the Sun, thus for planetary missions longer than a few hundred days, a greater radiation dose is likely to arise from cosmic rays. These energetic and often multiply ionized particles yield dense ion-pairs when they penetrate materials, and generate secondary radiation which can be more hazardous than the initial radiation. Low atomic weight materials such as hydrogen-rich polymers absorb nucleonic radiation better than metals and so from a sterilization point of view, the move from traditional building materials such as aluminium towards composites in spacecraft structures causes a slight reduction in the killing efficacy of cosmic radiation (Wilson *et al.*, 2001).

14.5 Cleanliness as a separate goal

Related to the issue of contamination of a spacecraft is the question of cleanliness. If a spacecraft is carrying an instrument that is able to detect the presence of organic molecules, then steps must be taken to ensure that those instruments do not have their data degraded unacceptably by the spacecraft's chemical inventory. Generally this goal is achieved by careful cleaning of sampling inlets and handling tools, ensuring that the path taken by a planetary sample is appropriately clean at all stages. This is a burgeoning field of study, made more challenging by the complex nature of current spacecraft sampling tools and the potential for migration of material on landers arriving at an atmosphere-bearing world. Protocols for non-biological organic contamination are being developed, similar to those of the COSPAR bioload regulations. Where these fields overlap is in those cases that might permit an otherwise permissible bioburden to be revived and to grow, and in doing so generate an intolerable level of organic matter through their metabolism and subsequent death.

---

[11] The event of October 1989 as monitored by GOES-7 in geostationary orbit.

14.6 Sample return

To date no mission has been classed as being Category V according to the current COSPAR protocols, but it is likely that sample-return from Mars will occur within a decade or so. The potential for damage to the returned sample's integrity by contamination or degradation by handling is considerable. A precautionary stance is being adopted, with the SSB recommending that samples from Mars should have their containment integrity verified and proven during their return leg, and upon arrival all Martian samples should be treated as hazardous until proven otherwise (Rummel, 2001). Establishing the non-presence of a hazardous agent is not strictly possible, and to bring the level of certainty to a credible level will require as yet unknown combinations of analytical techniques. The overlap of disciplines (biochemistry, spacecraft engineering, law, among others) needed to ensure the safe and fruitful exploration of our Solar System makes planetary protection a vibrant and expanding field for present researchers, and one that has wide applicability to all denizens of the Earth, and perhaps, other worlds.

Part II

Previous atmosphere/surface vehicles
and their payloads

This part of the book provides a basic description, key data and a drawing for all planetary atmospheric or surface vehicles launched, or attempted, from the earliest examples to 2007. Key references concerning the design, payload and results of each craft or mission are given in each case so that the reader may find more detailed information elsewhere. For the payload experiments, the names in parentheses indicate the Principal Investigators (PIs) or otherwise-titled responsible experimenters. Details of the particular experiments and the results obtained (if any) can in most cases be found by searching publications authored (or co-authored) by those named.

The many vehicles are divided into six categories, reflecting the way in which they encounter an atmosphere or surface.

- Destructive impact probes (where the mission is intended to end with the vehicle being destroyed on impact with the surface). These probes are discussed only very briefly, since they are not landers yet do play a role in planetary surface exploration.
- Atmospheric entry probes (where the vehicle's design is driven by its mission in the atmosphere).
- Pod landers (where the vehicle is designed to land initially in any orientation).
- Legged landers (where the vehicle is provided with landing gear).
- Payload delivery penetrators (where the vehicle decelerates in the sub-surface to emplace a payload).
- Small-body surface missions (where the vehicle operates in a low surface gravity environment). These can include many operations that are possible in low gravity, and various types of surface element.

The diagrams in this part of the book were drawn using information gleaned from a variety of sources. While researching specific details for spacecraft, J. Garry and the authors were glad to receive help from the following people: Charles Sobeck, Bernard Bienstock, Corby Waste, Pat Flannery, Marty Tomasko, Marcie Smith, Dan Maas, Doug Lombardi, Satish Krishnan and Debra Lueb.

147

Finding accurate detailed information about the flight models of certain spacecraft has been difficult, not least because the hardware concerned is no longer available on Earth to view! In all of the diagrams, hardware items have been drawn only when they can be unambiguously identified in photographs or technical illustrations. Any errors are therefore of the authors' own making. Note that in some cases thermal blanketing has been omitted for clarity. The general style is that the lander or probe is shown in two side views (90° apart) in the upper left, its accommodation on the carrier in the upper right, and a larger, labelled perspective view in the lower part.

By way of a 'global overview', Figure II.1 shows a launch timeline of planetary landers and atmospheric entry probes. The first launches of such craft were, perhaps surprisingly, not to the Moon, but those of Venera 1 and its 'twin'; however, their inclusion here is somewhat marginal (see Section 16.1.1).

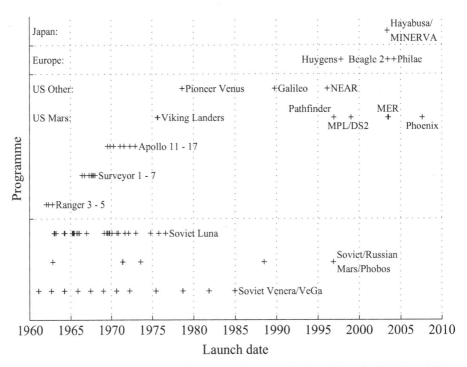

Figure II.1. Launch timeline of landers and atmospheric probes. Included are all launches, or launch attempts to 2007, carrying one or more craft able to operate on the surface of another world or in its atmosphere. NEAR-Shoemaker is included despite not having been designed to operate on the surface of an asteroid. Excluded are destructive impact probes and the tests of the Apollo LM and Soviet LK performed in Earth orbit.

Table II.1. *Successes and failures of launch attempts of planetary landers and entry probes, in terms of the goals of the landers and entry probes at their destination. Some missions that failed during cruise, i.e. before the lander or probe was deployed, still produced useful data (e.g. Phobos 2, lost in Mars orbit). Some useful descent data were returned by landers that failed during EDL (e.g. Mars 6), and the atmospheric probes Venera 4, 5, 6 were all successful despite not reaching the surface*

Outcome	Number of launches
Launch failure (all of which were Soviet Luna attempts in the period 1963–1975)	8
Failed to leave Earth orbit (all of which were Soviet/Russian lunar, Venus or Mars missions)	13
Failed during cruise (often for propulsion, thermal or electrical reasons)	13
Lost during, or very soon after, entry, descent or landing	13
Currently en route (Philae)	1
Success	41
Total	**89**

Particularly evident are the evolving programmes of the 'space race' era, with many launches to the Moon by the USA and Soviet Union, beginning with Rangers 3–5 in 1962 and ending in 1976 with Luna 24. The large number of launches in part reflects the high failure rate, in terms of launch failures, spacecraft failing to leave Earth orbit, failures during cruise, and crash-landings. Table II.1 presents a breakdown of the successes and failures. The Soviet Union took advantage of all bar four of the Venus launch windows from 1961 to 1984 (the 1983 window being used to launch orbiters instead), as illustrated by the clear periodicity of the launch dates. Many Venus and Mars projects have involved separate launches of multiple (usually two) spacecraft, in part to provide redundancy against failure and the long wait until a reflight can be attempted at the next launch window. For the Venera missions this proved to be a good approach, since one craft, but not both, were lost in each of the 1967, 1970 and 1972 windows. Twin missions are also able to yield complementary data, and the ground segment costs are less than they would be if the missions were not under way near-simultaneously. For the Moon, launch windows are frequent enough for the twin flight approach not to be necessary, while in other cases only a single craft has been launched, either for budgetary reasons or because redundancy is implemented by other means.

In parallel with (and in many cases as part of) these launches of landers and atmospheric probes, there have of course been equally vigorous programmes of flyby and orbiter missions. In many cases the lander or probe is delivered by a craft that carries a scientific payload of its own, for operation during cruise, flyby or in orbit around the target world. Quite often they also play a vital role as data relays for the lander or probe.

15

Destructive impact probes

The mission of a destructive impact probe ends successfully with a vehicle (or even just a passive projectile) being destroyed on impact with the surface of another world. The first destructive impact probe was Luna 2, which, along with the launcher's upper stage, impacted the Moon in 1959. Luna 2 hit the surface at $3315\,\mathrm{m\,s}^{-1}$ (Blagonravov, 1968), demonstrated by the loss of the radio signal. Rangers 6–9 impacted the Moon a few years later, obtaining (in the case of 7, 8, 9) close-up images of the lunar surface prior to impact at 2620–$2680\,\mathrm{m\,s}^{-1}$ (e.g. Schurmeier *et al.*, 1965; Hall, 1977). The craters made by these impacts were subsequently found in Lunar Orbiter and Apollo images. Discarded Apollo lunar module ascent stages and Saturn IVB rocket stages impacted the Moon and proved useful as artificial, well-characterised seismic sources (Latham *et al.*, 1970, 1978).

Many years later, Lunar Prospector ended its successful mission by impacting the lunar surface at $1700\,\mathrm{m\,s}^{-1}$, in an attempt to detect water ice by means of telescopic observations of the ejecta plume from Earth. No plume was seen, however, but the exercise resulted in calculations of possible H_2O ejecta cloud propagation that may be applicable to future events (Goldstein *et al.*, 2001). The lunar orbiters Hiten and SMART-1 also ended their missions by impacting the lunar surface. NASA's LCROSS (Lunar CRater Observation and Sensing Satellite) is due to make another attempt to detect ice using the impact technique.

The destructive impact approach was employed in spectacular fashion by the Deep Impact mission, launched in 2004 (A'Hearn *et al.*, 2000). The spacecraft comprised a flyby stage and an impactor, which separated prior to arrival at the target comet. The 370 kg impactor, most of the mass of which was copper, impacted the comet nucleus at $10.2\,\mathrm{km\,s}^{-1}$ to study the cratering process and nature of the comet nucleus sub-surface material. The flyby stage observed the comet and the impact event; simultaneous Earth-based observations were also made. The impactor spacecraft was instrumented with a camera and employed closed-loop targeting. The destructive impact probe approach had been proposed earlier, in the context of the never-implemented Clementine II mission, which planned to impact instrumented probes onto asteroid targets (Hope *et al.*, 1997). Such an approach is one of the concepts proposed for mitigation of a threatening near-Earth object (NEO), where the momentum of the impactor is used to deflect the NEO's orbit slightly. A demonstration of NEO deflection has been under study by ESA as the Don Quijote mission.

In summary, destroying a spacecraft, rocket stage or other projectile by impacting it onto another world can be useful for one or more of the following reasons.

- Remote or *in situ* observations/measurements of the ejecta plume for physical and compositional measurements, either telescopically from Earth or by another spacecraft.
- Remote observations of the crater from another spacecraft, for crater scaling information and exposure of sub-surface material.
- Generation of an artificial seismic source for measurements elsewhere on the body.
- NEO mitigation.

It is also worth mentioning here the possibility of observing the glow from atmospheric entry probes. This has been proposed in the context of Huygens (Lorenz, 2002) and, as for some other missions, was only successful in establishing an upper limit on emission (Lorenz *et al.*, 2006). The telescopic study of emission from bodies re-entering the Earth's atmosphere has also been the subject of recent work (e.g. in connection with the Genesis and Stardust return capsules, and the analysis of the ill-fated Space Shuttle Columbia) – these observations allow characterization of the thermal and non-thermal emissions from the shock layer, for a body with known mass and velocity. An additional serendipitous investigation was the encounter of the Mars Exploration rover Opportunity with part of its heat shield on the Martian surface, allowing at least partial documentation of the depth of charring of the thermal protection material. There are substantial and interesting intersections of the entry protection engineering disciplines with those associated with meteors and meteorites.

16

Atmospheric entry probes

The system design of atmospheric probes is dominated by the atmospheric entry and descent/drift through the atmosphere, even if surface operations are possible (e.g. Venera 7, 8, Pioneer Venus Day Probe, Huygens). Common experiment types for such probes include entry accelerometry, radio science for tracking the probe's motion, sensors for atmospheric temperature, pressure and humidity, mass spectrometry, aerosol analysis, (spectro-) photometry and nephelometry.

16.1 First Soviet Venera and Mars entry probes

This section covers early (1961–65) Soviet entry probes to Venus and Mars designed and built by Korolev's OKB-1 design bureau (now RKK Energia), all of which failed during launch or cruise. In 1965 further development of the deep space and lunar probes was handed over to NPO Lavochkin's Babakin Space Centre (then called OKB-301). Very few published details exist concerning the entry probes.

16.1.1 1VA entry probes

The first launches of spacecraft targeted at an extraterrestrial atmosphere were those of Venera 1, lost *en route*, and its 'twin' that failed to leave Earth orbit. The Venera 1 spacecraft was not designed to transmit a signal from the Venusian atmosphere. These 1VA spacecraft should be classed as 'destructive entry probes' rather than atmospheric entry probes in the modern sense of the phrase. The 1VA spacecraft were somewhat similar to those of the two Mars 1M craft, which were lost in launch failures in October 1960 (Figure 16.1).

Target	Venus	
Objectives	Demonstration of systems required for interplanetary flight; measurements of the environment in interplanetary space and close to Venus; entry into the Venus atmosphere and impact with the surface	
Prime Contractor	OKB-1	
Launch site, vehicle	Baikonour, Molniya 8K78	
	Sputnik 7	Venera 1
Launch date	04/02/1961	12/02/1961
Arrival date	–	–
Landing site co-ordinates	–	–
End(s) of Mission(s)	Failed to leave Earth orbit	Lost during cruise
Mass	Mass of capsule alone unknown; total for 1VA spacecraft: 643.5 kg	
Payload experiments	Only a Soviet pennant in a capsule with thermal protection, in a box mounted on the side. It was possibly not even due to separate from the main spacecraft. It was not due to transmit a signal	
Delivery architecture	Delivery by flyby s/c, from which it was possibly not even due to separate	
Thermal aspects	Unknown	
Power aspects	Unknown	
Communications architecture	Probably not due to transmit a signal	
EDL architecture	Ablative aeroshell?	
Landing speed(s)	Unknown	
Active operations (deployments, etc.)	Unknown	
Key references	Marov and Grinspoon, 1998; Hunten *et al.*, 1983; Kurt, 1994; Maksimov, 1997; Johnson, 1979; Chertok, 1999; Varfolomeyev, 1998	

16.1.2 2MV and 3MV entry probes

The second and third generations of OKB-1 probes saw continued development of the engineering subsystems needed to fly a spacecraft in interplanetary space reliably. The Venus and Mars probes were both based on the same generic design. A cruise/flyby spacecraft carried instrumentation for cruise science and either an entry probe ('descent apparatus') for *in situ* atmospheric measurements, or a remote sensing payload ('special compartment') for flyby observations.

Failures during launch, Earth escape or cruise meant that none of these probes returned data from another world, however, they were the first that are known to have carried scientific experiments destined for the atmospheres of Venus and Mars. Spherical in shape, they were attached to the flyby spacecraft by metal straps. These were to release the entry probe shortly before arrival. The probes

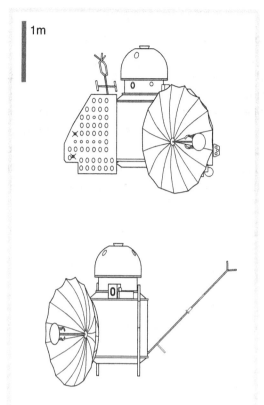

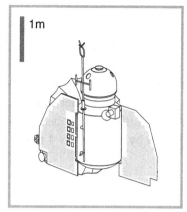

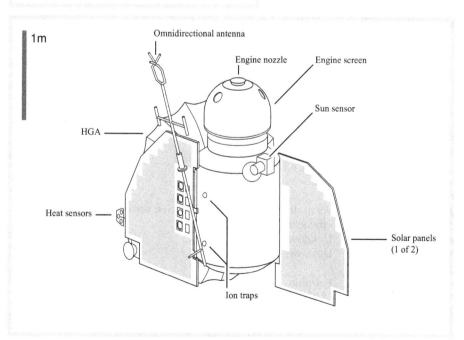

Figure 16.1 Venera IVA.

were designed to deploy a parachute system and transmit one-way data direct to Earth. Their design subsequently evolved into that of the successful Venera 4–8 probes.

Six 2MV probes were launched between August and November 1962; however, only Mars 1, carrying a remote sensing payload, left Earth orbit. Those carrying entry probes to Venus were type 2MV-1, while those carrying entry probes for Mars were type 2MV-3. The 2MV-2 and 2MV-4 craft were for Venus and Mars flybys, respectively. Interestingly, the design of the 2MV-3 entry probes assumed that the Martian atmosphere was much denser than the thin atmosphere later confirmed by the Mariner 4 flyby. It thus had a ballistic coefficient much too high

Target	Venus	Venus	Mars
Objectives	*In situ* measurements of the atmospheres and surfaces of Venus and Mars		
Prime contractor	OKB-1		
Launch site, vehicle	Baikonour, Molniya 8K78		
	Sputnik 19 (2MV-1)	Sputnik 20 (2MV-1)	Sputnik 24 (2MV-3)
Launch date	25/08/1962	01/09/1962	04/11/1962
Arrival date	–	–	–
Landing site co-ordinates	–	–	–
End(s) of mission(s)	Failed to leave Earth orbit		
Entry mass (kg)	~ 350		
Payload experiments	• T, P, density sensors (Mikhnevich)		
	• Chemical gas analysers (Florenskii, Andreichikov)		
	• Anti-coincidence gamma-ray counter using gas-discharge tubes (Lebedinskii, Krasnopolskii)		
	• Mercury-switch movement counter (Lebedinskii, Krasnopolskii)		
Delivery architecture	Delivery by flyby spacecraft		
Thermal aspects	Unknown		
Power aspects	Primary battery		
Communications architecture	One-way DTE		
EDL architecture	0.9 m diameter, spherical ablative aeroshell and 3-stage parachute system		
Landing speed(s)	Unknown		
Active operations (deployments, etc.)	Unknown		
Key references	Perminov, 1999; Kurt, 1994; Maksimov, 1997; Johnson, 1979; Semenov, 1994; Varfolomeyev, 1998		

to decelerate to a speed slow enough for safe parachute release, and would in fact have plummeted to the Martian surface at high speed.

Nine 3MV probes were launched between November 1963 and November 1965, to perform Mars and Venus (Venera) missions as well as interplanetary probe test (Zond) flights. Although one 3MV was lost in a launch failure, three failed to leave Earth orbit, and none returned data from Mars or Venus, the five remaining 3MV craft were partially successful in gaining valuable experience in the development and operation of spacecraft in interplanetary space. For example, Zond 3 returned images of the lunar far side, improving on those that had been returned by Luna 3 in 1959. Venera 3, and its flyby sibling Venera 2, were both lost shortly before arrival at Venus. Following the 3MV spacecraft, responsibility for the unmanned lunar and planetary probes was transferred to OKB-301, later the Babakin Space Centre of NPO Lavochkin.

Figure 16.2 shows a 2MV-1 or 2MV-3 craft (left) and a 3MV-3 (Venera 3) (right). The earlier 3MV-1 craft (e.g. Zond 1) were probably around 70 kg lighter, lacking modifications such as the large optical-baffle assembly seen on Venera 3.

Target	Venus		
Objectives	In situ measurements of the atmosphere and surface of Venus		
Prime contractor	OKB-1		
Launch site, vehicle	Baikonour, Molniya 8K78M		
	Cosmos 27 (3MV-1)	Zond 1 (3MV-1)	Venera 3 (3MV-3)
Launch date	27/03/1964	02/04/1964	16/11/1965
Arrival date	–	–	01/03/1966
Landing site co-ordinates	–	–	unknown
End(s) of mission(s)	Failed to leave Earth orbit	Lost during cruise	Lost during cruise (only 17 days before arrival)
Entry mass (kg)	~337?		337
Payload experiments	• T, P, density sensors (Mikhnevich) • G8-I & G8-II chemical gas analysers (Florenskii, Andreichikov (Surkov??)) • Anti-coincidence gamma-ray counter using STS-5 gas-discharge tubes (Lebedinskii or Avdiushin?) • Airglow photometer (Lebedinskii, Krasnopolskii) • Micro-organism detection experiment? (unflown proposal?)		

Table (Cont.)

Delivery architecture	Delivery by flyby spacecraft
Thermal aspects	Unknown
Power aspects	Primary battery
Communications architecture	One-way DTE
EDL architecture	0.9 m diameter, spherical ablative aeroshell and parachute system
Landing speed(s)	Unknown
Active operations (deployments, etc.)	Unknown
Key references	Marov and Grinspoon, 1998; Hunten *et al.*, 1983; Perminov, 1999; Kurt, 1994; Maksimov, 1997; Johnson, 1979; Semenov, 1994

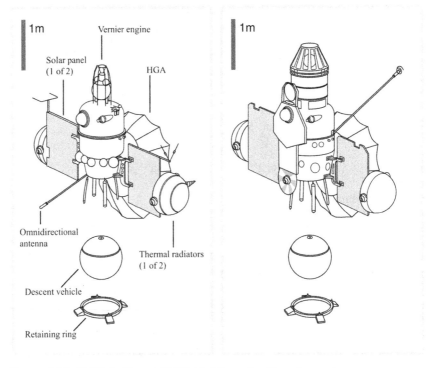

Figure 16.2 2MV (left) and 3MV (right) craft with entry probes.

16.2 Venera 4–8 (V-67, V-69, V-70 and V-72) entry probes

These fourth-generation Venus probes were used for four consecutive Venus launch windows from 1967 to 1972. In spite of the failure of three of the eight craft to leave Earth orbit (the result of ongoing problems with the launcher's upper stage) and the loss of the Venera 4, 5 and 6 probes before they reached the surface, these probes were the first successful planetary entry probes. They returned data on the atmospheric temperature and pressure profiles, composition, dynamics, light levels and surface composition, despite the highly restrictive data rate of 1 bit s^{-1}.

Following on from the 3MV entry probes, these probes were 10 cm larger in diameter than their predecessors, and became progressively more sophisticated and optimised to survive the Venusian temperature and pressure environment all the way down to the surface. Veneras 4–7 entered on the night side, Venera 8 on the day side. They eventually made way for a fifth generation of Veneras with greater payload capacity, surface capability and data rate. Venera 8 was still essentially an atmospheric probe; Venera 9 was to be a true lander. On Venera 8, however, to ensure that surface communication would still be possible if the probe did not come to rest in an upright position, an additional, ejected antenna was provided. Tethered to the probe, it contained a tilt switch to activate whichever face of the antenna landed uppermost. Venera 8's main transmitting antenna was also different. This was a result of the probe's planned landing near the dayside terminator – the Earth was much lower above the horizon and so an antenna having a beam pattern with higher gain at low elevations was chosen.

The first figure below (16.3) depicts the design of the Venera 4 probe (and its lost twin on Cosmos 167); Veneras 5 and 6 would have been almost identical in external appearance, with the possible exception of an aperture on 5 and 6 for the 'airglow photometer'. The second and third figures (16.4 and 16.5) depict Veneras 7 and 8, respectively (and of course their lost 'twins').

16.3 Pioneer Venus probes

Pioneer Venus involved two launches: an orbiter and a 'multiprobe bus' spacecraft to carry four entry probes, for release into the Venusian atmosphere prior to the destructive entry of the bus itself. To meet the objectives of both detailed measurements in the atmosphere and multiple measurements at different locations, one large probe and three smaller ones were flown, the Large Probe having seven times the payload capacity of each of the Small Probes (Day, North and Night). The bus also carried a small payload, returning mass spectrometer measurements down to 110 km altitude.

	Venera 4	Cosmos 167	Venera 5	Venera 6	Venera 7	Cosmos 359	Venera 8	Cosmos 482
Target	Venus							
Objectives	In situ measurements of the atmosphere and surface of Venus							
Prime contractor	NPO Lavochkin (formerly OKB-301)							
Launch site, vehicle	Baikonour, Molniya 8K78M							
Launch date	12/06/1967	17/06/1967	05/01/1969	10/01/1969	17/08/1970	22/08/1970	27/03/1972	31/03/1972
Arrival date	18/10/1967	–	16/05/1969	17/05/1969	15/12/1970	–	22/07/1972	–
Landing site co-ordinates	Entry at 19° N, 38° E	–	Entry at 3° S, 18° E	Entry at 5° S, 23° E	5° S, 351° E	–	10.70° S, 335.25° E	–
Mission durations	94 min TX, ended at 25 km altitude	–	53 min TX, ended at 18 km altitude	51 min TX, ended at 18 km altitude	35 min descent +23 min on surface	–	55 min descent +50 min on surface	– (still in Earth orbit!)
Entry mass (kg)	383	383	405	405	500	500	495	495

Payload experiments

Venera 4: 1 m diam
- MDDA T, P, density sensors (Avduevskii, Marov, Rozdestvensky, Mikhnevich) (TPV beta-particle densitometer & Pt resistance thermometers integrated in Mikhnevich's IS-164D unit)
- G-8 & G-10 chemical gas analysers (Florenskii (Surkov??))

- Doppler expt (Kerzhanovich)
- Radio altimeter using FMCW technique (? @ NII-17)

The Project Scientist was Mikhail Ya. Marov

Venera 5, 6: 1m diam
- MDDA & IS-164D T, P & VIP tuning-fork densitometer sensors (Avduevskiii, Marov, Rozdestvensky)
- Improved G-8 & G-10 chemical gas analysers (Florenskii (Surkov??))
- Doppler expt (Kerzhanovich)
- FO-69 Airglow photometer (Moroz)
- Radio altimeter using FMCW technique (? @ NII-17)

The Project Scientist was Mikhail Ya. Marov

Venera 7: 1m diam (egg-shaped)
- ITD T, P sensors (Avduevskii, Marov, Rozdestvensky)
- Doppler expt (Kerzhanovich)
- Radio altimeter (? @ NII-17)
- Possibly also beta particle densitometer (Mikhnevich), DOU-1M accelerometer (Avduevskii) and GS-4 gamma ray spectrometer (Vinogradov?, Surkov)

The Project Scientist was Mikhail Ya. Marov

Venera 8: 1m diam (egg-shaped).
- ITD T, P sensors (Marov)
- DOU-1M Accelerometer (Avduevskii)
- IOV-72 photometers (Selivanov, Avduevsky?, Ekonomov?)
- Doppler expt (Kerzhanovich)
- Radar altimeter (Natalovich, Tseitlin @ RNII KP)

Table (Cont.)

	• IAV-72 Ammonia analyser (Surkov)
	• GS-4 Gamma-ray spectrometer (Vinogradov?, Surkov)
	The Project Scientist was Mikhail Ya. Marov
Delivery architecture	Separation from flyby s/c 20 000–40 000 km from Venus
Thermal aspects	Pre-cooling to −10 °C before probe separation; pressurized main compartment; internal fan; lithium nitrate trihydrate (V-72 only)
Power aspects	28Ah Primary battery
Communications architecture	One-way DTE at 920 MHz, data rate ~1 bit s^{-1} using FSK
EDL architecture	Entry angles 43–65° at ~11.2 km s^{-1}. Spherical or egg-shaped aeroshell; pilot and main parachutes. Peak entry loads 400–500 g
Landing speed(s)	16.5 m s^{-1} (Venera 7)
Active operations (deployments, etc.)	Deployment of radar altimeter antennas, communications antenna and surface communications antenna (Venera 8 only)
Key references	Jastrow and Rasool, 1969; Hunten et al., 1983; Marov and Grinspoon, 1998

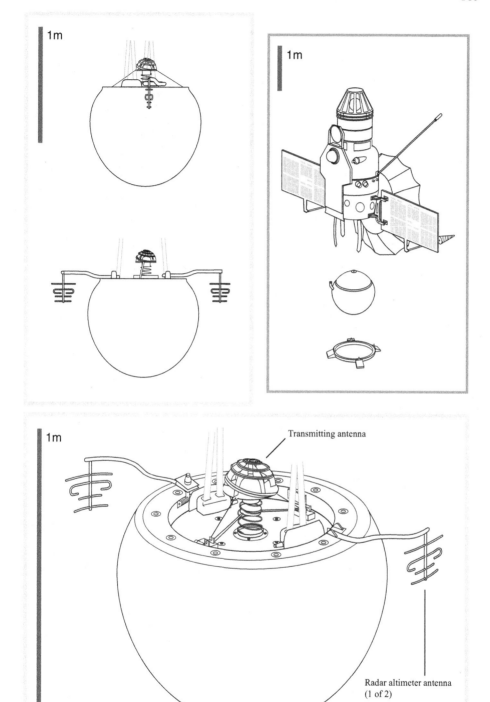

Figure 16.3 Venera 4, 5, 6.

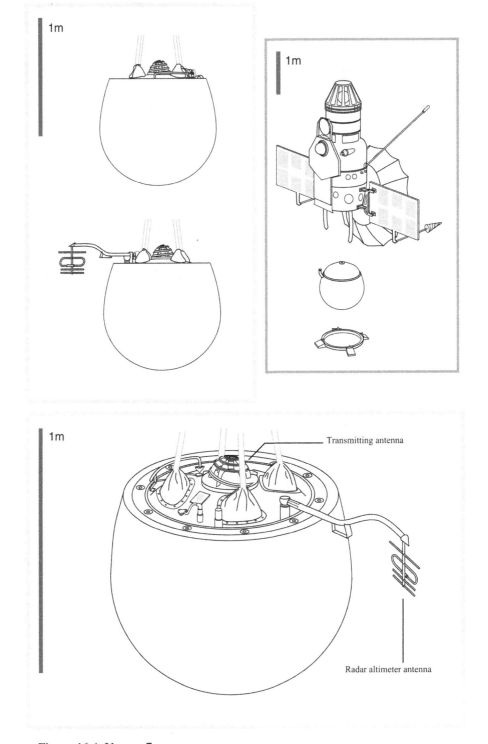

Figure 16.4 Venera 7.

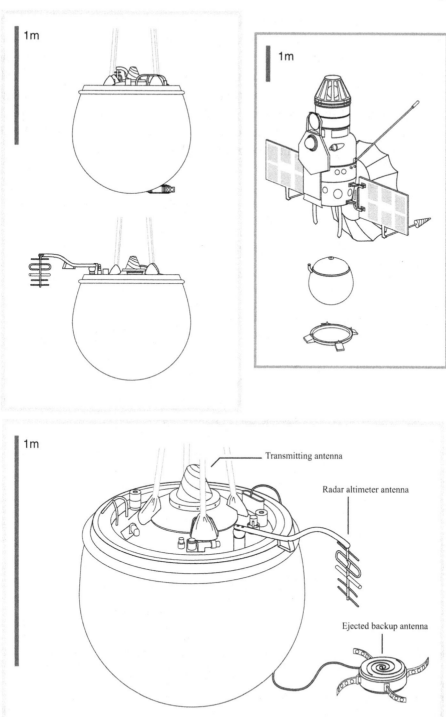

Figure 16.5 Venera 8.

The Small Probes carried no parachutes and retained their entry heat shields down to the surface, and the parachute of the Large Probe was jettisoned at 45 km altitude. This was to ensure rapid descent of the probes in the most hostile, lower part of the atmosphere. A particular challenge for the probes was the design of the (gas-filled) pressure vessels and the necessary hull penetrations that could withstand differential thermal expansion of the components. Optical windows of diamond and sapphire were used. All four probes were successful in their operation, two of the Small Probes even surviving for a time on the surface (Figures 16.6 and 16.7).

Pioneer Venus Large Probe

Target	Venus
Objectives	Atmospheric structure, dynamics, cloud structure and optical properties
Prime contractor	Hughes
Launch site, vehicle	ETR 36A Atlas-Centaur (on probe bus)
Launch date	08/08/1978
Arrival date	09/12/1978
Landing site co-ordinates	4.4° N, 304° E
End(s) of mission(s)	At impact
Entry mass (kg)	316.5
Payload experiments	• LAS (atmospheric structure) (Seiff)
	• LN (nephelometer) (Ragent, Blamont)
	• LCPS (cloud particle-size spectrometer) (Knollenberg)
	• LGC (gas chromatograph) (Oyama)
	• LIR (IR radiometer) (Boese)
	• LNMS (neutral particle-mass spectrometer) (Hoffman)
	• LSFR (solar-flux radiometer) (Tomasko)
	• DLBI (differential long baseline interferometer) (Counselman)
	• MPRO (atmospheric propagation) (Croft)
	• MWIN (Doppler tracking) (Kliore)
	• MTUR (atmospheric turbulence) (Woo?)
	Totals: 35 kg, 106 W.
	The Project Scientist was Lawrence Colin.
Delivery architecture	Release on approach by 15 rpm spin-stabilised probe bus carrying large probe and small probes. Release by spring, 23 days from encounter
Thermal aspects	0.78 m diameter 3-piece Ti pressure vessel with internal MLI and Be shelves. Pressurization with N_2
Power aspects	AgZn secondary battery 22 V, 40 A-hr
Communications architecture	One-way DTE: 40 W, 2.3 GHz, 256 bit s^{-1} (128 bit s^{-1} during entry blackout). Receiver for two-way Doppler only
EDL architecture	Entry at ~11.7 km s^{-1}, flight path angle −34°. 1.42 m diameter, 45° blunt half-cone, carbon-phenolic

Table (Cont.)

	Ablative aeroshell released after deployment of pilot chute, back cover and main parachute (conical ribbon); main parachute then released at 45 km. Peak entry load 280 g.
Landing speed(s)	9 m s^{-1}
Active operations (deployments, etc.)	EDL aspects only
Key references	Brodsky, 1979; Pioneer Venus, 1980; Fimmel *et al.*, 1983, 1995; Hunten *et al.*, 1983; Bienstock, 2004

Pioneer Venus Small Probes

Target	Venus		
Objectives	Atmospheric structure, dynamics, cloud structure and optical properties		
Prime contractor	Hughes		
Launch site, vehicle	ETR 36A, Atlas-Centaur (on probe bus)		
Launch date	08/08/1978		
Arrival date	09/12/1978		
	Day Probe	North Probe	Night Probe
Landing site co-ordinates	31.3° S, 317° E	59.3° N, 4.8° E	28.7° S, 56.7° E
End(s) of mission(s)	67 min after impact	at impact	2 s after impact
Entry mass (kg)	94 (each)		
Payload experiments	• SAS (atmospheric structure) (Seiff)		
	• SN (nephelometer) (Ragent, Blamont)		
	• DLBI (differential long base-line interferometer) (Counselman)		
	• SNFR (net flux radiometer) (Suomi)		
	• MPRO (atmospheric propagation) (Croft)		
	• MWIN (Doppler tracking) (Kliore)		
	• MTUR (atmospheric turbulence) (Woo)		
	Totals: 5 kg, 10 W.		
	The Project Scientist was Lawrence Colin		
Delivery architecture	Release and dispersion on approach by 48 rpm spin-stabilized probe bus carrying large probe and small probes, 19 days from encounter. Despin to 17 rpm by yo-yo mechanism		
Thermal aspects	0.46 m diameter 2-piece Ti pressure vessel with internal MLI and Be shelves. Pressurization with Xe		
Power aspects	AgZn secondary battery 22 V, 11 A-hr		
Communications architecture	One-way DTE: 10 W, 2.3 GHz, 64 bit s^{-1} during entry blackout and at high altitude; 16 bit s^{-1} below 30 km		
EDL architecture	Entry at $\sim$11.7 km s^{-1}, flight path angles from $-23°$ to $-71°$. 0.76 m diameter, 45° blunt half-cone, carbon-phenolic ablative aeroshell remained attached through out descent. Peak load 223–458 g		
Landing speed(s)	10 m s^{-1}		
Active operations (deployments, etc.)	Yo-yo despin mass release; door mechanisms for instrument deployment		
Key references	Brodsky, 1979; Pioneer Venus, 1980; Fimmel *et al.*, 1983, 1995; Hunten *et al.*, 1983; Bienstock, 2004		

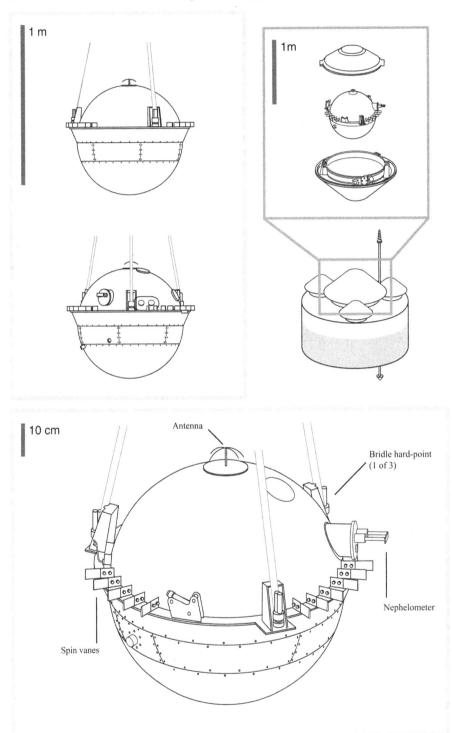

Figure 16.6 Pioneer Venus Large Probe.

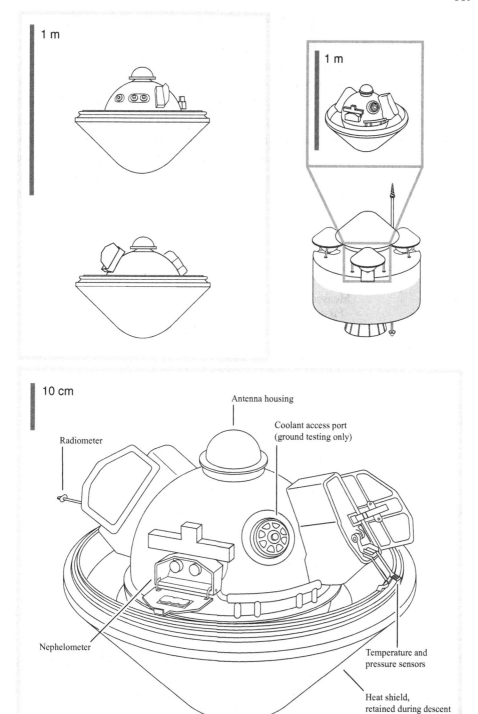

Figure 16.7 Pioneer Venus Small Probes.

16.4 VeGa AZ balloons

The two VeGa (Venera–Halley) spacecraft released balloons ('Aerostatic Zonds') into the atmosphere of Venus, as part of their deployment of the latest Venera-class landers to investigate the planet's atmosphere and surface. At the time of writing these remain the only balloons to have been sent to the atmosphere of another world. During the planning stage for this mission, larger French-led balloons had been considered but the inclusion of comet Halley in the mission plan for the flyby spacecraft forced the use of the smaller, Soviet balloons eventually flown. Both were successful, surviving for nearly two days in the nightside atmosphere while being tracked by radio telescopes on Earth (Figure 16.8).

Target	Venus	
Objectives	Measure winds, vertical heat flux and cloud particle density. First planetary balloon	
Prime contractor	NPO Lavochkin (formerly OKB-301)	
Launch site, vehicle	Baikonour, Proton 8K82K/11S824M (within VeGa 1 & 2 probes)	
	VeGa 1 AZ	VeGa 2 AZ
Launch date	15/12/1984	21/12/1984
Arrival date	11/06/1985	15/06/1985
Landing site co-ordinates	Floating altitude 54 km, latitude 8° N	Floating altitude 54 km, latitude 7° S
End(s) of mission(s)	46.5 h later	46.5 h later
Mass(es)	6.9 kg gondola, 2 kg helium, total 20.5 kg	
Payload experiments	• VLBI measurements of position & velocity (Sagdeev, Blamont, Preston) • Doppler expt (Kerzhanovich) • Meteocomplex (T, P sensors, vertical-wind anemometer, light level/lightning detectors) (Linkin, Blamont) • Nephelometer (Ragent, Blamont, Linkin) The Project Scientist was Roald Z. Sagdeev	
Delivery architecture	Release from lander back cover during descent.	
Thermal aspects	Float at benign altitude with temperatures ∼30–40 °C	
Power aspects	250 W-hr lithium battery (1 kg)	
Communications architecture	One-way DTE: 4.5 W at 1.667 GHz at 4 bit s^{-1} (initially for 270 s every 30 minutes); DVLBI measurements of position	

Table (Cont.)

EDI architecture	Released from VeGa lander at 61 km; parachute deployed at 55 km. Rise from 50 km to 54 km operating altitude. 3.54 m Teflon fabric balloon, filled with helium
Landing speed(s)	–
Active operations (deployments, etc.)	Jettison of parachute and inflation system. Ballast released at 50 km. Deployment of boom carrying temperature sensors and anemometer
Key references	Kremnev *et al.*, 1986; Sagdeev *et al.*, 1986; Aleksashkin *et al.*, 1988a,b; Vorontsov *et al.*, 1988; Hunten *et al.*, 1983; Marov and Grinspoon, 1998; Blamont *et al.*, 1993; Sagdeev *et al.*, 1986; TsUP, 1985; MNTK, 1985

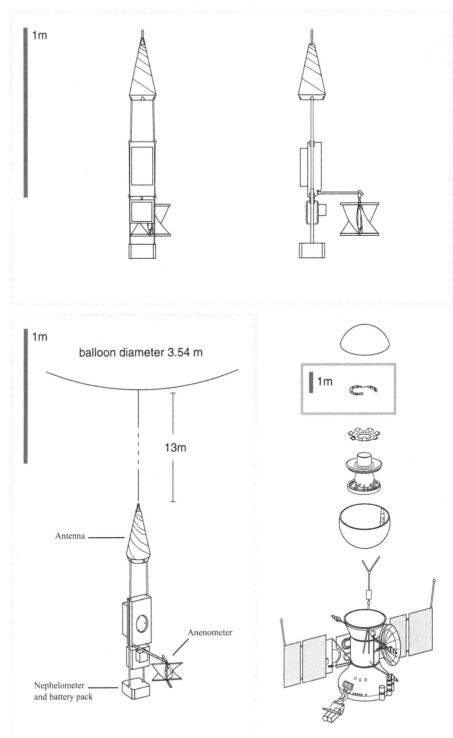

Figure 16.8 VeGa AZ balloons.

16.5 Galileo Probe

The Galileo probe (Figure 16.9) entered the atmosphere of Jupiter in December 1995. For more details see the Case Study, Chapter 22.

Target	Jupiter
Objectives	The probe's scientific objectives were to:
	• Determine the chemical composition of the Jovian atmosphere
	• Characterize the structure of the atmosphere to a depth of at least 10 bar
	• Investigate the nature of cloud particles and the location and structure of cloud layers
	• Examine the Jovian radiative heat balance
	• Study the nature of Jovian lightning activity
	• Measure the flux of energetic charged particles down to the top of the atmosphere
Prime contractor	Hughes
Launch site, vehicle	ETR, Shuttle with IUS (on board Galileo)
Launch date	18/10/1989
Arrival date	07/12/1995
Entry position	6.57° N
End(s) of mission(s)	Last transmissions received 61.4 min after entry interface
Entry mass (kg)	335
Payload experiments	• ASI atmospheric-structure instrument (Seiff)
	• NMS neutral mass spectrometer (Niemann)
	• NEP nephelometer (Ragent)
	• LRD lightning and radio-emissions detector (Lanzerotti)
	• HAD helium abundance detector (von Zahn)
	• NFR net-flux radiometer (Sromovsky)
	• EPI energetic-particles instrument (Fischer)
	• DWE Doppler wind experiment (Atkinson)
	Totals: 28 kg, 26 W. Ground-based Doppler tracking was also performed (Folkner).
	The Project Scientist was Richard Young
Delivery architecture	Release by Galileo on approach, 148 days before encounter
Thermal aspects	0.66 m diameter vented vessel with internal MLI and RHUs
Power aspects	$LiSO_2$ primary batteries, 22 A-hr; $Ca/CaCrO_4$ themal battery
Communications architecture	One-way redundant relay via Galileo, at 1387.0 and 1387.1 MHz, each 24 W and 128 bit s^{-1}. See Part 3 for more details
EDL architecture	Entry at $\sim$48 km s^{-1} (relative) or 60 km s^{-1} (inertial), relative flight path angle $-8.4°$. 1.25 m diameter, 45° blunt half-cone, carbon-phenolic ablative aeroshell released after deployment of pilot chute, back cover and main parachute. Peak entry load 250 *g*
Active operations (deployments, etc.)	Nephelometer arm deployment
Key references	Russell, 1992; Bienstock, 2004; Harland, 2000; Vojrodich *et al.*, 1983; Young *et al.*, 1996; Young, 1998

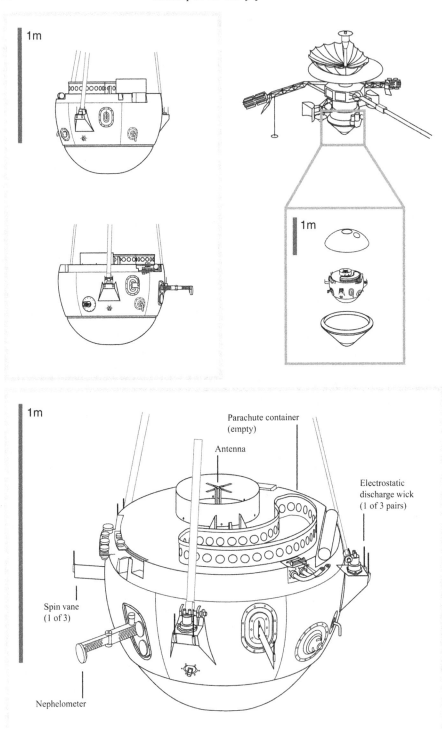

Figure 16.9 Galileo Probe.

16.6 Huygens

The Huygens probe entered the atmosphere of Titan in January 2005 and descended
to its surface. For more details see the Case Study, Chapter 23. (Figure 16.10)

Target	Titan
Objectives	• Determine atmospheric composition
	• Investigate energy sources for atmospheric chemistry
	• Study aerosol properties and cloud physics
	• Measure winds and temperatures
	• Determine properties of the surface and infer internal structure
	• Investigate the upper atmosphere and ionosphere
Prime contractor	Aerospatiale (now part of Thales Alenia Space)
Launch site, vehicle	ETR, Titan IVB (on board Cassini)
Launch date	15/10/1997
Arrival date	14/01/2005
Landing site co-ordinates	10.2° S, 192.365° W
End(s) of mission(s)	14/01/2005, after a descent of 149.5 min and 72 min of surface data.
Mass(es)	Entry mass: 318 kg, $\sim$200 kg without entry and descent subsystems
Payload experiments	• DISR descent imager/spectral radiometer (Tomasko)
	• DWE Doppler wind experiment (Bird)
	• GCMS gas-chromatograph mass spectrometer (Niemann)
	• HASI Huygens atmospheric-structure instrument (Fulchignoni)
	• ACP aerosol collector pyrolyser (Israël)
	• SSP surface science package (Zarnecki)
	Total mass 48.57 kg plus 2.95 kg balance mass. Radar altimetry data was also returned. Ground-based Doppler and VLBI tracking was also performed (Gurvits). The Project Scientist was Jean-Pierre Lebreton
Delivery architecture	Delivery by Cassini from Saturn orbit 20 days before Titan encounter. Ejection at 0.35 m s^{-1}, spinning at 7.5 rpm
Thermal aspects	Foam insulation; RHUs
Power aspects	LiSO$_2$ Primary batteries
Communications architecture	One-way relay via Cassini on two S-band channels; one carrier detected directly on Earth (for Doppler/VLBI)
EDL architecture	Entry at $\sim$6 km s^{-1}, flight path angle $-64°$. Aeroshell (peak load $\sim$20 g). Spin stabilised. Pilot chute pulls away back cover; main chute deployment; aeroshell separation; main chute separation and stabilisation chute deployment
Landing speed(s)	4.67 m s^{-1}
Active operations (deployments, etc.)	Deployment of HASI booms; ejection of DISR cover and GCMS cap; actuation of ACP aerosol collector
Key references	Wilson, 1997; *Space Sci. Rev.* 104(1), 2002; *Nature* 438(7069), 2005; Lorenz and Mitton, 2002; Harland, 2002; Hassan and Jones, 1997; Spilker, 1997.

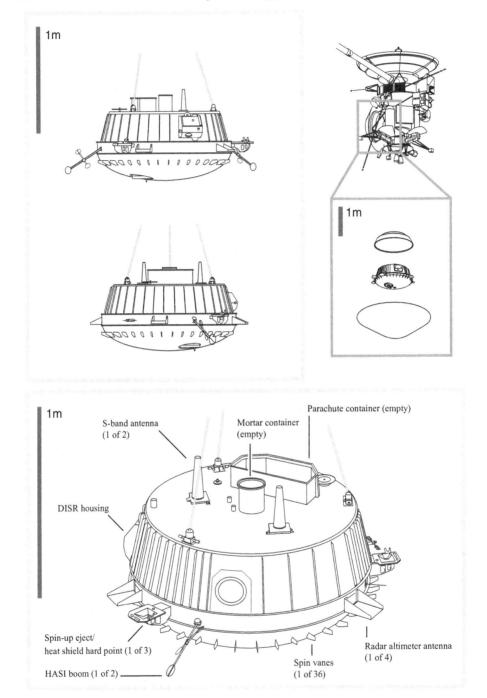

Figure 16.10 Huygens.

17

Pod landers

The landers covered in this chapter have the ability to survive an initial landing impact, which may send the vehicle rolling and/or bouncing across the surface, and then commence operations having come to rest in whatever orientation is finally reached. Most achieve this by means of airbags to cushion and dampen the initial impact and subsequent rolling/bouncing motion, followed by the opening out of a system of 'petals' to bring the lander itself to its proper orientation for surface operations. The Ranger seismometer capsules are the exception to this; their impact damping was provided by the balsa-wood shell and liquid-bath system surrounding the experimental equipment, and the orientation being achieved by means of the natural position of the equipment within its liquid bath.

Typical payload experiments for such landers include cameras, meteorological, geological, geophysical and environmental sensors for investigation of the landing site. While some can be body-mounted on the probe, others may require deployment by means of masts, arms or a rover. In the case of the Mars Exploration Rovers, the pod landing stage itself plays no further role once the rover has rolled off.

Pod landers are particularly suited to 'network science', where simultaneous seismological, meteorological or other geophysical measurements are made at multiple locations. Such a network was the aim of the NetLander mission[12] a network of four Mars landers to be carried on the CNES-led Mars Premier mission. The mission was cancelled in 2003 towards the end of Phase B of the project, however.

At the time of writing, ESA is planning the ExoMars mission, which will send a rover to Mars to search for evidence of life and to assess possible hazards for human exploration. The rover would be deployed in a similar fashion to the Mars Exploration Rovers, namely by means of a three-petalled

---

[12] The payload of each of the four NetLanders was as follows: ATMIS (Harri), ARES (Berthelier), GPR (Berthelier), MAGNET (Menvielle), NEIGE (Dehant), PANCAM (Jaumann), SEIS (Lognonné), SPICE (Spohn), MIC (Delory). The Project Scientist was Jean-Louis Counil.

pod lander. However, following roll-off of the rover, the landing stage may also incorporate long-term geophysical measurements, using payload heritage from a number of previous projects including NetLander, Beagle 2 and the cancelled Mercury Surface Element of the ESA-led BepiColombo mission.

17.1 Ranger Block 2 Seismo capsules

The three Ranger Block 2 missions – Rangers 3, 4 and 5 – were the first, albeit unsuccessful attempts to soft land and return data from an extraterrestrial surface. Having separated from the upper stage of the launcher, the Ranger cruise stage would have delivered the landers to the lunar surface on a direct descent trajectory. The landers comprised a 63.5 cm diameter balsa wood capsule mounted on a solid retro motor for descent braking. The spherical capsule enclosed and protected a 27.5 cm diameter seismometer experiment. Lunar planetary protection requirements of the time were met by baking of components at 125 °C for 24 h, clean assembly and bathing of the assembled spacecraft in ethylene oxide (Figure 17.1).

Target	The Moon		
Objectives	To achieve the first soft landing on the Moon and transmit lunar seismic data		
Prime contractor	Aeronutronic Division, Ford Motor Company		
Launch site, vehicle	ETR, Atlas-Agena B		
	Ranger 3	Ranger 4	Ranger 5
Launch date	26/01/1962	23/04/1962	18/10/1962
Arrival date	Missed by 37745 km on 28/01/1962	Impacted 6/04/1962	Missed by 724 km on 21/10/1962
Landing site co-ordinates	–	140.5° W, 15.5° S (lunar far side)	–
End(s) of mission(s)	Seismo capsules had an expected lifetime of 30–60 days Rangers 3 and 5 were tracked until 31/01/1962 and 30/10/1962, respectively.		
Masses	Lunar capsule system: 148.3 kg, incl. ∼98 kg retro motor		
Payload experiments	• Seismometer (3.36kg) (Press) • Capsule temperature & maximum deceleration measurement (Ewing)		
	The Project Scientist was Harold Washburn		

Table (Cont.)

Delivery architecture	Separation from cruise stage during approach at 21.4 km altitude
Thermal aspects	Insulation and 1.68 kg of water evaporating in an open-loop cooling system. 'Shower curtain' aluminized Mylar insulation surrounded the lander during cruise. Black 'saw-tooth' surface pattern added to white capsule for Rangers 4 and 5 to maintain a more even temperature during cruise
Power aspects	AgCd battery
Communications architecture	One-way DTE phase-modulated analogue signal at 960 MHz, 50 mW
EDL architecture	Spin-up to 300 rpm and firing of solid retro-rocket (Hercules BE-3). Separation of 63.5 cm diameter balsa wood capsule from retro upon burnout at 300 m altitude, then free-fall to surface
Landing speed(s)	Intended: $<56\,\mathrm{m\,s^{-1}}$; Ranger 4 impacted at $2.67\,\mathrm{km\,s^{-1}}$
Active operations (deployments, etc.)	Righting of seismometer achieved passively using freon liquid, which was then removed by means of a puncturing device
Key references	Hall, 1977; NASA, 1962, 1963; Wilson, 1982; Adamski, 1962

17.2 Luna 4–9, 13 (Ye-6 and Ye-6M) landers

After 11 previous attempts that ended in launch failure, loss during cruise or landing failure, the first successful landing on the Moon was achieved by 'automatic lunar station' Luna 9 in February 1966. Luna 9 was also the first of the series to be built by OKB-301, thus being designated a Ye-6M, as opposed to the previous 11 Ye-6 craft. A further Ye-6M became Luna 13, carrying an augmented payload. The landers were 0.58 m diameter pressurised vessels, with a system of four petals that opened to bring the lander upright. They also acted as the ground plane for the whip antennas. The two figures (17.2 and 17.3) show Luna 9 and Luna 13, respectively. The earlier Ye-6 landers are perhaps most easily distinguished from Luna 9 by their noticeably different camera turret.

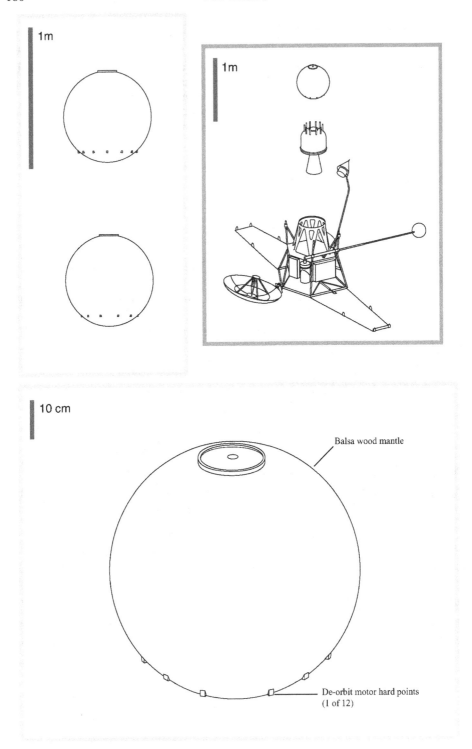

Figure 17.1 Ranger 3, 4, 5.

Target	The Moon						
Objectives	To achieve the first soft landing on the Moon and return imagery and other information on the lunar surface relevant to the design of subsequent missions						
Prime contractor	OKB-1 then NPO Lavochkin (formerly OKB-301)						
Launch site, vehicle	Baikonour, Molniya 8K78 or 8K78M						
	Sputnik 25	Luna 1963B	Luna 4	Luna 1964A	Luna 1964B	Cosmos 60	Luna 1965A
Lanch date	04/01/1963	03/02/1963	02/04/1963	21/03/1964	20/04/1964	12/03/1965	10/04/1965
Arrival date	–	–	Missed the moon	–	–	–	–
Landing site-co-ordinates	–	–	–	–	–	–	–
End(s) of Mission(s)	–	–	–	–	–	–	–
Lander mass (kg)			~82				

	Luna 5	Luna 6	Luna 7	Luna 8	Luna 9	Luna 13
Lanch date	09/05/1965	08/06/1965	04/10/1965	03/12/1965	31/01/1966	21/12/1966
Arrival date	Impacted 12/05/1965	Missed the Moon	Impacted 07/10/1965	Impacted 06/12/1965	03/02/1966	24/12/1966
Landing site co-ordinates	?	–	?	?	7° 8′ N 64° 32′ W	18° 52′ N 62° 04′ W
End(s) of Mission(s)	–	–	–	–	06/02/1966	28/12/1966
Lander mass (kg)		~82			105	112

Table (Cont.)

Payload experiments	Luna 9: the first Ye-6M: 0.58 m diam. capsule • Panoramic telephotometer (Selivanov) • Mirrors & calibration targets/tilt indicators • KS-17M radiation detector using SBM-10 tube(s) (Vernov?) www.laspace.ru says a gamma ray spectrometer was deleted from Luna 9 before launch. Keldysh (1980) mentions an LS seismometer (Sadovsky?) and an SG-57 3-axis boom-mounted magnetometer (Dolginov) – probably unflown Ye-6 payload candidates? The Ye-6 telephotometers were Volga instruments (Rosselevich) The Project Scientist was Aleksandr P. Vinogradov Luna 13: 0.58m diam. capsule. • Panoramic telephotometers (stereo pair) (Selivanov) • ID-3 infrared radiometer (Lebedinskii, Krasnopolskii) • GR-1 'gruntomer' penetrometer (Cherkasov) • RP 'plotnomer' radiation densitometer (Cherkasov) • KS-17MA radiation detector (Vernov?) • DS-1 'Yastreb' dynamograph (?) The Project Scientist was Aleksandr P. Vinogradov
Delivery architecture	Delivery by cruise/braking stage; descent direct from translunar trajectory; braking motor triggered by radar altimeter, followed by vernier engines for the final part of descent; separation of lander on contact with the surface
Thermal aspects	Open-cycle water cooling
Power aspects	Primary battery
Communications architecture	Two-way DTE, transmissions phase modulated at 183.538 MHz
EDL architecture	Airbag inflation and spring-loaded separation from descent stage on signal from 5 m contact probe. Jettison of airbags and opening of 4 petals
Landing speed(s)	Unknown
Active operations (deployments, etc.)	Opening of petals; deployment of whip antennas; deployment of plotnomer and gruntomer arms (Luna 13 only)
Key references	Vinogradov 1966,1969; Cherkasov et al., 1968a,b; Morozov et al., 1968; Siddiqi et al., 2000; Siddiqi, 2002

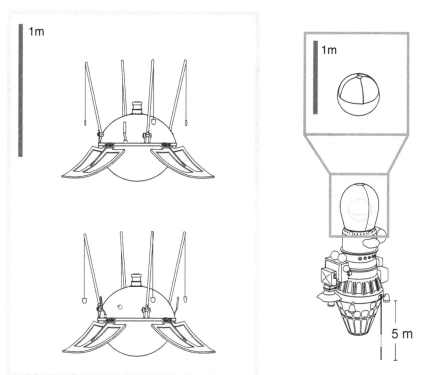

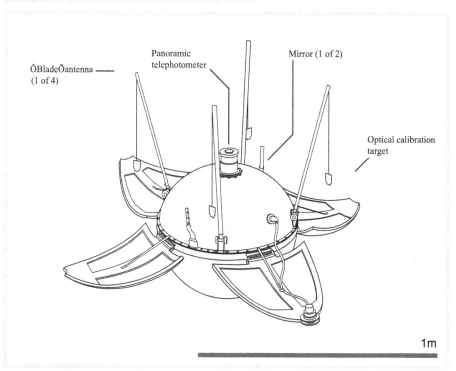

Figure 17.2 Luna 9.

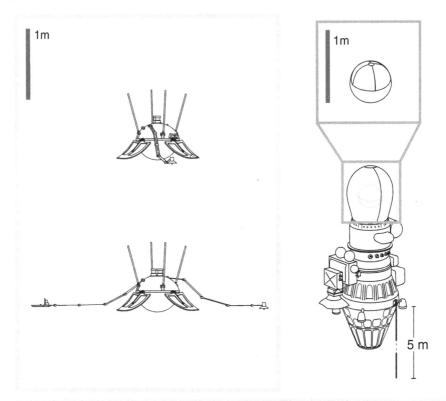

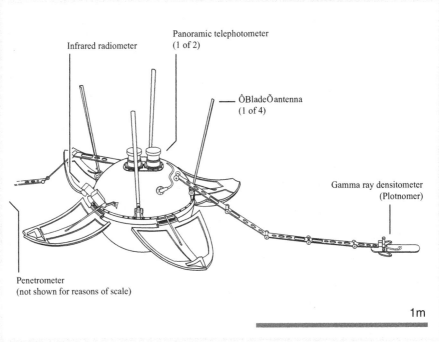

Figure 17.3 Luna 13.

17.3 Mars 2, 3, 6, 7 (M-71 and M-73) landers

These were the first attempts to soft land on Mars, following the failed 2MV-3 entry probe attempt in 1962 and the landers deleted for mass reasons from the Mars 69 mission. A lander with a Luna 9-style four-petalled opening/righting mechanism was encased in crushable material and delivered by parachute and braking rockets from the entry assembly (Figure 17.4).

Target	Mars			
Objectives	Soft landing on Mars and investigations of the atmosphere and surface			
Prime contractor	NPO Lavochkin (formerly OKB-301)			
Launch site, vehicle	Baikonour, Proton 8K82K/11S824			
	Mars 2	Mars 3	Mars 6	Mars 7
Launch date	19/05/1971	28/05/1971	05/08/1973	09/08/1973
Arrival date	27/11/1971	02/12/1971	12/03/1974	Missed by 1300 km
Landing site co-ordinates	44.2° S, 313.2° W	45° S, 158° W	23.9° S, 19.4° W –	
End(s) of mission(s)	Impact due to targeting error and thus wrong flight path angle	20 s after start of transmission from surface	0.3 s before impact	–
Masses	1210 kg descent module containing 350 kg landing capsule			
Payload experiments	Mars 2, 3:			

• Payload nearly identical to that of the 6 & 7 landers but the latter had significantly upgraded T, P sensors, mass spectrometer & telephotometry, and were able to transmit data during descent

The Project Scientist was Mikhail Ya. Marov

Mars 6, 7:
• panoramic telephotometers (stereo pair) (Selivanov)
• T, P sensors (Marov)
• density & wind sensors (Linkin)
• accelerometer (Cheremukhina)
• Doppler expt (Kerzhanovich)
• mass spectrometer (Istomin)
• 'automatic laboratory for activation analysis' (XRFS?) (Surkov)
• PROP-M tethered walking rover (4.5 kg):
  • penetrometer (Kemurdzhian)
  • densitometer (Surkov)

The Project Scientist was Mikhail Ya. Marov

Table (Cont.)

Delivery architecture	Separation from orbiter (Mars 2, 3) or flyby spacecraft (Mars 6, 7) on approach
Thermal aspects	Passive control with insulation
Power aspects	Primary batteries
Communications architecture	One-way VHF relay via orbiter or flyby spacecraft
EDL architecture	Solid rocket motor firing and separation; firing of spin-up motors on 120° aeroshell; drogue and main parachute deployment; entry shell separation; parachute separation and retrorocket firing; crushable material surrounding 4-petalled lander. Descent duration: 250 s (3), 240 s (6)
Landing speed(s)	$21 \, \mathrm{m \, s^{-1}}$ (Mars 3)
Active operations (deployments, etc.)	Opening of petals; deployment of whip antennas; deployment of PROP-M tethered walking rover, mast for temperature and wind sensors, and 'automatic laboratory for activation analysis' from one petal
Key references	Marov and Petrov, 1973; Kerzhanovich, 1977; Ivanov, 1977; Markov, 1989; Kieffer *et al.*, 1993; Perminov, 1999

17.4 Mars 96 Small Stations

The Mars 96 orbiter (which would have been named Mars 8 had it left Earth orbit successfully) carried four surface elements: two landers ('Small Stations') and two penetrators. These would have enabled a network of simultaneous measurements to be made. The mission was delayed by one launch window, before which it was known as Mars 94. Subsequent Mars missions were planned to carry balloons and rovers but were abandoned after the Mars 96 failure. The Small Stations (Figure 17.5) carried an international payload and were intended to operate for an extended period, with an extremely small power budget. The objectives of Martian network science are still being pursued through other proposed missions.

Target	Mars
Objectives	Exploration of the dynamics and structure of the atmosphere, the role of water and other materials containing volatiles, studies of atmospheric boundary layer processes, surface chemistry and geology, to obtain new information on the puzzle of an intrinsic magnetic field, and to study the interior of Mars by recording seismic activity
Prime contractor	IKI/NPO Lavochkin

Table (Cont.)

Launch site, vehicle	Baikonour, Proton 8K82K/11S824M (on Mars 96 orbiter)	
Launch date	16/11/1996	
Arrival date	(planned for 12/09/1997)	
	Small Station 1	Small Station 2
Landing site co-ordinates (planned)	37.6° N, 161.9° W	33° N, 169.4° W
End(s) of mission(s)	Nominal lifetime on Mars: 1 year	
Mass(es)	87 kg separation mass 33 kg lander including airbags 12 kg payload	
Payload experiments	• PANCAM panoramic camera (Linkin, Runavot) • DESCAM descent camera (Lipatov, Hua, Runavot) • OPTIMISM: • SIS seismometer (Lognonné) • MAG magnetometer (Menvielle, Musmann) & inclinometer • APX alpha-proton-X-ray spectrometer (Rieder, Economou) • MIS meteorology instrument system (Harri): • PTUW P,T, humidity & wind sensors (Harri, Linkin) • ODS optical depth sensor (Pommereau) • DPI descent phase instrument (accelerometer & T sensor) (Lipatov) • MOx Mars oxidant experiment (Lane) • MAPEX microelectronics and photonics experiment The Project Scientist was Viacheslav M. Linkin. The Finnish part was co-ordinated by Risto Pellinen. The French part was co-ordinated by Jacques Blamont	
Delivery architecture	Spin-up to 12 rpm and separation from orbiter 4–5 days before orbit insertion.	
Thermal aspects	RHUs, RTG heat and insulation to maintain payload temperature within ±55 °C. Heat power 8.5 W	
Power aspects	Two ^{238}Pu RTGs (total 440 mW) and NiCd secondary battery, plus Li battery for descent phase	
Communications architecture	Relay at 8 kbit s^{-1} via Mars 96 orbiter or MGS. Link from orbiter to lander was only for initiation of data transmission from lander, i.e. no commanding capability. Receiving frequency: 437.100 MHz, transmitting frequency: 401.5725 MHz	
EDL architecture	Entry at 5.75 km s^{-1}. 1 m diameter ablative aeroshell with wing-like stabiliser at rear. Parachute cover release and main parachute deployment, aeroshell separation, inflation of 2 airbags at 4–10 km altitude. Jettison of airbags	
Landing speed(s)	20 m s^{-1} vertical, 20 m s^{-1} horizontal	
Active operations (deployments, etc.)	Spring-loaded opening of petals, deployment of antenna/ sensor boom, deployment of MAG, APX and MOx booms	
Key references	Several papers in *Planet. Space Sci.* **46**(6–7), 1998, including Linkin *et al.*, 1998; Pellinen and Raudsepp, 2000	

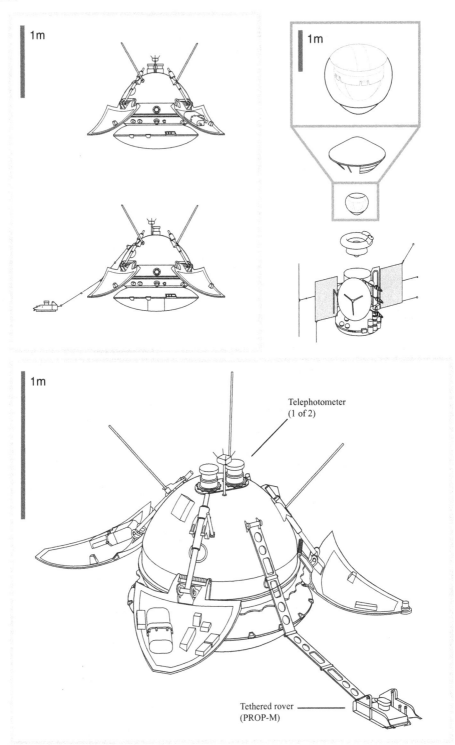

Figure 17.4 Mars 2, 3, 6, 7.

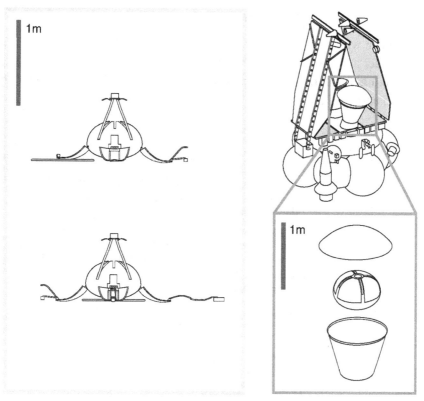

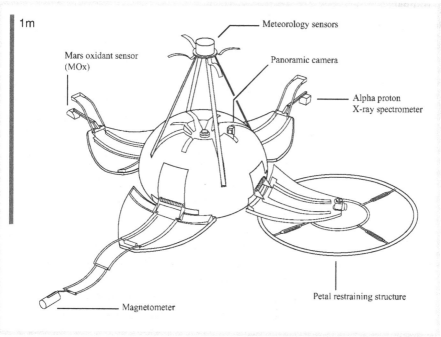

Figure 17.5 Mars 96 small stations.

17.5 Mars Pathfinder

Mars Pathfinder was highly successful, combining technological and scientific goals, and lander and rover elements, to become the first Martian surface mission since the Viking Landers. For more details see the Case Study, Chapter 24 (Figures 17.6 and 17.7).

Target	Mars
Objectives	Technological: demonstrating the feasibility of low-cost landings on and exploration of the Martian surface
	Scientific: atmospheric entry science, geological characterization of the landing site, meteorology, and long-range and close-up surface imaging, with the general objective being to characterize the Martian environment for further exploration
Prime contractor	JPL
Launch site, vehicle	ETR, Delta 2
Launch date	04/12/1996
Arrival date	04/07/1997
Landing site co-ordinates	19.33° N, 33.55° W
End(s) of mission(s)	27/09/1997
Mass(es)	Entry mass 585.3 kg. Landed mass 410 kg (incl. 99 kg airbag system, 264 kg lander + 10.5 kg rover)
Payload experiments	Sagan Memorial Station:

Sagan Memorial Station:
- IMP Imager for Mars Pathfinder (Smith, Keller)
- ASI/MET atmospheric structure instrument/meteorology package (Seiff)
- Windsock investigation (Sullivan)
- Magnetic properties investigation (Knudsen)
- Celestial mechanics radio science (Folkner)

The Project Scientist was Matthew Golombek

Sojourner:
- Rover imaging cameras
- APXS alpha-proton-X-ray spectrometer (Rieder)
- WAE wheel abrasion experiment (Ferguson)
- MAE materials adherance experiment (Landis)
- Surface material properties (Moore)

Table (Cont.)

Delivery architecture	Separation from cruise stage on approach
Thermal aspects	Lander: all sensitive electronics in thermally-isolated box (foam insulation) (no RHUs). Rover: electronics box aerogel insulated plus 3 RHUs
Power aspects	Lander: 2.5 m^2, ~177 W solar array, 27 V 50 A-hr Ag–Zn secondary battery
	Rover: 16.5 W GaAs/Ge solar array plus 150 W-hr lithium thionyl chloride (LiSOCl$_2$) primary battery
Communications architecture	X-band two-way DTE at 6 kbits s^{-1} to 70 m DSN via high-gain antenna. Also low-gain antennas. Sojourner-Lander: UHF at 9.6 kbits s^{-1}
EDL architecture	Entry at 7.26 km s^{-1} (inertial) or 7.48 km s^{-1} (relative), with a flight path angle of $-13.6°$ 2.65 m diameter, 70° blunt half cone, ablative SLA-561 aeroshell. Spin-stabilised. Max. deceleration: 16 *g*. Parachute deployment at 7.9 km altitude. Lander lowered from back shell and airbags inflated (each face of the tetrahedral lander having six 1.8 m diameter spherical lobes on a 1 m 'billiard rack' grid). Radar-altimeter triggered braking rockets at 98 m. Bridle cut, with motion continuing for 2 min after initial impact. Airbags deflated / retracted and petals opened
Landing speed(s)	Initial impact at 14 m s^{-1}
Active operations (deployments, etc.)	EDL aspects including airbag retraction and petal opening. Deployment of camera mast, met boom, HGA and rover deployment ramps
Key references	*J. Geophys. Res.* **102**(E2), 1997, and **104**(E4), 1997; *Science* **278**(5344), 1997; Shirley, 1998; Mishkin, 2004; Spencer *et al.*, 1999; http://mars.jpl.nasa.gov/

17.6 Beagle 2

The goal of carrying a mass-spectrometer-based gas analyser and other payload in the small mass and cost budget available to this mission produced a craft design with the highest payload/gross mass ratio for a planetary lander. The inclusion of a tethered mobile element and a multi-tool arm gave a significant 'reach' to the payload, and offered an affordable solution to the problem of analysing rocks / sands / rock interiors with a large number of sensor heads. Extremely high

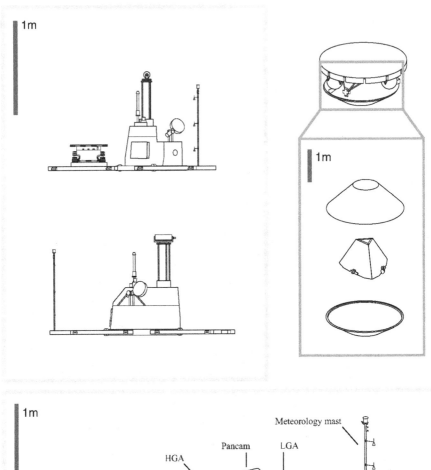

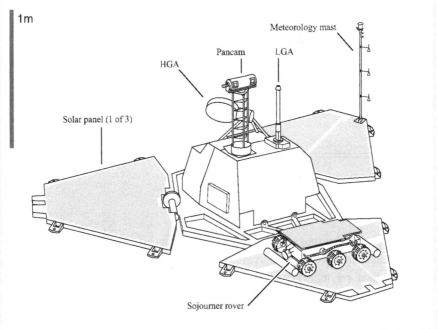

Figure 17.6 Mars Pathfinder.

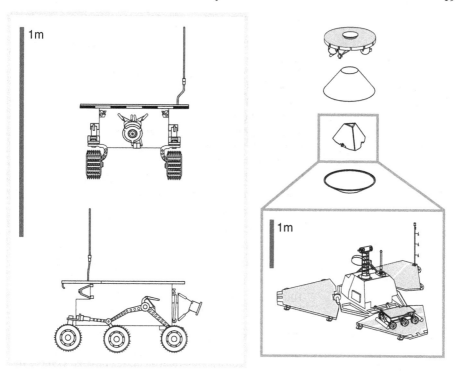

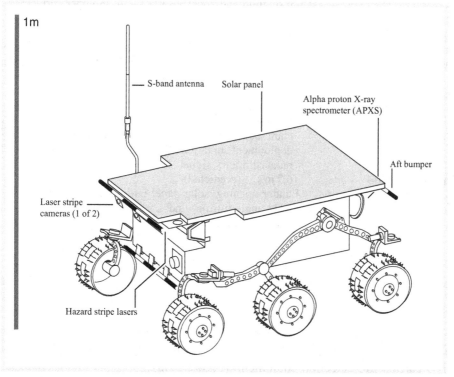

Figure 17.7 Sojourner.

Target	Mars
Objectives	To search for evidence of past or present life on Mars
Prime contractor	EADS Astrium
Launch site, vehicle	Baikonour, Soyuz-Fregat (on Mars Express)
Launch date	02/06/2003
Arrival date	25/12/2003
Landing site co-ordinates	11.53°N, 90.53°E (post-separation prediction)
End(s) of mission(s)	Intended primary mission: 180 days
Mass(es)	Separation/entry mass: 68.84 kg (incl. 35.4 kg EDLS and 33.2 kg lander)
	Probe support equipment on Mars Express: 4.88 kg
Payload experiments	• SCS stereo camera system (Coates)
	• GAP gas analysis package (Pillinger, Wright)
	• MIC microscope (Thomas)
	• MBS Mössbauer spectrometer (Klingelhöfer)
	• XRS X-ray fluorescence spectrometer (Fraser)
	• ESS environmental sensors suite (Sims, Zarnecki)
	Total mass: 9 kg
	The Project Scientist was Colin Pillinger
Delivery architecture	Spin-up and separation from orbiter on approach, at 14 rpm and $0.3\,\mathrm{m\,s^{-1}}$
Thermal aspects	Active control by electrical heating; passive control by insulation and solar absorber unit on upper surface of lander base
Power aspects	Li-ion secondary battery and GaAs solar array
Communications architecture	Two-way UHF relay via Mars Express or Mars Odyssey. Lander-orbiter 2–128 kbps at 401.6 MHz, orbiter-lander 2–8 kbps at 437.1 MHz
EDL architecture	Entry at $5.4\,\mathrm{km\,s^{-1}}$ (relative), $5.63\,\mathrm{km\,s^{-1}}$ (inertial), flight path angle $-16.5°$. 0.924 m diameter, 60° blunt half cone, Norcoat Liège aeroshell. Spin-stabilised. Pilot chute deployment, aeroshell separation, main chute deployment, inflation of 3 ammonia-filled gas bags on radar altimetry signal. Jettison of gas bags
Landing speed(s)	$16.7\,\mathrm{m\,s^{-1}}$ (predicted)
Active operations (deployments, etc.)	Lander opening, solar panel unfolding. Robotic arm, wide-angle mirror/wind sensor boom deployment, sampling mole, rock corer/grinder
Key references	Pillinger, 2003; Pillinger et al., 2003; Sims, 2004a,b; http://www.beagle2.com/; Pullan et al., 2004; Bonnefoy et al., 2004; Wright et al., 2003

standards of planetary protection were applied to this mission, both in terms of sterilisation and cleanliness. This was due to the fact that it was destined for Mars AND carried life-detection equipment. (Figure 17.8)

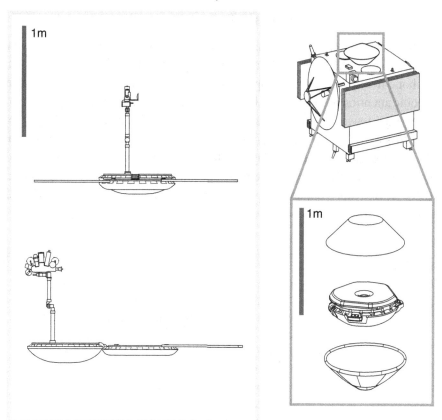

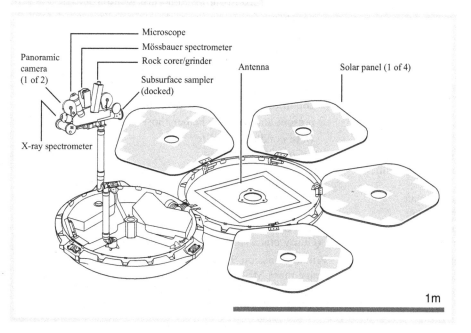

Microscope

Mössbauer spectrometer

Panoramic
camera
(1 of 2)

Rock corer/grinder

Antenna

Solar panel (1 of 4)

Subsurface sampler
(docked)

X-ray spectrometer

1m

Figure 17.8 Beagle 2.

17.7 Mars Exploration Rovers

'Follow the water' emerged as the mantra for NASA's Mars programme in the late 1990s. The intent of the Mars Exploration Rover (MER) missions was essentially to act as robotic field geologists, to map the rocks and soils around their landing sites with specific attention to minerals and formations that might indicate the presence or history of liquid water. These rovers were considerably larger than Sojourner. Although loss of solar power due to dust deposition on the arrays was expected to limit their lifetime to a few tens of days, both rovers are still operating at the time of writing, over 1.5 Martian years since their arrival, and have traversed a combined total of over 16.2 km. Of particular note are the number and quality of images returned (Figure 17.9). For more details, see the case study, Chapter 27.

Target	Mars	
Objectives	• Search for and characterize a variety of rocks and soils that hold clues to past water activity	
	• Determine the distribution and composition of minerals, rocks and soils surrounding the landing sites	
	• Determine what geologic processes have shaped the local terrain and influenced the chemistry	
	• Perform 'ground truth' of surface observations made by Mars orbiter instruments	
	• Search for iron-bearing minerals, identify and quantify relative amounts of specific mineral types that contain water or were formed in water	
	• Characterize the mineralogy and textures of rocks and soils and determine the processes that created them	
	• Search for geological clues to the environmental conditions that existed when liquid water was present and assess whether those environments were conducive to life	
Prime contractor	JPL/Cornell	
Launch site, vehicle	ETR, Delta 2 (7925)	
	Spirit (MER-A)	Opportunity (MER-B)
Launch date	10/06/2003	07/07/2003
Arrival date	04/01/2004	25/01/2004
Landing site co-ordinates	14.57° S, 175.47° E (Gusev crater)	1.95° S, 354.47° E (Meridiani Planitia)
End(s) of mission(s)	Primary mission 90 days; still operating into 2009	
Mass(es)	Entry mass MER-A 827 kg, MER-B 832 kg. Backshell & parachute 209 kg, heat shield 78 kg. Landed mass 540 kg incl. lander platform 348 kg and 185 kg rover.	

Table (Cont.)

Payload experiments	• Athena (Squyres):
	• Mini-TES (Christensen)

Payload experiments
- Athena (Squyres):
  - Mini-TES (Christensen)
  - Pancam panoramic camera (Bell)
  - APXS alpha-particle-X-ray spectrometer (Rieder)
  - Mössbauer spectrometer (Klingelhöfer)
  - Microscopic imager (Herkenhoff)
  - RAT rock abrasion tool (Gorevan)
  - Magnet arrays (Madsen)
  - Sundial
- Descent camera
- Navigation camera
- Two hazard cameras

The Project Scientist is Steven Squyres

Delivery architecture — Separation from cruise stage on approach

Thermal aspects — Warm electronics box; thermal control via gold paint, aerogel insulation, heaters, thermostats and radiators

Power aspects — Solar cells (gallium indium phosphide/gallium arsenide/germanium) and batteries; peak is 100 W near local noon early in mission. Power varies with time of day, season and rover tilt; dust accumulation periodically cleared by wind gusts

Communications architecture — Two-way DTE with high or low-gain antennas or by UHF relay ($128\,\mathrm{kbits\,s^{-1}}$) via Mars Global Surveyor, Mars Odyssey or Mars Express orbiters

EDL architecture — Incorporates IMUs in rover and backshell. Entry at $5.55\,\mathrm{km\,s^{-1}}$ (relative) or 5.65(A), $5.72\mathrm{(B)}\,\mathrm{km\,s^{-1}}$ (inertial), flight path angle $-11°$. 2.65 m diameter, 70° blunt half-cone, ablative SLA-561 aeroshell. Spin-stabilised. Max deceleration 6.3 *g*. Parachute deployment at 8.6 km altitude, $131\,\mathrm{m\,s^{-1}}$. Heatshield separation 20 s later. Lander lowered from backshell on bridle 10 s later. Radar altimeter acquires ground at 2.4 km altitude. Descent images acquired for DIMES (descent image motion estimation system). Airbags (Pathfinder configuration but strengthened) inflated 8 s before landing, 284 m altitude. Retro-rockets and TIRS (transverse impulse rocket system) fired 6 s before landing, 134 m altitude. Bridle cut 3 s before landing, 10 m altitude. Initial landing 354 s after entry. Bounces and rolls up to 1 km.

Landing speed(s) — $\sim 14\,\mathrm{m\,s^{-1}}$

Active operations (deployments, etc.) — Rover 'stand-up', solar array and HGA deployment, rover operations (av. speed $1\,\mathrm{cm\,s^{-1}}$, max $5\,\mathrm{cm\,s^{-1}}$). Robotic arm

Key references — *J. Geophys. Res.* **108**(E12), 2003; *Science* **305**(5685), 2004 and **306**(5702), 2004; Squyres, 2005; http://mars.jpl.nasa.gov/; http://athena.cornell.edu/. See also *Mars Exploration Rover Landings Press Kit*, NASA, 2004

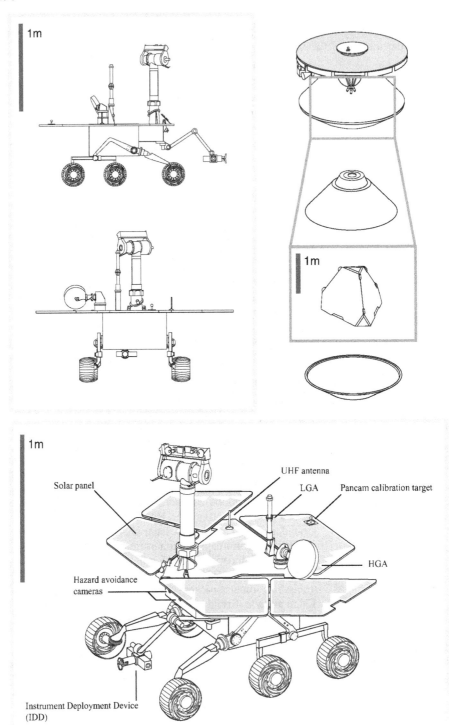

Figure 17.9 Mars Exploration Rovers.

18

Legged landers

These landers use a system of legs to cushion the landing and provide a stable platform for surface operations. With the exception of the Venera landers and the forthcoming Mars Science Laboratory (MSL) rover, all legged landers have three or four legs with footpads, and retro-thrusters perform final braking before landing. This was not required for the Veneras, whose terminal velocity at the surface was low enough ($\sim 8\,\mathrm{m\,s^{-1}}$) such that the landing gear alone was able to provide sufficient damping. The landing gear was toroidal and we thus consider it as effectively a single 'leg'. Mars Science Laboratory is due to make use of the rover's wheels as landing gear. A key feature of legged landers is that they must be the right way up for landing – beyond some tolerable limit such landers would topple over and fail. This attitude control must be performed during descent, usually by thrusters. Only for sufficiently thick atmospheres, such as that of Venus, can aerodynamic stabilisation be used.

Beyond those described here, future possible legged landers include robotic and crewed lunar landers from the US, robotic lunar landers from China and Japan, and a Mars sample-return mission.

18.1 Surveyor landers

The Surveyor landers performed soft landings on the Moon, largely as reconnaissance of the surface for the later Apollo landings. For more details see the Case Study, Chapter 21 (Figure 18.1).

18.2 Apollo lunar modules

To date the largest lunar or planetary landers, and the only successfully implemented crewed landers, the Apollo LM delivered twelve astronauts and their

199

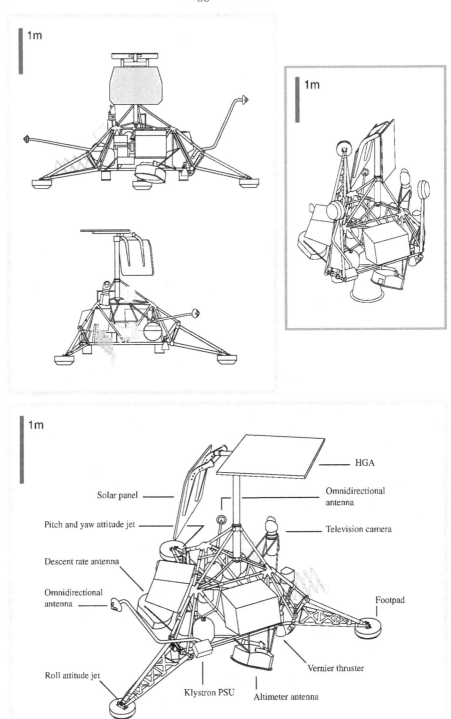

Figure 18.1 Surveyor.

	Surveyor 1	Surveyor 2	Surveyor 3	Surveyor 4	Surveyor 5	Surveyor 6	Surveyor 7
Target	The Moon						
Objectives	• To accomplish soft landings on the Moon • To provide basic data in support of the manned Apollo programme • To perform operations on the lunar surface which contribute new scientific knowledge about the Moon						
Prime contractor Launch site, vehicle	Hughes Aircraft Company ETR, Atlas-Centaur						
Launch date	30/05/1966	20/09/1966	17/04/1967	14/07/1967	08/09/1967	07/11/1967	07/01/1968
Arrival date	02/06/1966	Impacted 23/09/1966	20/04/1967	Impacted 17/07/1967	11/09/1967	10/11/1967	10/01/1968
Landing site co-ordinates[1]	43.32° W, 2.46–2.50° S	SE of crater Copernicus	23.32° W, 3.06° S	1.50° W, 0.43° N (from in-flight tracking)	23.20° E, 1.42° N	1.37° W, 0.46° N	11.44° W, 40.97° S
End of mission (last data return)	07/01/1967	–	04/05/1967	–	17/12/1967	14/12/1967	21/02/1968
Mass(es)	At injection: 995–1040 kg; at touchdown: 294–306 kg						
Payload experiments	• TV camera (Shoemaker) • Strain gauges (Christensen, Batterson) • SMSS soil mechanics surface sampler (3, 4, 7) (Shoemaker, Scott) • Alpha-scattering instrument (5, 6, 7) (alpha-proton spectrometer) (Turkevich) • Temperature sensors • Mirrors						

Table (Cont.)

	• Magnets
	• Descent camera (1 and 2 only; not used)
	The Project Scientist was Leonard Jaffe
Delivery architecture	Approach direct from translunar trajectory
Thermal aspects	Passive control via white paint, high IR-emittance thermal finish and polished aluminium. Electronics compartments were equipped with insulating blankets, conductive heat paths, thermal switches and electric heaters
Power aspects	Solar array and main battery; auxiliary, non-rechargeable battery carried on Surveyors 1–4 only.
Communications architecture	Two-way DTE: S-band (2295 MHz downlink, 2113 MHz uplink) via planar array HGA or either of two omni-directional antennas
EDL architecture	Descent from trans-lunar trajectory. Solid retro motor firing and jettison followed by vernier liquid bipropellant system for final descent. 3-legged landing gear and crushable blocks
Landing speed(s)	Actual: 1.4–$4.2\,\mathrm{m\,s^{-1}}$ (designed to withstand $4.6\,\mathrm{m\,s^{-1}}$)
Active operations (deployments, etc.)	Solar array and HGA articulation, TV camera movement, SMSS and alpha-scattering experiment deployment, restart of propulsion system
Key references	NASA TR 32–1265, 1968; NASA SP-184, 1969; Le Croissette, 1969; Heiken et al., 1991

[1] Inertial co-ordinates from on-surface tracking; inertial co-ordinates from in-flight tracking, and selenographic co-ordinates (Surveyor Project Final Report, NASA TR 32–1265), differ slightly due to random and systematic errors.

equipment to the lunar surface and returned them safely to lunar orbit. Three test missions were flown prior to Apollo 11, and on Apollo 13 the LM acted as a 'lifeboat' for the crew. Apollos 15, 16 and 17 carried a Lunar Roving Vehicle to extend the crew's operational range. Also deployed were experiments that would continued operating after the astronauts' departure – acting essentially as independent, though manually deployed, landers in their own right. The initial EASEP (Early Apollo Surface Experiments Package) on Apollo 11 was followed by the ALSEP (Apollo Lunar Surface Experiments Package) on subsequent missions. (Figure 18.2)

18.3 Luna 17, 21 (Ye-8) landers and the Lunokhods

The later Luna landers used the same basic design for both rover deployment (Ye-8) and sample return (Ye-8-5 and Ye-8-5M) (Figures 18.3 and 18.4).

18.4 Luna 15, 16, 18, 20 (Ye-8-5) landers

These landers performed the first successful robotic sample return missions (Figure 18.5).

18.5 Luna 23, 24 (Ye-8-5M) landers

These upgraded sample return craft carried a modified drill to obtain a deep drill core with preserved stratigraphy. Unlike Luna 16 and 20, no telephotometer cameras were carried (Figure 18.6).

18.6 Soviet LK lunar lander

The Soviet Union built and flew its lunar lander in Earth orbit three times (analogous to the Apollo 5 flight) (Figure 18.7).

18.7 Venera 9–14 (4V-1) and VeGa (5VK) landers

The later Venera landers facilitated investigation of the Venusian atmosphere and surface. Upgrades were made with each pair of launches; the VeGa landers also carried balloons. Veneras 9–14 entered on the day side, VeGa 1, 2 on the night side (Figures 18.8–18.11).

18.8 Viking landers

The Viking project, comprising two soft landers and two orbiters, was a massive and ambitious project. The primary mission objectives were to obtain high

	5	9	10	11	12	13	14	15	16	17
Target	The Moon									
Objectives	Land a two-man crew and associated equipment on the Moon, and return them to lunar orbit									
Prime Contractor	Grumman									
Launch site, vehicle	ETR, Saturn V (Saturn 1B for Apollo 5)									
Launch date	22/01/1968	03/03/1969	18/05/1969	16/07/1969	14/11/1969	11/04/1970	31/01/1971	26/07/1971	16/04/1972	07/12/1972
Arrival date	–	–	–	20/07/1969	19/11/1969	–	05/02/1971	30/07/1971	21/04/1972	11/12/1972
Landing site co-ordinates	–	–	–	0.67408 N, 23.47297 E	3.01239 S, 23.42157 W	–	3.64530 S, 17.47136 W	26.13222 N, 3.63386 E	8.97301 S, 15.50019 E	20.19080 N, 30.77168 E
End of Surface mission[1]	–	–	–	21/07/1969	20/11/1969	–	06/02/1971	02/08/1971	24/04/1972	14/12/ 1972
Mass(es)	14,696 kg at launch,									

Payload experiments

Apollo 11: EASEP (laser retroreflector, passive seismometer, dust detector), soil mechanics, solar-wind collector, photography, field geology and sample collection

Apollo 12: ALSEP (passive seismometer, suprathermal ion detector, cold cathode ion gauge, dust detector, magnetometer, solar wind spectrometer, radio science), soil mechanics, solar wind collector, photography, field geology and sample collection

Apollo 14: ALSEP (laser retroreflector, passive seismometer, active seismometer, suprathermal ion detector, charged-particle experiment, cold-cathode ion gauge, dust detector, radio science), soil mechanics, solar-wind collector, magnetometer, photography, field geology and sample collection

Apollo 15: ALSEP (laser retroreflector, passive seismometer, magnetometer, solar-wind spectrometer, suprathermal ion detector, cold-cathode ion gauge, dust detector, heat flow experiment, radio science), soil mechanics, solar wind collector, magnetometer, photography, field geology and sample collection

Apollo 16: ALSEP (passive seismometer, active seismometer, magnetometer, heat-flow experiment, radio science), soil mechanics, solar-wind collector, magnetometer, cosmic-ray detector, far-UV camera/spectroscope, photography, field geology and sample collection.

Apollo 17: ALSEP (seismic-profiling experiment, gravimeter, mass spectrometer, ejecta and meteorites experiment, heat-flow experiment, radio science), soil mechanics, cosmic-ray detector, neutron probe, surface electrical properties, gravimeter, photography, field geology and sample collection

Delivery architecture	Separation from CSM in lunar orbit
Thermal aspects	Environmental control system for cabin
Power aspects	Ascent stage: 4×400 Ah AgZn batteries Descent stage: 2×296 Ah AgZn batteries
Communications architecture	2-way S-band DTE. ALSEP: 2-way S-band DTE
EDL architecture	Powered, piloted descent and landing
Landing speed(s)	~0.9 m s^{-1}
Active operations (deployments, etc.)	Many (HGA, astronaut operations, LRV and EASEP/ALSEP deployment, launch of ascent stage)
Key references	Apollo Preliminary Science Reports; Heiken *et al.*, 1991; Kelly, 2001; Cortright, 1975; Beattie, 2001. See also the series of repackaged material in the Apollo 11–17 NASA Mission Reports books, from Apogee Books

[1] The ALSEP stations were shut down on 30/09/1977.

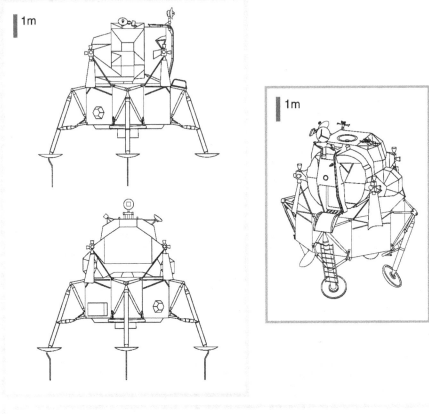

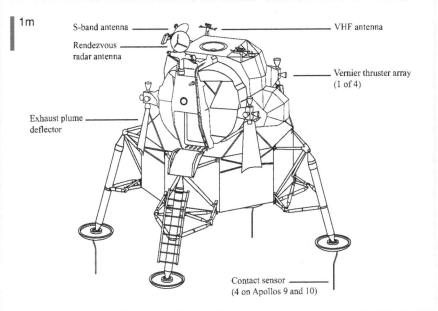

Figure 18.2 Apollo LM.

Target	The Moon		
Objectives	Investigations at multiple locations of the lunar surface material and environment; celestial mechanics		
Prime contractor	NPO Lavochkin (formerly OKB-301)		
Launch site, vehicle	Baikonour, Proton 8K82K/11S824		
	Luna 1969A	Luna 17	Luna 21
Launch date	19/02/1969	10/11/1970	08/01/1973
Arrival date	–	17/11/1970	15/01/1973
Landing site co-ordinates	–	38.26° N 35.19° W[1] (Sinus Iridium)	26°03′ N 30°22′ E (Le Monnier)
End(s) of mission(s)	–	09/1971 (total odometry 10.54 km)	10/05/1973 (total odometry 37 km)
Masses		5700 kg launch	5700 kg launch
		1900 kg landing	1900 kg landing
		756 kg rover	836 kg rover

Payload experiments   Luna 17/Lunokhod 1:

- Cameras (two TV & four panoramic telephotometers) (Selivanov)
- RIFMA XRFS (Kocharov)
- RT-1 X-ray telescope (?)
- PrOP odometer/penetrometer (Kemurdzhian)
- RV-2N radiation detector (Chuchkov)
- TL-1 laser retroreflector (Kokurin)

The Project Scientist was Aleksandr P. Vinogradov

Luna 21/Lunokhod 2:

- Cameras (three TV & four panoramic telephotometers) (Selivanov)
- RIFMA-M XRFS (Kocharov)
- X-ray telescope (?)
- PrOP odometer/penetrometer (Kemurdzhian)

Table (Cont.)

- RV-2N-LS radiation detector (Chuchkov)
- TL-2 laser retroreflector (Kokurin)
- AF-3L UV/visible astrophotometer (Zvereva?)
- SG-70A magnetometer (Dolginov)
- Rubin 1 photodetector (?)

The Project Scientist was Aleksandr P. Vinogradov

Delivery architecture	Powered descent from lunar orbit
Thermal aspects	Polonium RHU at rear; air circulation inside main pressurised compartment, with open-cycle water cooling; radiator on upper-surface lid closed during lunar night
Power aspects	Solar cells on lid (Si on Lunokhod 1, GaAs on Lunokhod 2); secondary batteries
Communications architecture	Two-way DTE
EDL architecture	Descent engine, retros, 4 landing legs
Landing speed(s)	$2\,\mathrm{m\,s^{-1}}$ (Luna 21) Max. $5\,\mathrm{m\,s^{-1}}$
Active operations (deployments, etc.)	Deployment of ramps from descent stage; rover chassis; articulated lid, HGA retroreflector cover, odometer/penetrometer assembly and magnetometer boom (Lunokhod 2 only)
Key references	Vinogradov, 1971; Barsukov, 1978; Heiken et al., 1991

[1] Suggested by Stooke (2005), with the rover now parked some 2 km further North at 38.29°N, 35.19°W.

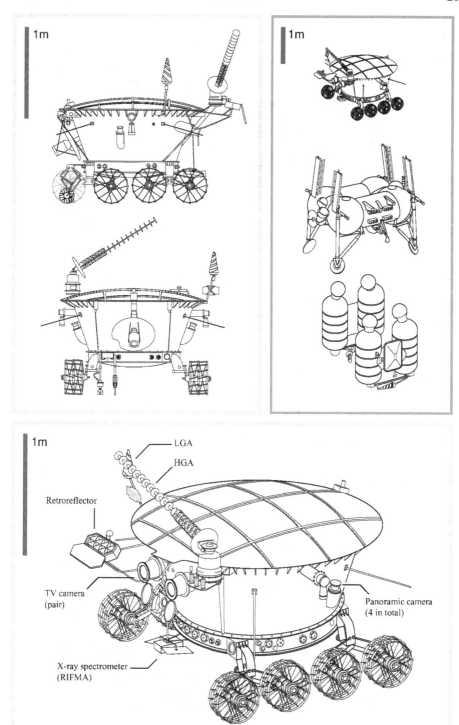

Figure 18.3 Lunokhod 1.

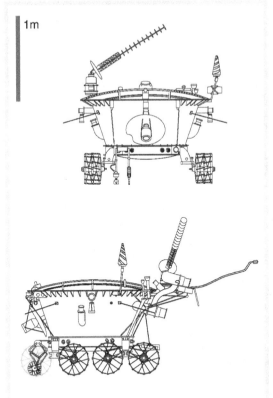

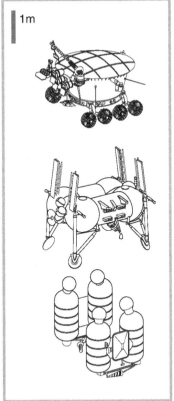

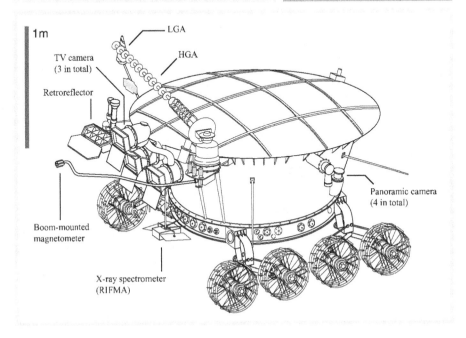

Figure 18.4 Lunokhod 2.

	Luna 1969B	Luna 15	Cosmos 300	Cosmos 305	Luna 1970A	Luna 16	Luna 18	Luna 20
Target	The Moon							
Objectives	To collect a sample of lunar surface material and bring it to the Earth							
Prime contractor	NPO Lavochkin (formerly OKB-301)							
Launch site, vehicle	Baikonour, Proton 8K82K/11S824							
Launch date	14/06/1969	13/07/1969	23/09/1969	22/10/1969	06/02/1970	12/09/1970	02/09/1971	14/02/1972
Arrival date	–	Impacted 21/07/1969	–	–	–	20/09/1970	Impacted 11/09/1971	21/02/1972
Landing site co-ordinates	–	?	–	–	–	0°41' N 56°18' E	3°34' N 56°30' E	3°32' N 56°33' E
End(s) of mission(s)	–	–	–	–	–	?	?	?
Mass(es)	Launch: 5667 kg (15); 5725 kg (16, 18, 20) 1880 kg on landing 520 kg ascent stage 35 kg return capsule entry mass							
Payload experiments	• GZU ground-sampling device • Return vehicle • Radiation detector(s) (Vernov?) • Panoramic telephotometers (stereo pair, with lamps) (Selivanov) The Project Scientist was Aleksandr P. Vinogradov							

Table (Cont.)

Delivery architecture	Powered descent from lunar orbit. Vertical launch of ascent stage for direct return to Earth from particular landing longitude avoided the need for trajectory correction. Separation on command from Earth of re-entry capsule from ascent stage
Thermal aspects	Water cooling of descent-stage instrument compartment
Power aspects	Secondary batteries (AgZn 14 A-hr on return stage; 4.8 A-hr in re-entry capsule)
Communications architecture	Two-way DTE for main s/c at 922 and 768 MHz, plus backup at 115 and 183 MHz. Ascent stage: 101.965 MHz and 183.537 MHz only. 121.5 and 114.167 MHz beacon on re-entry capsule.
EDL architecture	Descent engine, retros, 4 landing legs. Spherical re-entry capsule 0.5 m diameter, re-entry speed $\sim$11 km s^{-1}. Peak load 315 g. Parachute system (1.5 m^2 pilot followed by 10 m^2 main). System of 2 inflated balloons for capsule orientation after landing
Landing speed(s)	4.8 m s^{-1} (Luna 16)
Active operations (deployments, etc.)	Deployment of drill arm; activation of drill; delivery of sample to return capsule; launch of ascent stage
Key references	Vinogradov, 1974; Barsukov & Surkov, 1979; Heiken et al., 1991; Grafov et al., 1971

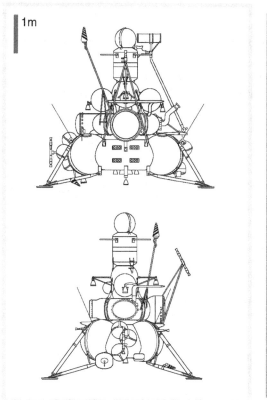

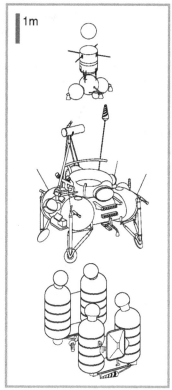

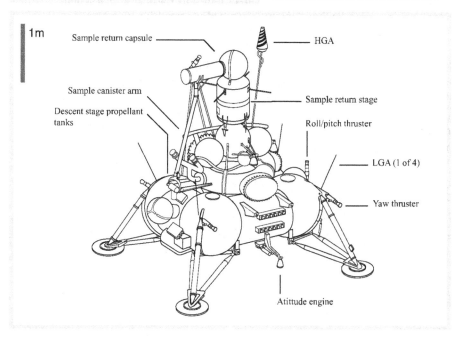

Sample return capsule —

Sample canister arm

Descent stage propellant tanks

HGA

Sample return stage

Roll/pitch thruster

LGA (1 of 4)

Yaw thruster

Atittude engine

Figure 18.5 Luna 16, 20.

Target	The Moon		
Objectives	To collect a sample of lunar surface material and bring it to the Earth		
Prime contractor	NPO Lavochkin (formerly OKB-301)		
Launch site, vehicle	Baikonour, Proton 8K82K/11S824 or 8K82K/11S824M		
	Luna 23	Luna 1975A	Luna 24
Launch date	28/10/1974	16/10/1975	09/08/1976
Arrival date	06/11/1974	–	18/08/1976
Landing site co-ordinates	12°41′ N 62°17′ E	–	12°45′ N 62°12′ E
End(s) of mission(s)	?	–	?
Mass(es)	5795 kg launch 514.8 kg ascent stage 34 kg return capsule entry mass		
Payload experiments	• GZU LB-09 upgraded ground sampling device • Return vehicle • Radiation detector(s) (Vernov?)		
	The Project Scientists were Aleksandr P. Vinogradov and Valerii L. Barsukov		
Delivery architecture	Powered descent from lunar orbit. Vertical launch of ascent stage for direct return to Earth from particular landing longitude avoided the need for trajectory correction. Separation on command from Earth of re-entry capsule from ascent stage		
Thermal aspects	Water cooling of descent stage instrument compartment		
Power aspects	Secondary batteries (AgZn 14 A-hr on return stage; 4.8 A-hr in re-entry capsule)		
Communications architecture	Two-way DTE for main s/c at 922 and 768 MHz, plus backup at 115 and 183 MHz. Return stage: 115 and 183 MHz only		
EDL architecture	Descent engine, retros, 4 landing legs		
Landing speed(s)	$11 \, \mathrm{m \, s^{-1}}$ (23) versus max $5 \, \mathrm{m \, s^{-1}}$		
Active operations (deployments, etc.)	Deployment of drill; activation of drill; delivery of sample to return capsule; launch of return stage		
Key references	Barsukov, 1980; Heiken *et al.*, 1991		

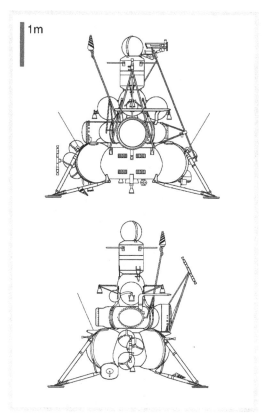

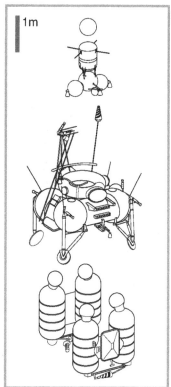

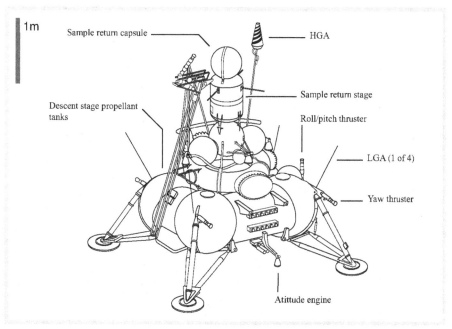

Figure 18.6 Luna 24.

Target	The Moon, but only tested in Earth orbit
Objectives	Test of the LK (T2K type, minus landing gear) in Earth orbit
Prime contractor	Yangel
Launch site, vehicle	Baikonour, Soyuz 11A511L
	Cosmos 379 (T2K no. 1) / Cosmos 398 (T2K no. 2) / Cosmos 434 (T2K no. 3)
Launch date	24/11/1970 / 26/02/1971 / 12/08/1971
End(s) of mission(s)	Each mission lasted a few days, including orbital manoeuvres to simulate descent, ascent and rendezvous. The spacecraft decayed from orbit in 1983, 1995 and 1981, respectively
Mass(es)	~5700 kg launch
Payload experiments	Test flights
Delivery architecture	Lunar version would have separated from Block D after braking manoeuvre and performed powered descent
Thermal aspects	Unknown
Power aspects	Batteries
Communications architecture	Two-way DTE
EDL architecture	The lunar LK had 4 legs with footpads and hold-down thrusters; the T2K lacked the legs and crew ladder and carried other modifications needed for the test flights
Landing speed(s)	Unknown
Active operations (deployments, etc.)	HGA articulation, planned astronaut activities, launch of ascent stage
Key references	Semenov, 1996; Johnson, 1995; Siddiqi, 2000

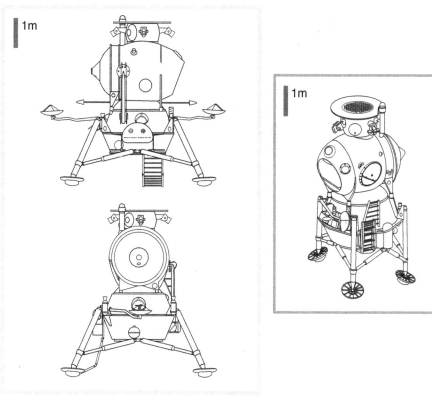

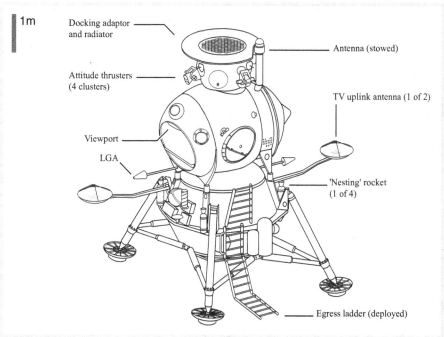

Figure 18.7 LK Lander.

	Venera 9	Venera 10	Venera 11	Venera 12	Venera 13	Venera 14	VeGa 1	VeGa 2
Target	Venus							
Objectives	Studies of the Venus atmosphere and surface							
Prime contractor	NPO Lavochkin (formerly OKB-301)							
Launch site, vehicle	Baikonour, Proton 8K82K/11S824 or 8K82K/11S824M							
Launch date	08/06/1975	14/06/1975	09/09/1978	12/09/1978	10/10/1981	04/11/1981	15/12/1984	21/12/1984
Arrival date	22/10/1975	25/10/1975	25/12/1978	21/12/1978	01/03/1982	05/03/1982	11/06/1985	15/01/1985
Landing site co-ordinates	31.01° N, 291.64° E	15.42° N, 291.51° E	14° S, 299° E	7° S, 294° E	7.5° S, 303.5° E	13° S, 310° E	8.10° N, 175.85° E	7.14° S, 177.67° E
Duration of transmission from surface:	53 min	65 min	95 min	110 min	127 min	57 min	20 min	?
Masses (kg)		Entry: 1560 Landing: 660		Entry: ~1700 Landing: ~760		Entry: ~1700 Landing: 760		Entry: 1750 Landing: ~750

Payload experiments Venera 9,10:
- T, P sensors (Marov)
- accelerometer (Avduevskii)
- IOV-75 vis/IR photometers (Avduevskii, Marov, Ekonomov, Moshkin??)
- MNV-75 backscatter & multi-angle nephelometers (Marov)
- MAV-75 P-11 mass spectrometer (Istomin)
- Doppler expt (Kerzhanovich)
- Panoramic telephotometers (two, with lamps) (Selivanov)
- ISV-75 anemometer (Avduevskii, Marov)
- GS-12V Gamma-ray spectrometer (Surkov)
- RP-75 Densitometer (Surkov)

The Project Scientist was Mikhail Ya. Marov

Venera 11,12:
- T, P sensors (Marov, Avduevskii)
- Bizon(-M?) accelerometry (Marov, Avduevskii, Cheremukhina)
- Colour panoramic telephotometers (two) (Selivanov)
- Sigma gas chromatograph (Gel'man)
- MKh-6411 mass spectrometer (Istomin)
- IOAV scanning spectrophotometer (Moroz, Ekonomov, Moshkin)
- Backscatter nephelometer (Marov)
- Doppler expt (Kerzhanovich)
- BDRA-1V Chemical composition of aerosols by XRFS (Surkov)
- Gamma-ray spectrometer (Surkov)
- GZU VB-01 drill (Barmin) + BDRP-1V soil XRFS (Surkov)
- Groza electrical/acoustic activity expt (Ksanfomaliti)
- PrOP-V penetrometer (Kemurdzhian)
- MSB small solar batteries (Lidorenko)

The Project Scientists were Valerii L. Barsukov and Mikhail Ya. Marov

Venera 13,14:
- ITD T, P sensors (Avduevskii)
- Bizon-M accelerometry (Avduevskii, Cheremukhina)
- TFZL-077 colour panoramic telephotometers (two) (Selivanov)
- Kontrast geochemical indicator (Florenskii)
- IOAV-2 scanning spectrophotometer & UV photometer (Moroz)
- MNV-78-2 nephelometer (Marov)

Table (Cont.)

- MKh-6411 mass spectrometer (Istomin)
- Sigma-2 gas chromatograph (Mukhin)
- Doppler expt (Kerzhanovich)
- BDRA-1V chemical composition of aerosols by XRFS (Surkov)
- VM-3(R?) hygrometer (Surkov)
- GZU VB-02 drill (Barmin) + Arakhis-2 / BDRP-2V soil XRFS (Surkov)
- Groza-2 electrical/acoustic activity expt (Ksanfomaliti)
- PrOP-V penetrometer (Kemurdzhian)
- MSB small solar batteries (Lidorenko)

The Project Scientist was Valerii L. Barsukov

VeGa 1, 2:
- Meteocomplex T, P sensors (Linkin, Blamont, Kerzhanovich)
- Sigma-3 gas chromatograph (Mukhin)
- LSA particle size spectrometer (Mukhin)
- IFP aerosol analyser (Mukhin)
- VM-4 hygrometer (Surkov)
- ISAV-A nephelometer/scatterometer (Moroz)
- Malakhit-V mass spectrometer (Surkov, Thomas, Israël)
- ISAV-S UV spectrometer (Bertaux, Moroz)
- GZU VB-02 drill (Barmin) + BDRP-AM25 soil XRFS (Surkov)
- GS-15-STsV gamma ray spectrometer (Surkov)
- PrOP-V penetrometer (Kemurdzhian)
- MSB small solar batteries (Lidorenko).

Also VeGa AZ.
The Project Scientist was Valerii L. Barsukov

Delivery architecture	Separation from orbiter (Venera 9,10) or flyby spacecraft (Venera 11–14, VeGa 1,2) on approach
Thermal aspects	Thermal louvre on entry shell. Pre-cooling to −10°C. Pressure vessel insulated with KG-25 (a high-temperature polyurethane foam) and PTKV-260. lithium nitrate trihydrate thermal accumulator
Power aspects	Batteries
Communications architecture	One-way VHF relay via orbiter (Venera 9, 10) or flyby s/c (others), data rate to orbiter/flyby s/c 256 bits s^{-1} (9/10) or ~3kbits s^{-1} (11/12) on each of two channels
EDL architecture	Entry angles 18–21°; 2.4 m diameter spherical aeroshell; drogue & lead-away parachutes on upper lid; drag parachute; jettison of lower shell; jettison of drag parachute & deployment of main parachute(s); jettison of main parachute(s); final descent using aerodynamic braking disc; toroidal landing gear. Descent duration: ~60.5 min. Entry loads 150–180g.
Landing speed(s)	8 m s^{-1} (9 and 10), 8 m s^{-1} (12), 7.5–7.6 m s^{-1} (13 and 14)
Active operations (deployments, etc.)	Ejection of camera covers (9–14), deployment of colour charts (11–14), deployment of densitometer arm (9, 10), deployment of PrOP-V (11–14 and VeGa 1, 2); soil sampling drill operation (11–14, VeGa 1, 2)
Key references	Keldysh,1979; Hunten *et al.*,1983; Marov and Grinspoon, 1998; *Cosmic Res.* **14**(5), 1976; **17**(5), 1979; **21**(2), 1983; **25**(5), 1987; *TsUP*, 1985; *MNTK*, 1985

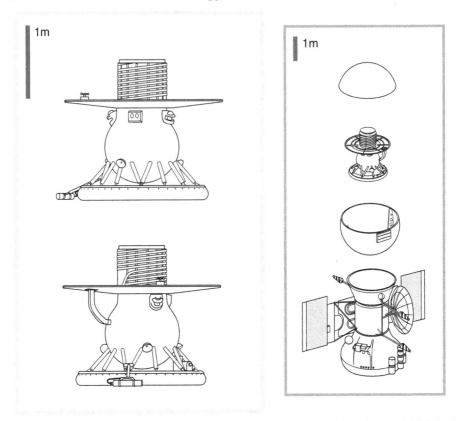

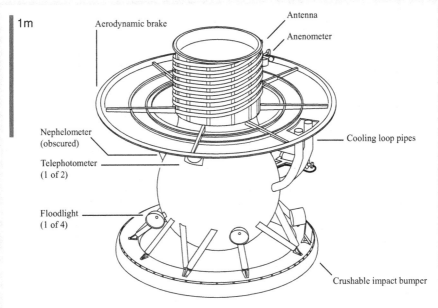

Figure 18.8 Venera 9, 10.

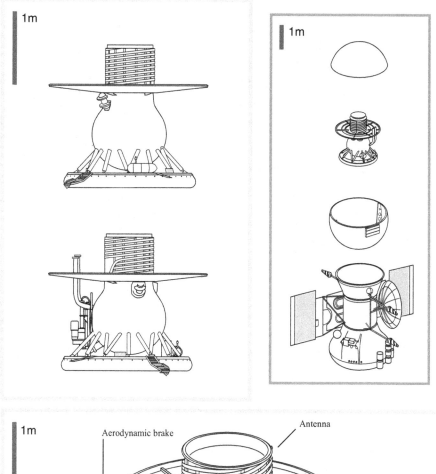

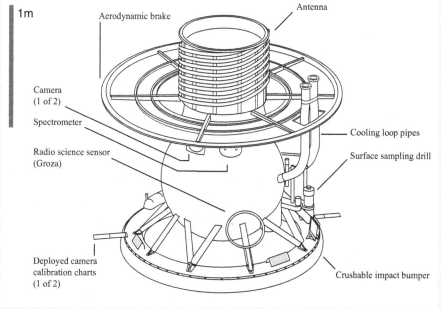

Figure 18.9 Venera 11, 12.

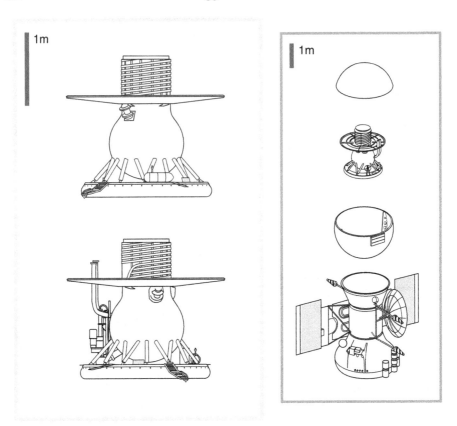

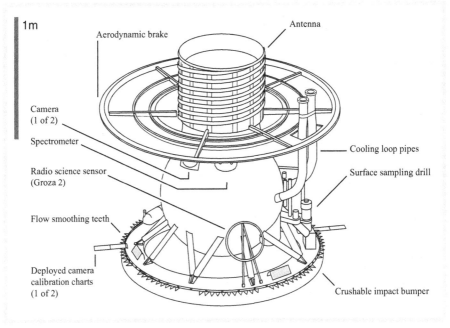

Figure 18.10 Venera 13, 14.

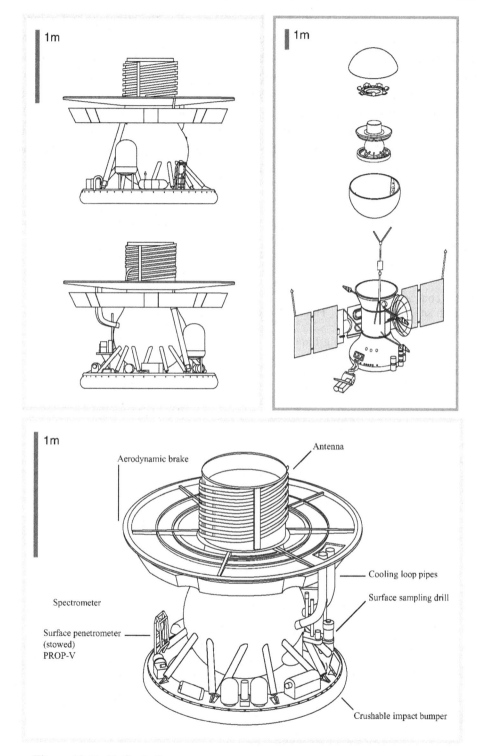

Figure 18.11 VeGa 1, 2.

resolution images of the Martian surface, characterize the structure and composition of the atmosphere and surface, and search for evidence of life. Viking Landers 1 and 2 soft-landed successfully, and during their surface operations lasting over 3 and over 6 years, took hundreds of pictures, made a series of meteorological measurements that has not been rivalled, manipulated the ground with a soil scoop, and analysed the soil: a remarkable achievement for the 1970s. The project required considerable technological investment, particularly in parachute technology and in instrumentation – many Viking developments have yet to be improved upon (Figure 18.12).

Target	Mars	
Objectives	Soft land on Mars; search for life; meteorological, environmental and seismological monitoring; compare orbital and surface data	
Prime contractor	Martin Marietta	
Launch site, vehicle	ETR, Titan IIIE-Centaur D1 (with Viking Orbiters)	
	Viking Lander 1 (Mutch Memorial Station)	Viking Lander 2 (Soffen Memorial Station)
Launch date	20/08/1975	09/09/1975
Landing date	20/071976	03/09/1976
Landing site co-ordinates	22.48° N, 47.94° W (Chryse Planitia)	47.97° N, 225.86° W (Utopia Planitia)
End(s) of mission(s)	13/11/1982	11/04/1980
Mass(es)	1185 kg entry mass, 663 kg lander wet mass, 612 kg at landing. See Ezell and Ezell (1984) for detailed breakdown	
Payload experiments	• RPA (retarding potential analyser) (Nier)	
	• Atmospheric structure (Nier)	
	• NMS (neutral mass spectrometer) (Nier)	
	• Facsimile cameras (Mutch)	
	• GEX/LR/PR (exobiology) (Klien)	
	• GCMS (Biemann)	
	• XRFS (Toulmin)	
	• P, T and wind sensors (Hess)	
	• 3-axis seismometer (Anderson)	
	• Magnetic properties (Hargraves), physical properties (Shorthill)	
	• Radio science (Michael) 91 kg on lander, 9 kg on aeroshell. The Project Scientist was Gerald Soffen	

Table (Cont.)

Delivery architecture	Separation from orbiter 1500×32800 km; small hydrazine motors for de-orbit burn
Thermal aspects	Thermal control was effected by optical coatings, fibrous insulation and a gas bellows-activated thermal switch which conducted waste heat from the RTG into the lander. The RTG was covered with a wind shield to prevent excessive convective cooling. A water circulation loop was used to remove waste heat during prelaunch operations
Power aspects	90 W electrical power from 2 RTGs with NiCd secondary batteries
Communications architecture	Two-way 76 cm pointable DTE 2.2 GHz dish, 20 W at 500 bits s^{-1} UHF (381 MHz) two-way relay via orbiter, 30 W at 4 and 16 kbit s^{-1} through 8-element crossed dipole
EDL architecture	Incorporated a full IMU guidance package. Entry at 4.42–4.48 km s^{-1} (relative), flight path angle $-17.6°$. 3.54 m diameter, 70° blunt half cone, ablative SLA-561 aeroshell – peak deceleration 8.4 g at 27 km altitude. Further, it performed a lifting entry, flying at a nominal angle of attack of 11°. 16.2 m DGB parachute deployed at 5.9 km by radar altimeter. Parachute jettison at 1.4 km and 54 m s^{-1}. Throttlable 276–2840 N descent engines using hydrazine
Landing speed(s)	2.4 m s^{-1}
Active operations (deployments, etc.)	Sample acquisition and soil mechanics experiments using robot arm
Key references	Ezell and Ezell, 1984; *J. Geophys. Res.* **82**(28), 1977; Mutch, 1978; Martin Marietta, 1976; Kieffer *et al.*, 1993; Burgess, 1978; Corliss, 1975; Holmberg *et al.*, 1980; Cooley and Lewis, 1977

18.9 Mars Surveyor landers

These legged Mars landers arose from the US 'Mars Surveyor' programme in the wake of the success of Mars Pathfinder. At that stage the programme envisaged launching one Surveyor lander and one Surveyor orbiter with each Mars launch window, starting in '98/'99 with the Mars Surveyor '98 lander (later renamed Mars Polar Lander), launched a few weeks after Mars Climate Orbiter (which was in part reflying payload lost with Mars Observer in 1993). Mars Polar Lander also carried the second New Millennium technology mission, the Deep Space 2 Mars Microprobes. Before the failure of both these missions on arrival at Mars, the plan was to launch a Mars Surveyor '01 lander in the same launch window as the Mars Surveyor '01 orbiter (launched as planned as Mars Odyssey). Further

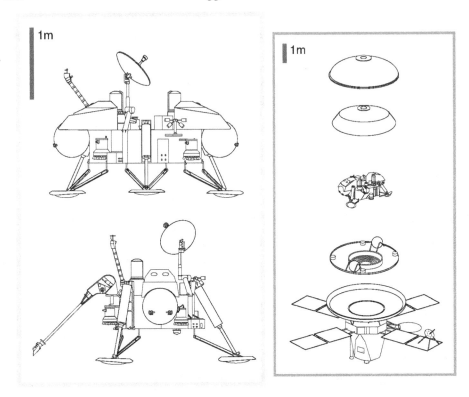

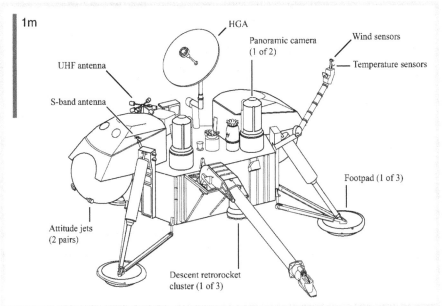

Figure 18.12 Viking Lander.

Target	Mars
Objectives	Study high-latitude Martian surface environment; search for H_2O and CO_2 in soil; monitor weather. More specifically, to:

- Record local meteorological conditions near the Martian south pole, including temperature, pressure, humidity, wind, surface frost, ground ice evolution, ice fogs, haze, and suspended dust
- Analyse samples of the polar deposits for volatiles, particularly H_2O and CO_2
- Dig trenches and image the interior to look for seasonal layers and analyse soil samples for water, ice, hydrates, and other aqueously deposited minerals
- Image the regional and immediate landing site surroundings for evidence of climate changes and seasonal cycles
- Obtain multi-spectral images of local regolith to determine soil types and composition

Prime contractor	Lockeed Martin Astronautics
Launch site, vehicle	ETR, Delta II (Delta 7425)
Launch date	03/01/1999
Arrival date	03/12/1999
Landing site co-ordinates	76° S, 195° W (South Polar layered terrain)
End(s) of mission(s)	Mission expected to end after ∼ 3 months when batteries freeze as days get shorter in late summer. Contact lost prior to entry – never regained
Mass(es)	Launch mass 583 kg, incl. 64 kg cruise/descent propellent, 82 kg cruise stage, 140 kg aeroshell and two 3.6 kg DS-2 Mars Microprobes. Landed dry mass 290 kg
Payload experiments	• MVACS Mars Volatiles and Climate Surveyor (Paige)

- • SSI surface stereo imager (Smith, Keller) (clone of Pathfinder camera, dual lenses focusing onto single CCD chip, with filters between 0.4 and 1.1 μm for mineralogy and atmospheric science.)
- • RA robotic arm (2 m long, to insert temperature probe, perform soil mechanics analyses, and to deliver surface samples to TEGA), carrying the RAC robotic arm camera (Keller)
- • MET meteorological package (1.2 m mast with windspeed/ direction sensor, temperature sensors, and TDL absorption cells to measure water vapour content and carbon dioxide and water isotope ratios. Pressure sensors mounted on spacecraft. Secondary 0.9 m submast for saltation layer windspeed and temperature measurements. RAATS robotic arm atmospheric temperature sensor, and STP soil temperature probe, also on the arm) (Crisp, May, Harri)

Table (Cont.)

- TEGA thermal & evolved gas analyser (set of 8 one-shot ovens for differential scanning calorimetry (DSC) of surface samples, coupled to oxygen detector and TDL H_2O/CO_2 absorption cell to determine ice concentration, adsorbed volatiles and volatile-bearing minerals) (Boynton)
- MARDI Mars descent imager (nested downlooking images 1.25 mrad per pixel. 1000×1000 pixels, panchromatic electronically shuttered CCD. 9 km at start of descent (7.5 m resolution) to 9 m at end (9 mm)) (Malin)
  - Mars microphone (Friedman)
- LIDAR (light detection and ranging) (400 nJ, 100 ns pulses at 2.5 kHz, 0.88 μm GaAlAs diode to probe lowest $\sim$3 km for dust and ice haze) (Linkin)
- Mars magnetic properties experiment (Knudsen)
- Cruise stage also carried the DS-2 Mars Microprobes (Section 19.2 and Chapter 25)

The Project Scientist was Richard Zurek.

Delivery architecture	Separation from cruise stage on approach
Thermal aspects	Thermally regulated interior component deck (min. $-30\,°C$) Very cold ambient environment
Power aspects	16 A h NiH secondary batteries $+ 2.9\,m^2$ GaAs solar arrays. Nominal power 200 W
Communications architecture	Via UHF antenna: two-way relay via Mars Climate Orbiter, or one-way relay (Mars-to-orbit) via Mars Global Surveyor (128 kbits s^{-1}). Two-way DTE via articulated X-band medium gain antenna (2.1–12.6 kbits s^{-1}) or low-gain antenna
EDL architecture	Entry at 6.9 km s^{-1}. 2.4 m diameter, $70°$ blunt half cone, ablative SLA-561 aeroshell. Active attitude control, using thrusters to minimise angle of attack. Max. deceleration 12 g. Parachute deployment at 8.8 km altitude, 430 m s^{-1}. Heatshield separation at 7.5 km altitude, 250 m s^{-1}. Landing legs deployed 70–100 s before landing. Back shell/parachute jettisoned and descent engines fired, controlled using four-beam Doppler radar. Final 40 m of descent controlled using gyros and accelerometers. Cruise/ descent propulsion system: 64 kg hydrazine in 2 diaphragm tanks; regulated He pressurization; 12 266 N descent engines (3 groups of 4)
Landing speed(s)	Nominally 2.4 m s^{-1} under thruster control for last 12 m of descent
Active operations (deployments, etc.)	Deployment of solar arrays, camera boom and meteorology masts. Sample acquisition via robotic arm and scoop. Articulated medium-gain antenna
Key references	*J. Geophys. Res.* **106**(E8), 2001; Casani *et al.*, 2000; Warwick, 2003; Backes *et al.*, 2000. See also *Mars Polar Lander/Deep Space 2 Press Kit*, NASA, 1999.

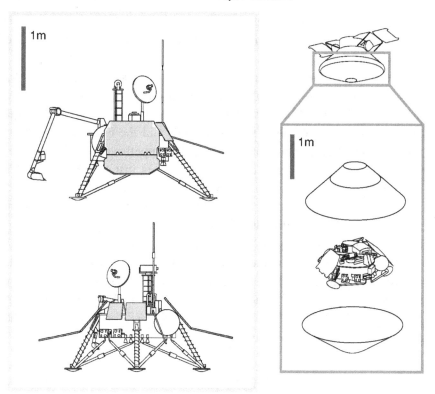

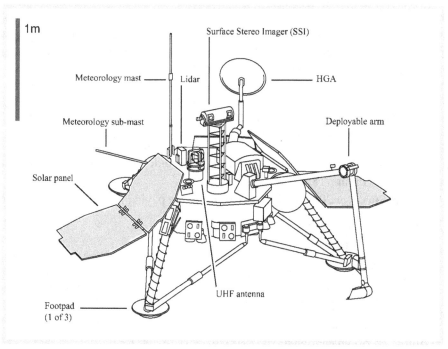

Figure 18.13 Mars Polar Lander.

Mars Surveyor landers were planned for at least 2003 and 2005, with greater international participation (e.g. from Italy), in parallel with a series of smaller 'Mars Micromissions'.

However, following the MPL failure, work was stopped on the '01 lander and it was put into storage. Already at an advanced stage of construction, it was eventually resurrected as the 2007 Phoenix lander, with a modified payload that drew upon elements from both MPL and the original '01 lander. Phoenix is the first 'Mars Scout' mission (i.e. a competitively selected PI-led mission, like the Discovery series of missions; in this instance the mission is led by Peter Smith of the University of Arizona). The planned 2003 Surveyor lander was dropped in favour of the twin Mars Exploration Rovers (drawing significant heritage from Pathfinder's successful EDL system and, indeed, from the experience gained with operations of Sojourner), with the 2005 opportunity being used to send the Mars Reconnaissance Orbiter. The 2011 surface mission is planned to be a large rover, the Mars Science Laboratory, which evolved from the initial concept of a 'Mars Smart Lander' and 'Long-Range Science Rover'. Further Scout missions are planned for 2013 and beyond, with sample return at some later date.

18.9.1 Mars Polar Lander (MPL)

This, the first Mars Surveyor lander, was targeted at the volatiles and surface environment of the South Polar layered terrain. Also riding on the same cruise stage were the two Deep Space 2 Mars Microprobes. All three elements were lost during entry, descent or landing (Figure 18.13).

18.9.2 Phoenix

Derived from the Mars Surveyor 2001 lander platform, with a payload drawing on those of both the '01 lander and Mars Polar Lander, Phoenix is the first Mars Scout mission (Figure 18.14).

The Mars Surveyor 2001 lander had the following payload when it was cancelled:

- APEX Athena Precursor Experiment (Squyres)
  - Mini-TES (Christensen)
  - Pancam (Bell)
  - Mössbauer spectrometer (Klingelhöfer)
  - Capture magnet (Madsen)
- Marie Curie (Sojourner-class) rover, with:
  - Rover imaging cameras
  - Athena APXS alpha-proton-X-ray spectrometer (Rieder)

Target	Mars
Objectives	To understand the near-surface chemistry and geology of a polar landing site, with particular attention to H_2O, organics and meteorology
Prime contractor	Lockeed Martin Astronautics
Launch site, vehicle	ETR, Delta II (2925)
Launch date	4 August 2007
Arrival date	25 May 2008
Landing site co-ordinates	Northern plains
End(s) of mission(s)	Digging phase to last 3 months, then weather station mode Total duration ~150 sols
Mass(es)	Comparable to Mars Surveyor 2001 lander (328 kg landed dry mass)
Payload experiments	• SSI surface stereo imager (Smith)
	• Robotic arm & RAC robotic arm camera (Keller)
	• MET meteorology suite (LIDAR, T sensor, P sensor) (Carswell, Michelangeli, Whiteway)
	• TEGA thermal & evolved gas analyser (Boynton)
	• MARDI Mars descent imager (Malin)
	• Mars microphone (Friedman)
	• MECA microscopy, electrochemistry and conductivity analyser (Hecht, Meloy?)
	• WCL wet chemistry lab
	• Microscopy station (optical & AFM)
	• Material patch plates
	• TECP thermal & electrical conductivity probe on RA
	• magnetic properties experiment (Madsen)
	The project scientist is Peter Smith
Delivery architecture	Separation from cruise stage on approach
Thermal aspects	Thermally regulated interior component deck (min. $-30\,°C$). Very cold ambient environment
Power aspects	Secondary batteries + solar arrays
Communications architecture	Two-way UHF relay via Mars Odyssey or Mars Reconnaissance Orbiter

Table (Cont.)

EDL architecture	Entry at $5.7\,\mathrm{km\,s^{-1}}$. $2.64\,\mathrm{m}$ diameter, $70°$ blunt half cone, ablative SLA-561 aeroshell. Active attitude control, using thrusters to minimise angle of attack. Max. deceleration $7\,g$. Parachute deployment at $13\,\mathrm{km}$ altitude, $<504\,\mathrm{m\,s^{-1}}$. Heatshield separation at $12\,\mathrm{km}$ altitude, $<286\,\mathrm{m\,s^{-1}}$. Landing legs deployed $\sim182\,\mathrm{s}$ before landing. Back shell/parachute jettisoned at $740\,\mathrm{m}$ and descent engines fired, controlled using four-beam Doppler radar and hazard detection and avoidance system. Final $40\,\mathrm{m}$ of descent controlled using gyros and accelerometers. Cruise/descent propulsion system: $64\,\mathrm{kg}$ hydrazine in 2 diaphragm tanks; regulated He pressurization; 12 266 N descent engines (3 groups of 4)
Landing speed(s)	Nominally $1.6\,\mathrm{m\,s^{-1}}$ under thruster control for last $12\,\mathrm{m}$ of descent
Active operations (deployments, etc.)	Deployment of solar arrays, camera boom and meteorology masts. Sample acquisition via robotic arm and scoop. Articulated medium gain antenna
Key references	Smith *et al.*, 2004; *J. Geophys. Res.* **113**(E12), 2008; *Science* **325**(5936), 2009. See also http://mars.jpl.nasa.gov/ and http://phoenix.lpl.arizona.edu/

- - MEEC Mars experiment on electrostatic charging (Ferguson)
  - WAE wheel abrasion experiment (Ferguson)
- Robotic arm & RAC robotic arm camera (Keller)
- MARDI Mars descent imager (Malin)
- MECA Mars environmental compatibility assessment (Meloy):
  - WCL wet chemistry lab
  - Microscopy station
  - Material patch plates
  - Electrometer on RA
- MIP Mars *in situ* propellant production precursor (Kaplan)
- MARIE Martian radation environment experiment (Badhwar)
- Sundial

  The Project Scientist was Steve Saunders.

18.10 Mars Science Laboratory

Currently planned for 2011, the Mars Science Laboratory is a large, long-range rover equipped with a sophisticated and diverse payload. MSL is also due to use a new 'sky crane' landing technique, landing on its own wheels rather than encased in a lander platform.

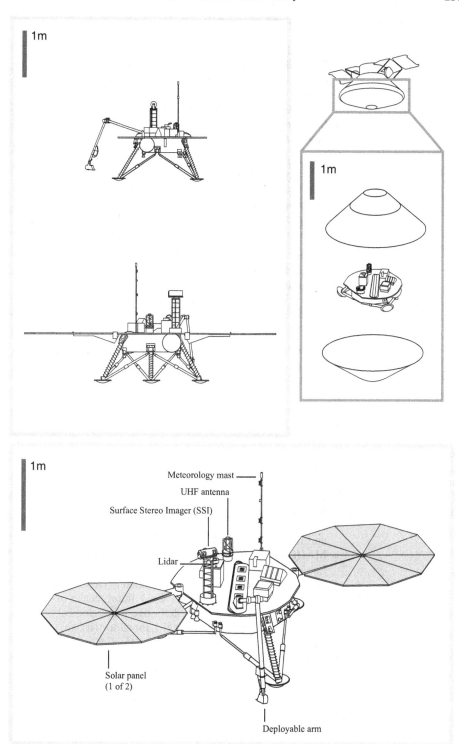

Figure 18.14 Phoenix.

Target	Mars
Objectives	Biological objectives:

Biological objectives:
- Determine the nature and inventory of organic carbon compounds
- Inventory the chemical building blocks of life (carbon, hydrogen, nitrogen, oxygen, phosphorous, and sulfur)
- Identify features that may represent the effects of biological processes

Geological and geochemical objectives:
- Investigate the chemical, isotopic, and mineralogical composition of the Martian surface and near-surface geological materials
- Interpret the processes that have formed and modified rocks and soils

Planetary process objectives:
- Assess long-timescale (i.e., 4-billion-year) atmospheric evolution processes
- Determine present state, distribution, and cycling of water and carbon dioxide

Surface radiation objective:
- Characterize the broad spectrum of surface radiation, including galactic cosmic radiation, solar proton events, and secondary neutrons

Prime contractor	JPL
Launch site, vehicle	ETR, Atlas V 541
Launch date	August 2011
Arrival date	Autumn 2012
Landing site co-ordinates	To be determined. Landing ellipse 20–40 km in length, 3–5 times smaller than previous missions
End(s) of mission(s)	Nominal lifetime of 1 Martian year
Mass(es)	Launch mass ~2800 kg. Rover mass ~775 kg
Payload experiments	• Mast camera (Malin)

Payload experiments
- Mast camera (Malin)
- ChemCam: laser induced remote sensing for chemistry and micro-imaging (Wiens)
- MAHLI: Mars handlens imager for the Mars Science Laboratory (Edgett)
- APXS alpha-particle-X-ray spectrometer (Gellert)
- CheMin: an X-ray diffraction/X-ray fluorescence (XRD/XRF) instrument for definitive mineralogical analysis in the MSL Analytical Laboratory (Blake)
- RAD radiation assessment detector (Hassler)
- Mars descent imager (Malin)
- SAM: sample analysis at Mars with an integrated suite consisting of a gas chromatograph mass spectrometer, and a tunable laser spectrometer (Mahaffy)
- Meteorological package (Spanish Ministry of Education and Science)

Table (Cont.)

	• UV sensor (Spanish Ministry of Education and Science)
	• Pulsed neutron source & detector (Roskosmos)
	• MEDLI MSL Entry, Descent and Landing Instrumentation
Delivery architecture	Separation from cruise stage on approach
Thermal aspects	Details not yet available
Power aspects	Secondary battery with RTG or solar arrays
Communications architecture	Two-way relay via orbiters
EDL architecture	Despin and cruise balance mass jettison. Precision landing techniques, by means of lifting body, guided entry, parachute descent, separation of parachute/backshell assembly, then 'skycrane' technique, lowering the rover on a tether from the retro assembly
Landing speed(s)	Not yet available
Active operations (deployments, etc.)	Rover operations (range 5–20 km over 1 Martian year), robotic arm operations
Key references	http://mars.jpl.nasa.gov/

19

Payload delivery penetrators

Payload delivery penetrators are bullet-shaped vehicles designed to penetrate a surface and emplace experiments at some depth. The basic technology for these has existed for several decades based largely on military heritage (e.g. Simmons, 1977; Murphy *et al.*, 1981a; Bogdanov *et al.*, 1988), however only in the mid 1990s did proposals for their use in Solar System exploration begin to be adopted for actual flight. In the US, Mars penetrators were studied for several years (and, indeed, field tested) as part of a possible post-Viking mission, while in the Soviet Union planetary penetrator work seems to have started in the mid 1980s.

Impact speeds range from about 60 to $300\,\mathrm{m\,s^{-1}}$. The resulting impact load experienced by penetrators as they decelerate in geological materials routinely exceeds $500\,g$, and terrestrial systems in the military field can be rated at $10\,000\,g$ or even $100\,000\,g$, although the choice of components at these levels is severely limited (being more suited to the relatively simple job of triggering a detonator than making planetary science measurements). Additional impact damping may be included in the form of crushable material (e.g. honeycomb or solid rocket motor casing), sacrificial 'cavitator' spikes protruding ahead of the penetrator's tip (e.g. Luna-Glob high-speed penetrator concept, with speeds exceeding $1.5\,\mathrm{km\,s^{-1}}$) and gas-filled cavities (e.g. the Mars 96 penetrators).

Masses have ranged from the tiny DS-2 Mars Microprobes at $2.5\,\mathrm{kg}$ each (excluding aeroshell) to $62.5\,\mathrm{kg}$ each for the Mars 96 penetrators.

Penetrators may consist of a single unit, or a slender forebody and a wider aftbody linked by an umbilical tether, the two parts separating during penetration to leave the aftbody at the surface. Expected forebody penetration depths have ranged from $\sim 0.5\,\mathrm{m}$ for the Mars Microprobes (impacting at $\sim 190\,\mathrm{m\,s^{-1}}$) up to 4–6 m for Mars 96, with 1–3 m expected for the single-body Lunar-A penetrators, which do not need to retain any components at the surface ($13\,\mathrm{kg}$ each, $140\,\mathrm{mm}$ diameter, impacting at $\sim 285\,\mathrm{m\,s^{-1}}$).

Power is usually provided by primary batteries or radioisotope thermoelectric generators (RTGs), although solar arrays have been high-*g*-tested successfully. The DS-2 Mars Microprobes' nominal lifetime was only a few hours, while the Lunar-A penetrators are expected to have enough power for about a year. Transmission of data back to Earth is usually by means of an omnidirectional antenna and a relay spacecraft.

Experiments flown on (or proposed for) penetrators include the following.

- Accelerometry/gravimetry/tiltmeter
- Thermal sensors (temperature profile, thermal conductivity/diffusivity, heat flow)
- Imaging
- Magnetometer
- Permittivity/conductivity sensors
- Seismometer
- Spectrometers (γ-ray, neutron, α/proton/X-ray, X-ray fluorescence, etc.)
- Sample collection for evolved gas analyser/mass spectrometry/spectroscopic analysis
- Penetrators with combined sampling and pyrotechnic return
- Explosive charge
- Meteorological sensors (not applicable to atmosphereless bodies of course!)

Sadly, neither the Mars 96 penetrators nor the DS-2 Mars Microprobes completed their missions – Lunar-A now has the task of demonstrating penetrator technology on another world for the first time, although at the time of writing no launch date has been set. Table 19.1 gives key references for penetrator missions and proposals.

Table 19.1. *Penetrator missions, studies and key references*

Mission	Reference
Mars 96 penetrators	Surkov and Kremnev, 1998
DS-2 Mars Microprobes	Smrekar *et al.*, 1999; Smrekar *et al.*, 2001
Lunar-A penetrators	Mizutani *et al.*, 2001
Vesta/Mars-Aster penetrators	ESA, 1988; Surkov, 1997
CRAF/Comet Nucleus Penetrator	Boynton and Reinert, 1995
Luna-Glob high-speed penetrators	Galimov *et al.*, 1999
Luna-Glob large penetrators/ Polar Station	Surkov *et al.*, 1999; Surkov *et al.*, 2001
BepiColombo Mercury surface element-penetrator option	Pichkhadze *et al.*, 2002
BepiColombo Mercury surface element-hard lander/penetrator option	ESA, 2000
Polar Night lunar penetrators	Mosher and Lucey, 2006

19.1 Mars 96 penetrators

These were the first attempted planetary penetrators, lost when Mars 96 failed to leave Earth orbit. They carried an international payload and employed an innovative two-stage inflatable entry and descent braking device (Figure 19.1).

Target	Mars	
Objectives	• Imaging of the Martian surface	
	• Data on meteorology of the planet	
	• Chemistry of rocks	
	• Water content in Martian rocks	
	• Seismic activity of Mars	
	• Physical and mechanical characteristics of Martian regolith	
	• Magnetic field and magnetic properties of rocks	
Prime contractor	NPO Lavochkin	
Launch site, vehicle	Baikonour, Proton 8K82K/11S824M (on Mars 96 orbiter)	
Launch date	16/11/1996	
Arrival date	(Would have been 12/09/1997)	
	Penetrator 1	Penetrator 2
Landing site co-ordinates (planned)	36° N, 161° W	36° N, 125° W
Nominal lifetime on Mars	1 Earth year	
Mass(es)	62.5 kg each, incl. 17 kg braking system	
Payload experiments	• PTV-1 television camera (Surkov, Chumak)	
	• MEKOM meteorological instrumentation (Marov)	
	• PUI pressure & humidity transmitter (Harri, Polkko)	
	• P, T, humidity & wind sensors (Marov, Manuilov)	
	• Optical sensor (Esposito, Maki)	
	• PEGAS gamma ray spectrometer (Surkov, Moskaleva)	
	• ANGSTREM XRFS (Surkov, Dunchenko)	
	• ALPHA alpha-proton spectrometer (Wänke, Rieder)	
	• NEUTRON-P neutron spectrometer (Surkov, Scheglov)	
	• GRUNT accelerometers (Khavroshkin, Tsyplakov)	
	• TERMOZOND T sensors (Okhapkin)	
	• KAMERTON seismometer (Khavroshkin)	
	• IMAP-7 magnetometer (Zhuzgov)	
	Total mass 4.5 kg. The Project Scientist was Yuri A. Surkov	
Delivery architecture	Separation from Mars 96 orbiter after orbit insertion, spinning around longitudinal axis. $30\,\mathrm{m\,s^{-1}}$ deorbit burn	
Thermal aspects	Active heating radiator and passive insulation	
Power aspects	0.5 W RTG + 150 W-h Li battery	

Table (Cont.)

Communications architecture	UHF relay at 8 kbits s^{-1} via Mars 96 orbiter or MGS
EDL architecture	Entry at 5.6 km s^{-1}. Separation of deorbit motor; inflation of entry shield; inflation of additional descent brake. Impact penetration loads damped by fluid reservoir shock absorber
Landing speed(s)	60–80 m s^{-1} (planned)
Active operations (deployments, etc.)	Deployment of mast carrying antenna, camera, magnetometer, meteorological and wind sensors; extension and retraction of the TERMOZOND sensors
Key references	Surkov, 1997; Surkov and Kremnev, 1998

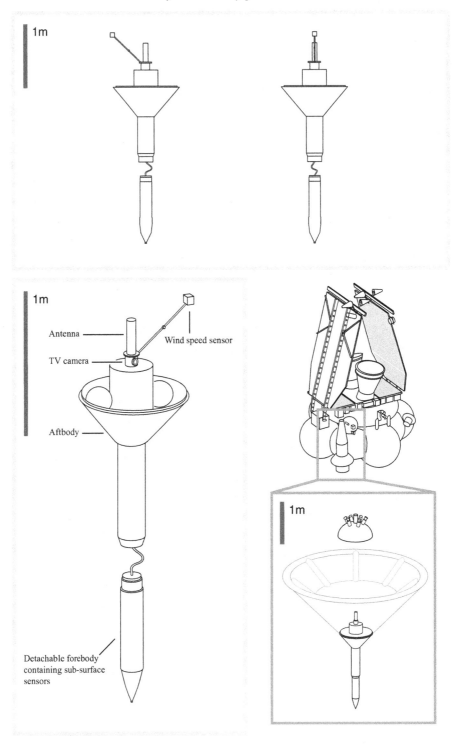

Figure 19.1 Mars 96 Penetrators.

19.2 Deep Space 2 Mars Microprobes

The Deep Space 2 (DS-2) Mars Microprobes were two small penetrators attached to the cruise stage of Mars Polar Lander. Forming part of NASA's New Millennium programme, they were designed to demonstrate penetrator technology and search for sub-surface water. Sadly, no signals were received from either probe after separation, and several possible failure modes have been postulated. See Chapter 25 for a more detailed case study (Figure 19.2).

Target	Mars
Objectives	Penetrator technology demonstration; sub-surface water detection; atmospheric density profile measurement
Prime contractor	JPL
Launch site, vehicle	ETR, Delta II (Delta 7425) (on Mars Polar Lander)
Launch date	03/01/1999
Arrival date	03/12/1999
	Scott Amundsen
Landing site co-ordinates	75° S, 196° W (south polar layered terrain), i.e. 60 km NW of MPL; statistically 2 km from each other
End(s) of mission(s)	Expected lifetime (limited by battery energy) of 1–3 days
Mass(es)	3.6 kg with entry shell; 2.4 kg post-impact
Payload experiments	• Atmospheric descent accelerometer (Catling, Magalhães)
	• Impact accelerometer (Lorenz, Moersch)
	• Evolved water experiment (Murray, Zent, Yen)
	• Soil conductivity experiment (Morgan, Presley)
	The Project Scientist was Suzanne Smrekar
Delivery architecture	Separation from Mars Polar Lander during approach, 10 min before impact. Separation $\Delta V < 0.3 \, m \, s^{-1}$, no spin
Thermal aspects	Cold-tolerant electronics. No heaters
Power aspects	$LiSOCl_2$ primary batteries
Communications architecture	One-way UHF relay via Mars Global Surveyor
EDL architecture	Aeroshell – peak deceleration of 12.4 g, at 44 km altitude. Aeroshell shatters at impact, forebody/aftbody separation during penetration
Landing speed(s)	140–210 m s^{-1} (design envelope); 180–200 m s^{-1} (predicted)
Active operations (deployments, etc.)	Sample acquisition via drill in forebody; pyrotechnic closure of sample oven door
Key references	Smrekar *et al.*, 1999, 2001; Casani *et al.*, 2000; Braun *et al.*, 1999b. See also *Mars Polar Lander/Deep Space 2 Press Kit*, NASA, 1999.

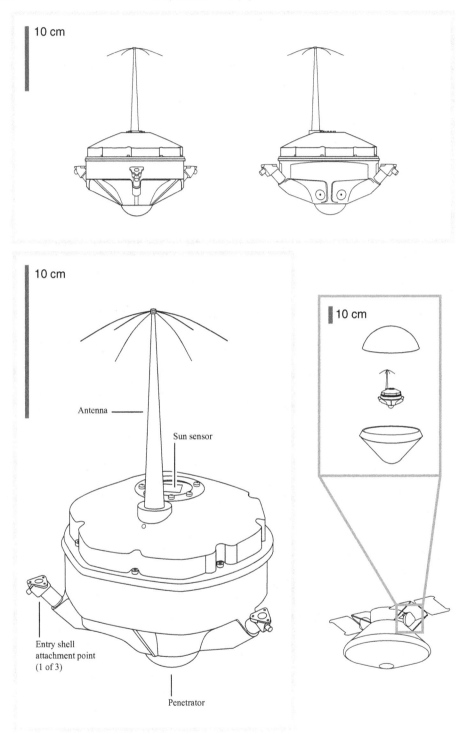

Figure 19.2 Deep Space 2 Mars Microprobes.

19.3 Lunar-A penetrators

The Lunar-A project was originally intended for launch in the late 1990s, carrying two (originally three) penetrators for lunar seismology. However, the mission has been subject to multiple delays; at the time of writing no new launch date has been set (Figure 19.3).

Target	The Moon
Objectives	Lunar seismology and heat flow measurements
Prime contractor	ISAS/JAXA + NTS
Launch site, vehicle	Unknown
Launch date	Undetermined
Arrival date	~1 year after launch
Landing site co-ordinates	One penetrator will be targeted in the vicinity of the Apollo 12 and 14 landing sites on the lunar near-side, with the other aimed close to the antipodal point on the far side
End(s) of mission(s)	Each penetrator has a battery life of approximately 1 year
Mass(es)	Each penetrator has a mass of 14 kg, plus ~31 kg for deorbit and attitude control
Payload experiments	• Seismometer
	• T & thermal conductivity sensors
	• Accelerometry
	• Tiltmeter
	The Project Scientist is Hitoshi Mizutani (retired)
Delivery architecture	Separation from Lunar-A orbiter after orbit insertion. Deorbit
Thermal aspects	Passive thermal control (protected underground by regolith)
Power aspects	$LiSOCl_2$ primary batteries
Communications architecture	Two-way relay via Lunar-A orbiter
EDL architecture	Rhumb-line manoeuvre; separation of de-orbit motor/attitude control stage; impact penetration
Landing speed(s)	$285 \, m \, s^{-1}$ (planned)
Active operations (deployments, etc.)	None
Key references	Mizutani *et al.*, 2001, 2003

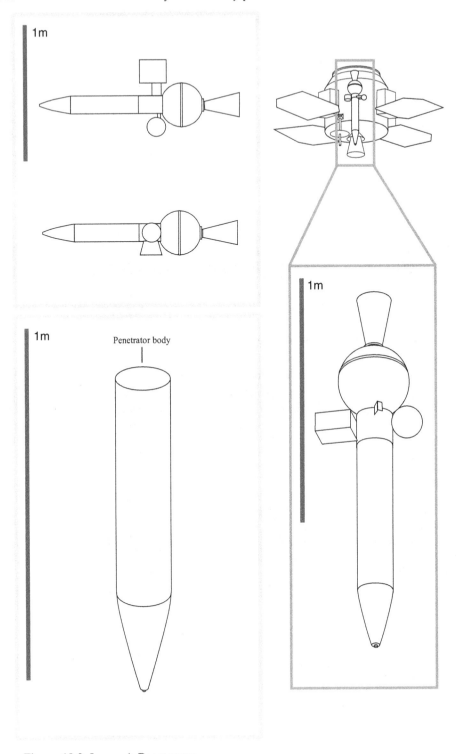

Penetrator body

Figure 19.3 Lunar-A Penetrators.

20

Small body surface missions

Missions to small bodies differ from those to larger worlds because the low surface gravity means that an orbiter (or rendezvous) spacecraft can approach close enough to perform a surface mission while hovering (with little or no thrust) and the speed of a landing can be very low. This blurs the distinction between orbiters and landers, and may enable orbiter spacecraft to survive landing, as shown by the landing of NEAR on asteroid Eros. Low gravity also means that a landing vehicle may risk being lost entirely on rebound from the surface, or ejected by outgassing in the case of a comet nucleus. Anchoring systems may thus be required. On the positive side, the low gravity also makes it easy to achieve mobility by jumping, and to perform 'touch and go' surface-sampling manoeuvres (e.g. Yano *et al.*, 2003; Sears *et al.*, 2004). Most small bodies are highly irregular, and their gravitational fields can be challenging environments in which to navigate. Dust thrown up from the surface (whether from natural cometary activity or the action of a spacecraft) is another hazard. Many small bodies, particularly comets, are in elliptical orbits and so experience wide variations of temperature and solar power production with time and surface location.

20.1 Phobos 1F

The Phobos project involved two large Mars orbiters, Phobos 1 and Phobos 2 (Sagdeev *et al.*, 1988; TsUP, 1988). Their main target was Phobos itself, the larger of Mars' two moons, being $26.1 \times 22.2 \times 18.6$ km^3 in size and probably a captured asteroid. Amongst the payloads were two types of lander, a stationary lander (on both) and a hopping rover (on Phobos 2 only). Sadly, Phobos 1 was lost during cruise and Phobos 2 was lost in Martian orbit, before the low altitude (~ 50 m) hovering phase during which the landers were to be deployed to the surface. However, at the time of writing, Russia has a Phobos sample return project under way, named Phobos-Grunt (Marov *et al.*, 2004).

20.1.1 Phobos 1, 2 DAS

Phobos 1 and 2 each carried a 'long-lived autonomous station' (Russian abbreviation: DAS) to enable the study of Phobos' orbit (which is thought to be slowly spiralling in towards Mars) and libration by means of radio science and Sun-sensor data. Other *in situ* instruments were also included (Figure 20.1).

Target	Phobos	
Objectives	Measurement of the orbit of Phobos and its libration	
Prime contractor	NPO Lavochkin (formerly OKB-301)	
Launch site, vehicle	Baikonour, Proton 8K82K / 11S824F (on Phobos 1 & 2)	
	Phobos 1	Phobos 2
Launch date	07/07/1988	12/07/1988
Arrival date	Deployment was due March/April 1989	
Landing site co-ordinates	–	–
End(s) of mission(s)	DAS lifetime on surface designed to be $\geq$3 months	
Mass(es)	67 kg, 20.6 kg payload	
Payload experiments	• TV camera (Blamont, Kerzhanovich) • ALPHA-X alpha-proton-X-ray spectrometer (Hovestadt; Mukhin) • LIBRATION sun sensor (a.k.a. STENOPEE?) (Blamont, Linkin) • Seismometer (Khavroshkin, Tsyplakov, Linkin) • RAZREZ anchor penetrometer (Khavroshkin, Tsyplakov) • Celestial mechanics experiment (Linkin, Preston, Blamont) The Project Scientist was Viacheslav M. Linkin	
Delivery architecture	Separation from Phobos orbiter at 2.2 m s^{-1} during low altitude flyover	
Thermal aspects	Details unknown (thermal blanket on electronics boxes)	
Power aspects	Solar arrays, secondary battery	
Communications architecture	Two-way DTE at 1672 MHz; transmission rate 4–20 bits s^{-1}	
EDL architecture	Spin-stabilisation at $\leqslant$2 rad s^{-1} during descent; detection of surface with contact probes; firing of hold-down thrusters and anchoring harpoon (100 m s^{-1}); deployment of upper section followed by deployment and pointing of solar arrays and antenna	
Landing speed(s)	0–4 m s^{-1} vertical, 2 m s^{-1} horizontal	
Active operations (deployments, etc.)	Descent and landing aspects; deployment of ALPHA-X sensor heads; solar array pointing mechanism	
Key references	Sagdeev *et al.*, 1988; TsUP, 1988	

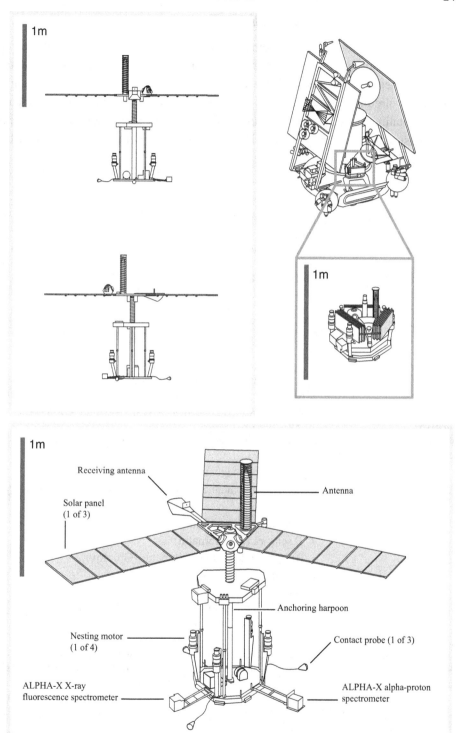

Figure 20.1 Phobos DAS.

20.1.2 Phobos 2 PROP-F

PROP-F was a mobile surface element due to be deployed from the Phobos 2 orbiter. It was due to land and perform a sequence of physical and compositional properties measurements at multiple locations. Motion was actuated by a single 'foot' in its circular base, and it used a system of four control rods (one static pair and one rotating pair) to bring itself upright at each new site, ready to perform the next measurement sequence. Sadly Phobos 2 was lost before PROP-F could be deployed (Figure 20.2).

Target	Phobos
Objectives	*In situ* measurements of physical and compositional properties of Phobos' surface material at several locations
Prime contractor	VNIITransMash
Launch site, vehicle	Baikonour, Proton 8K82K/11S824F (on Phobos 2)
Launch date	12/07/1988
Arrival date	PROP-F was due to be deployed in March/April 1989
Landing site co-ordinates	–
End(s) of mission(s)	PROP-F not yet deployed when Phobos 2 lost on 27/03/1989. The envisaged total time of operation was 4 h, during which ~ 7 measurement cycles were expected.
Mass(es)	50 kg
Payload experiments	• ARS-FP automatic X-ray fluorescence spectrometer (Surkov)
	• Ferroprobe magnetometer (Dolginov)
	• Kappameter magnetic permeability/susceptibility sensor (Dolginov)
	• Gravimeter (Ksanfomaliti)
	• T sensors (Ksanfomaliti)
	• BISIN conductometer/tiltmeter (Gromov)
	• Mechanical sensors (Kemurdzhian):
	• Penetrometer
	• UIU accelerometer
	• Sensors on hopping mechanism
	Total 7 kg.
	The Project Scientist was Aleksandr L. Kemurdzhian
Delivery architecture	Separation from Phobos orbiter during low altitude flyover, with velocity relative to orbiter of 3 m s^{-1} horizontal component and 0.45 m s^{-1} downward vertical component. Free fall to surface.
Thermal aspects	Unknown

Table (Cont.)

Power aspects	Battery
Communications architecture	One-way (?) relay via Phobos 2 orbiter
EDL architecture	Free fall to surface. Time to come to rest after initial impact and subsequent bouncing ~ 45 min. Ejection of damper and righting of lander using control rods
Landing speed(s)	Initial impact: <1 m s^{-1} vertical, ~ 3 m s^{-1} horizontal
Active operations (deployments, etc.)	Hopping manoeuvre, movement of rotating arms to bring itself upright, penetrometer operational sequence
Key references	Kemurdzhian *et al.*, 1988, 1989a,b, 1993

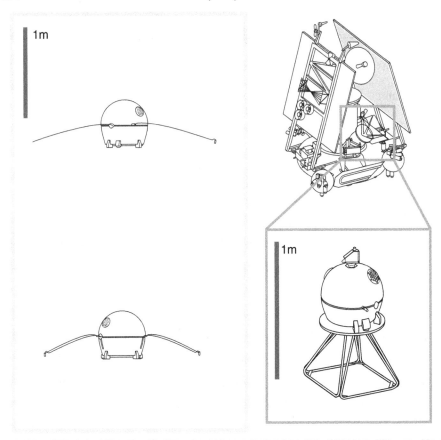

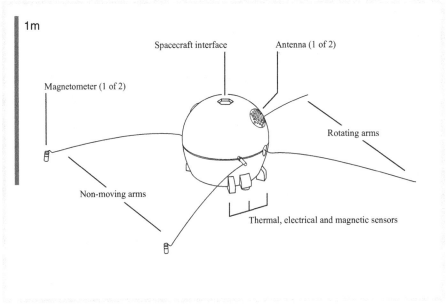

Figure 20.2 Phobos 2 PROP-F.

20.2 NEAR Shoemaker

The Near-Earth Asteroid Rendezvous (NEAR-Shoemaker) mission was a NASA Discovery mission launched in February 1996 to study asteroid Eros (Russell, 1998; Bell and Mitton, 2002). No aspect of its design was required to support a landing manoeuvre, since it was able to meet all its mission goals from orbit around the asteroid. However on 12 February 2001, in a grand finale to the mission, NEAR was commanded to descend to the surface, where it landed softly (at a speed of around 1.9 m s^{-1}) and continued to operate (Dunham *et al.*, 2002; see also Section 5.6), thus performing the first ever controlled landing on an asteroid. During descent, high resolution imagery was telemetered, and while on the surface valuable additional data was acquired from the gamma ray spectrometer and magnetometer instruments. Surface operations continued for 14 days.

20.3 Rosetta Lander Philae

Philae is a comet nucleus lander launched in 2004 as part of ESA's Rosetta mission, after a 1-year delay and change of the target comet. See Chapter 26 for a more detailed case study (Figure 20.3).

Target	Comet 67P/Churyumov-Gerasimenko
Objectives	To land on a comet nucleus and measure the following:
	• The elemental, molecular, mineralogical, and isotopic composition of the comet's surface and subsurface material
	• Characteristics of the nucleus such as near-surface strength, density, texture, porosity, ice phases and thermal properties; texture measurements will include microscopic studies of individual grains
Prime contractor	DLR/MPAe
Launch site, vehicle	Kourou, Ariane 5 G+ (on Rosetta orbiter)
Launch date	02/03/2004
Arrival date	Nominal lander delivery: November 2014
Landing site co-ordinates	To be determined based on orbiter data
End(s) of mission(s)	First science sequence 120 h (possible with battery power only), followed by ∼3 month long-term mission to 2 AU, then extended mission until lander overheats
Mass(es)	97.4 kg mass ejected from orbiter
Payload experiments	• APX alpha-particle-X-ray spectrometer (Rieder, Klingelhöfer)
	• ÇIVA comet nucleus infrared and visible analyser (Bibring)

Table (Cont.)

- ÇIVA-P panoramic cameras (5 single + 1 stereo pair)
- ÇIVA-M visible/IR microscope
- COSAC cometary sampling and composition experiment (evolved gas analyser) (Rosenbauer, Goesmann)
- CONSERT comet nucleus sounding experiment by radiowave transmission (Kofman)
- MUPUS multi-purpose sensors for surface and sub-surface science (Spohn)
  - TM thermal IR radiometer
  - ANC-M,T accelerometer and temperature sensor in harpoon anchors
  - PEN thermal probe
- Ptolemy evolved gas analyser (Pillinger, Wright)
- ROMAP Rosetta Lander magnetic field investigation and plasma monitor (Auster)
- ROLIS Rosetta Lander imaging system (Mottola)
- SESAME surface electrical, seismic and acoustic monitoring experiments (Möhlmann, Seidensticker)
  - CASSE cometary acoustic surface sounding experiment (Möhlmann, Seidensticker)
  - PP permittivity probe (Laakso, Schmidt),
  - DIM dust impact monitor (Apathy)
- SD2 sampling, drilling and distribution system (Finzi)

Total mass 26.7 kg
The Project Scientists are Helmut Rosenbauer, Hermann Böhnhardt and Jean-Pierre Bibring

Delivery architecture	Separation at low speed (5–52 cm s^{-1}) from the Rosetta orbiter during low altitude (few km) manoeuvre
Thermal aspects	'Warm' compartment inside lander body, and 'cold' balcony for external instruments
Power aspects	Primary ~1 kWh LiSOCl$_2$ battery plus secondary 100 Wh Li-ion battery with body-mounted Si solar array
Communications architecture	Two-way S-band relay via Rosetta orbiter
EDL architecture	Ejection velocity from orbiter causes lander to fall towards surface. Momentum wheel attitude stabilisation. Landing legs unfolded. Hold-down thruster, anchoring harpoon and foot screws activated on contact with surface. Impact damping performed partly by damping mechanism in cardanic joint between landing gear and lander body
Landing speed(s)	⩽1.2 m s^{-1} (planned)

Table (Cont.)

Active operations (deployments, etc.)	Descent and landing aspects including unfolding legs, hold-down thruster and anchoring harpoon. Movement of lander body on landing gear via cardanic joint and rotation device. Deployment of CONSERT antennas, APX, sampling drill, magnetometer boom and MUPUS boom
Key references	Biele *et al.*, 2002; Biele, 2002; Biele and Ulamec, 2004; Ulamec *et al.*, 2006; ESA SP-1165 (in preparation); *Space Sci. Rev.* **128**(1–4), 2007.

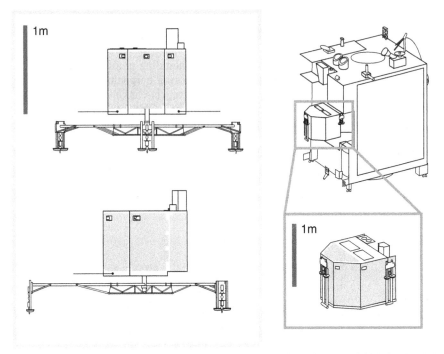

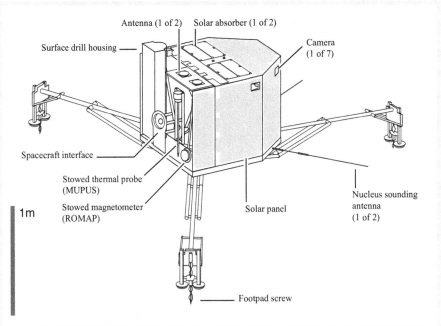

Figure 20.3 Rosetta Lander Philae.

20.4 Hayabusa (MUSES-C) and MINERVA

Hayabusa (called MUSES-C prior to launch) is a Japanese near-Earth asteroid sample-return mission that includes a small mobile surface element. It carries a horn-shaped sampling mechanism that, after contact with the surface, fires small projectiles to collect the ejecta for return to Earth in a re-entry capsule. The MINERVA hopping rover – at 591 g the lightest ever planetary vehicle – was due to be deployed during descent to the surface. Hayabusa had also been due to carry a Small Science Vehicle (SSV) or 'MUSES-CN' wheeled Nanorover from JPL (PI: Yeomans) (Jones, 2000). It was, however, cancelled in November 2000 for mass and budgetary reasons. It would have carried the following payload: multi-band camera (Smith), near-IR point reflectance spectrometer (Clark), AXS alpha-X-ray spectrometer (Economou), and a laser ranging system. Figure 20.4 shows MINERVA and its accommodation on Hayabusa.

Target	25143 Itokawa (1998 SF36) (previously 4660 Nereus then 10302 (1989ML))
Objectives	Technology demonstration (electric propulsion autonomous navigation microgravity, sampling, direct atmospheric re-entry), *in situ* asteroid science
Prime Contractor	ISAS/JAXA + NTS
Launch site, vehicle	Kagoshima, M-V-5
Launch date	09/05/2003
Arrival date	Having arrived at Itokawa on 12/09/2005 for remote sensing operations, MINERVA separated from Hayabusa on 12/11/2005; Hayabusa released a target marker and contacted the surface three times and stayed 30 minutes on 19/11/2005 and once on 25/11/2005
Landing site co-ordinates	MINERVA is believed not to have encountered the asteroid surface, due to deployment at a time when Hayabusa was moving away
End(s) of mission(s)	Contact with MINERVA ended shortly after deployment, yet MINERVA had already successfully imaged Hayabusa and measured heat radiated by the asteroid surface. Hayabusa mission still under way at time of writing
Mass(es)	Hayabusa: launch wet mass 530 kg, incl. 50 kg chemical propellant and 65 kg of xenon
	MINERVA: 591 g, plus ~ 0.6 kg of support equipment on Hayabusa
Payload experiments	Hayabusa:
	• AMICA asteroid multiband imaging camera (ONC-T (telescopic camera) + 8-colour filters & polarisers) (Saito)
	• NIRS near-IR spectrometer (Abe)

Table (Cont.)

- XRS X-ray spectrometer (Okada, Kato)
- LIDAR (Mukai)
- ONC-W (wide-view cameras)
- Three target markers
- LRF (Laser range-finders)
- Surface sampling device (Yano)

MINERVA hopper (Yoshimitsu):
- Three colour CCD cameras (one plus stereo pair)
- Thermal sensors
- Mechanical sensors
- Sun sensors

Delivery architecture	Separation of MINERVA from Hayabusa during low altitude manoeuvre and free fall to surface
Thermal aspects	MINERVA: passive control, plus sleep mode if too hot or cold, and hopping to avoid extremes
Power aspects	MINERVA powered by battery and Si solar array (max 2.2 W at 1 AU) with two 2.3 V, 50 F condensers
Communications architecture	Hayabusa 2-way DTE; MINERVA two-way relay via orbiter at 9600 bits s^{-1}
EDL architecture	Intended method for Hayabusa: descent at 12 cm s^{-1} and release of target marker at ~ 40 m; braking to 3 cm s^{-1} then free fall from rest at 17 m
	MINERVA: separation from Hayabusa at a few tens of m altitude (at low speed) and passive free fall
Landing speed(s)	~ 0.1 m s^{-1}
Active operations (deployments, etc.)	MINERVA: hopping mechanism (turntable, rotor, brake) Hayabusa: deployment of solar arrays, MINERVA, target markers, sampling mechanism and sample return capsule
Key references	Yano *et al.*, 2002; Kawaguchi et al., 2003; Yoshimitsu *et al.*, 2001, 2003; *Science* **312** (5778), 2006

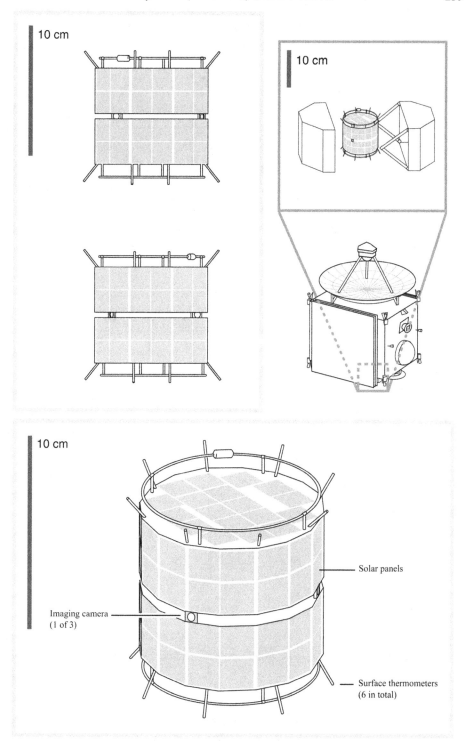

Figure 20.4 MINERVA on Hayabusa.

Part III
Case studies

Each of the missions or spacecraft in this part has been selected for description in greater detail because they have faced and overcome an unusual challenge in their design and/or mission. Collectively, the seven case studies cover: atmospheric probes and surface/sub-surface missions; worlds with and without atmospheres; low and high gravity environments, and both static and mobile elements.

21

Surveyor landers

The Surveyor spacecraft were a series of seven lunar soft-landing vehicles launched by the USA in the period 1966–1968. They were a second generation of lunar spacecraft, following the Ranger series that ran from 1961 to 1965, and paved the way for the later soft landings required for Apollo. The main aims of the Surveyor project were to accomplish a soft landing on the Moon, provide basic data in support of Apollo, and perform scientific operations on the lunar surface for an extended period. The Ranger 3, 4, 5 soft landing attempts having failed, Surveyor was to achieve the USA's first soft landings on another world. Orbital surveys by the Lunar Orbiter spacecraft complemented the *in situ* investigations by Surveyor.

Industrial studies for the project that became Surveyor began in mid 1960, with the Hughes Aircraft Company being chosen as prime contractor, under NASA JPL. The first launch was initially planned for late 1963 but a series of technical and programmatic issues forced an accumulated delay of nearly three years, by which time development of the Apollo landers was already well under way, and the Soviet Union had already made the first successful soft landing with Luna 9.

The main challenge for Surveyor was designing one of the first systems for performing a soft landing on another planetary body, with the associated terminal guidance and control problems of braking the spacecraft to land intact, and the then great uncertainty regarding the lunar surface's physical properties. The spacecraft was also required to survive the cold of the fortnight-long lunar night.

The launch vehicle chosen was the Atlas Centaur, which itself was still under development. This resulted in changes to the mass available for Surveyor and its payload. In the event, the first four Surveyors were classed as engineering test models. The final three carried more payload, though still somewhat less than had originally been envisaged (see Corliss, 1965 for information on other instruments developed). The mass on separation from the Centaur upper stage ranged from 995 to 1040 kg, and that at touchdown from 294 to 306 kg, of which up to 32.2 kg was scientific payload. After separation, the spacecraft locked onto the Sun and Canopus and performed a mid-course correction (using its vernier engines) before descending to the lunar surface directly (i.e. without first entering lunar orbit). The total flight time from launch to landing ranged from 63.5 to 66.5 h.

21.1 Design

Perhaps the first point to note regarding the tripod-shaped Surveyors is that they were delivered directly from the launcher, rather than separated from a carrier vehicle (such as a cruise stage, flyby craft, orbiter or descent stage). The Moon's proximity and lack of atmosphere make this architecture a possible solution, and indeed it was also used by the Soviet Ye-8 series. Its use for missions to worlds beyond the Moon has in practice been precluded, for example by requirements for both an entry shield and mid-course correction capability, the need for relay communications, and often the presence anyway of an orbital element to the mission. (Such a 'lander-only' mission architecture may also be appropriate for some minor-body missions, however.)

The Surveyor spacecraft employed a distinctive open structure of tubular aluminium, onto which the spacecraft subsystems and payload were mounted. The landing gear comprised three hinged landing legs with shock-absorbers and hinged footpads, backed up by three crushable blocks mounted on the underside. The footpads and crushable blocks used aluminium honeycomb to ensure damping of the landing loads. Electronic equipment was housed in two thermally controlled compartments attached to the spaceframe.

A vertical mast carried the solar array and a planar high-gain antenna, both of which were articulated. Two deployable low-gain antennas were also incorporated. Radio communications operated in the S-band, and the transmitters could feed either low power (100 mW) or high power (10 W) to any of the three antennas.

Reading an account of the history of the project (Koppes, 1982), one can speculate that the open structure may have arisen as a result of the prime contractor's highly granular division of the project, into about a hundred 'units' or 'control items', rather than the now conventional set of major subsystems (structure, thermal, power, communications, propulsion, etc.). While this may have limited the scope for design optimisation, there was ample scope for flexibility from one mission to the next. Variations occurred in both payload instrumentation and engineering subsystems.

Propulsion for braking and descent was provided by a main retro-rocket using solid propellant (a Thiokol TE-364), complemented by a three-nozzle throttlable vernier propulsion system using liquid bipropellant (monomethyl hydrazine hydrate and MONO-10 oxidiser). Two radar systems were used, the first to initiate firing of the main retro motor, the second forming part of a closed-loop control system with the vernier engines in the final stages of descent. The first radar was mounted in the nozzle of the retro motor and, its job having been done, was jettisoned before the retro fired. The retro motor was nominally planned to burn out after ~40 s at around 10 km altitude, after which it too was jettisoned and the vernier motors took over the braking for the final part of the descent. Attitude measurement was done using Sun and Canopus sensors, and gyroscopes. Attitude control was achieved using a nitrogen cold-gas system. See Chapter 5 for more detail on the descent phase.

Power was provided by a 0.855 m^2 solar array, which generated up to 85 W, and silver–zinc rechargeable batteries. On Surveyor 1–4 an additional, 'auxiliary' primary battery was installed to ensure operation until shortly after landing.

Thermal control was achieved by a mixture of passive and active control. Passive control was achieved by a combination of white paint, high IR-emittance thermal finish and polished aluminium, while the electronics compartments were equipped with insulating blankets, conductive heat paths, thermal switches and electric heaters.

The landing sites were equatorial on the near-side (as for Apollo), with the exception of the more scientifically-driven Surveyor 7, which touched down in the southern highlands to sample contrasting terrain. The landing accuracy with respect to the intended target location ranged from 2.4 km (Surveyor 7) to 28.8 km (Surveyor 5). Landings generally occurred shortly after local sunrise, allowing the maximum period of time for surface operations before sunset.

The three main payload experiments were the TV camera (10.6 kg), soil mechanics surface sampler (SMSS, 9.2 kg) and alpha-scattering instrument (12.4 kg), although only the camera was carried on all seven missons. Additional experiments and engineering sensors included strain gauges and temperature sensors distributed throughout the spacecraft, mirrors, magnets and photometric targets for the camera. A descent camera was carried, but not used, on Surveyors 1 and 2 only. Its mounting position was used instead for the SMSS on Surveyor 3.

21.2 Flight performance

Of the seven missions, Surveyors 2 and 4 failed. Surveyor 2 crashed into the Moon at high speed as a consequence of the failure of one of the vernier thrusters during the mid-course correction. The resulting thrust imbalance caused the spacecraft to tumble. Operations continued, however, and although the situation could not be corrected, engineering information on the functioning of the spacecraft was gained prior to impact. Surveyor 4 operated nominally until contact was lost 2.5 minutes before touchdown, as the retro engine was completing its 40 s burn. Surveyor 5 managed to recover from a helium pressurant leak in the vernier system and land successfully, as detailed in Chapter 5.

The data from Surveyor 3 showed that it touched down on the lunar surface three times before landing, because the engines did not shut down as intended. The spacecraft moved 20 m between the first and second touchdowns and about 11 m between the second and third. A final translation movement of about 30 cm occurred following the third touchdown. The engines were finally shut down prior to the third touchdown. This behaviour is thought to be the result of the radar beams traversing the lip of a crater during the final part of the descent.

During surface operations of the successful missions, a total of 87,674 TV images were returned (in either 200 and 600-line modes, some with colour or polariser filters). Many of the image frames were composited together manually to form panoramas. Targets for observations apart from the lunar surface (both undisturbed and disturbed) included the Earth, laser emissions from Earth, the solar corona, Mercury, Venus and Jupiter, and stars to 6th magnitude. The SMSS instruments on Surveyors 3 and 7 performed 1898 and 4397 mechanism movements, respectively, and carried out a total of 51 bearing, trenching or impact tests. The alpha-scattering instrument was deployed and operated successfully on Surveyors 5, 6 and 7, being lowered to the ground by a winch mechanism. On Surveyor 7, the availability of the SMSS allowed it to be repositioned to examine a different location. On several of the missions, thrusters were fired to look at plume impingement and dust contamination of the spacecraft. Surveyors 1, 5 and 7 were all successful in surviving at least one lunar night to be reactivated after sunrise.

As part of the surface mechanical properties investigation, Surveyor 6 performed a 'hop' manoeuvre, moving 2.4 m away from its original landing area. This manoeuvre, the first launch and controlled movement across the lunar surface, provided excellent views of the surface disturbances produced by the initial landing and the effects of firing rocket engines close to the lunar surface. Photography obtained after the hop contributed to the soil-mechanics investigation.

A 'Surveyor Block II' series of missions, carrying more payload and to act as scouts and target markers for particular Apollo landing sites, was studied briefly but never implemented. Another unflown proposal included a rover.

22

Galileo probe

The Galileo mission (e.g. O'Neill, 2002; Bienstock, 2004; Hunten *et al.*, 1986) was conceived early in the 1970s. In 1975 initial work started at NASA Ames for a Jupiter orbiter and probe for launch in 1982 on the Space Shuttle, with Jupiter arrival in 1985 after a Mars flyby *en route*. The project was transferred to JPL, and was approved by Congress in 1977. Development difficulties with the Space Shuttle led to a slip, and over the following years political pressures from various NASA centres led to several redesigns and different upper stages. Eventually, Galileo was set for a May 1986 launch on the Shuttle with a powerful Centaur upper stage. The Challenger disaster, however, interrupted the Shuttle launch schedule, and a re-examination of safety considerations ruled out the Centaur upper stage with its volatile cryogenic propellants. The revised mission, with a two-stage inertial upper stage (IUS) solid propellant upper stage would launch (after yet more delays) on October 18, 1989.

The low energy of the launcher then required Galileo to make one Venus and two Earth flybys to reach Jupiter. Although this trajectory afforded two asteroid flybys, the thermal design reworking needed to protect the spacecraft in the inner solar system led inadvertently to the failure of the high-gain antenna deployment mechanism, which drastically reduced the downlink performance during the scientific mission.

The release date of the probe was driven by trade-off of the higher fuel penalty (for the orbiter to retarget from the probe entry trajectory to the orbit insertion trajectory) at a later release, against the poorer accuracy of an earlier release. In fact, the coast after release was some 148 days, close to the battery limit of 150 days, with entry on 7 December 1995.

The entry point was driven largely by the need to minimize the entry speed, which is extremely high given Jupiter's deep gravity well. Jupiter's rapid rotation provides an opportunity to mitigate the entry speed, in that the receding (evening) limb is moving at some 11 km s^{-1} (the problem is analogous to landing an aircraft on an aircraft carrier – it is easier if the carrier is moving away from the aircraft, such that the velocities are subtracted).

22.1 Equipment

Probe data was received by an articulated 1.1 m diameter L-band high-gain antenna on the orbiter, equipped with dual feeds for the redundant (parallel) L-band transmissions from the probe at 1387.0 and 1387.1 MHz. The two channels had opposite circular polarizations. Transmit RF power was 23 W. The receivers on the orbiter were specified to acquire the probe signal within 50 s, with a minimum acquisition signal strength of 31 dB Hz^{-1} and tracking threshold of 26 dB Hz^{-1}.

The 1.25 m diameter entry shield (comprising a carbon–phenolic sphere–cone forebody of 152 kg and a hemispherical nylon phenolic aft cover) had a mass of 220 kg. The shield thickness varied from 14.6 cm at the nose to about 5 cm at the edge (e.g. Green and Davy, 1981).

The dominant energy transfer from the atmosphere to the heat shield at these high speeds (48 km s^{-1}) is radiative – the shock layer was expected to reach 14 000 K and heat loads at the nose to reach some 42 kW cm^{-2}.

Resistive sensors (called ARADs – Analog Resistance Ablation Detectors) embedded in the heat shield recorded its ablation – some 4.1 cm of material was removed at the nose, falling to about 2.5 cm at the edge, corresponding to a loss in mass of some 79 ± 4 kg (Milos, 1997; Milos *et al.*, 1999a). It is believed the aft cover lost about 8.5 kg of material. Together with some pyrolysis loss (i.e. outgassing from unablated material) the total mass drop was some 88.9 kg.

The nominal energy requirement for the mission until entry +48 minutes was 16.3 A-hr, and a margin of 1.7 A-hr was carried. Some degradation during storage and cruise was expected and the total battery capacity at manufacture was 21 A-hr (about 730 W-hr, or 2.6 MJ). The battery was made with three modules of D-size lithium/sulphur dioxide ($LiSO_2$) cells, each module with 13 cells with bypass diodes. Additionally four thermal batteries were carried for pyro actuation.

Note that unlike the Pioneer Venus probes the Galileo descent module was not a pressure vessel; individual units were protected with hermetically sealed housings as necessary. Views of the probe's interior are shown in Figure 22.1 and Figure 22.2, and the main characteristics of the experiments in Table 22.1.

The parachute system (deployed nominally at Mach 0.9 and a dynamic pressure of 6000 Pa) comprised a pilot parachute thrown through the wake at 30 m s^{-1} by a mortar; separation nuts then fired to release the aft cover, to which the pilot chute was attached. The cover pulled the main chute and stripped its bag, with full inflation completed 1.75 s after mortar actuation (Rodier *et al.*, 1981). Then, 10.25 s

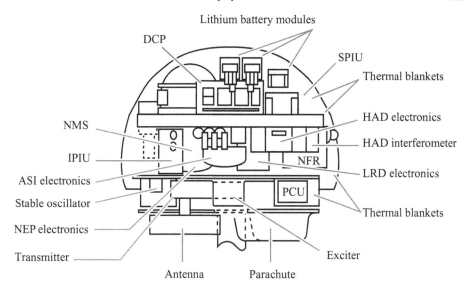

Figure 22.1. Layout of the Galileo Probe equipment. Note that this view is inverted – the parachute and antenna are of course pointed upwards relative to the local Jovian gravity during descent.

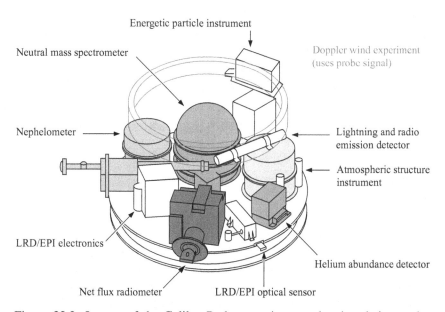

Figure 22.2. Layout of the Galileo Probe experiments, showing their exterior access. Accommodating sensor requirements for field of view, exposure to airflow, etc. is a sometimes challenging task for the space probe designer.

after mortar firing, the aeroshell was allowed to fall away – 1.5 s later, with the aeroshell 30 m away, marked the 'official' start of the descent sequence. The sequence is shown in Figure 22.3.

Table 22.1. *Galileo Probe science instruments and their main resource requirements*

Instrument	Mass (kg)	Power (W)	Volume (l)	Data rate (bits s^{-1})
Atmospheric structure (ASI)	4.0	6.3	3.1	18
Nephelometer (NEP)	4.8	13.5	4.6	10
Helium abundance (HAD)	1.4	1.1	2.3	4
Net flux radiometer (NFR)	3.0	10.0	4.6	16
Neutral mass spectrometer (NMS)	12.3	29.3	8.6	32
Lightning and RF emissions/ energetic particles (LRD/EPI)	2.5	2.3	2.9	8
Total	28.0	62.5	26.1	88

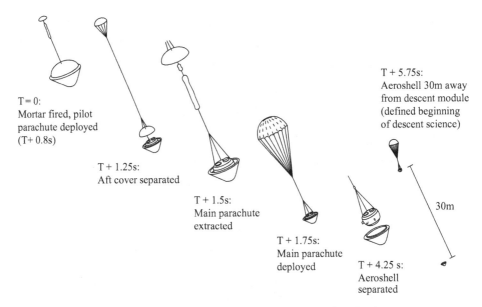

Figure 22.3. Parachute deployment sequence of the Galileo Probe.

Both parachutes were 20° conical ribbon chutes (chosen largely for attitude stability) made with Dacron. The pilot and main chutes had projected diameters of 0.74 and 2.5 m; the corresponding drag areas were 0.51 m^2 and 4.97 m^2, and fabric masses of 0.36 and 3.7 kg respectively. Kevlar was used for the main parachute riser and bridle.

22.2 Flight performance

The probe had been designed for a 3-year interplanetary cruise, but endured one twice as long. A probe checkout was conducted on 15 March 1995, verifying battery performance. On 12 April an 8 cm s^{-1} trajectory correction was applied to line the probe up with the entry corridor. Formal release readiness reviews were conducted, and the probe activated on 5 July. The umbilical was severed by explosive guillotine on 11 July, ready for spin-up the following day (the entire orbiter, which had a dual-spin architecture wherein part usually remained three-axis stabilised, spun at 10.5 rpm) and on 13 July, the probe was released by the firing of explosive bolts. Separation ΔV of 0.3 m s^{-1} was introduced by springs. The orbiter performed a 61 m s^{-1} deflection manoeuvre on July 27.

The probe was powered up by timer 6 h before entry, the entry interface being defined as 450 km above the 1 bar level. About 3 h prior to entry, at 5 Jupiter radii (and within the Io plasma torus) the probe collected data on energetic particles (which penetrated through the heat shield). Data was stored in solid-state memory for subsequent transmission during the descent.

The entry site was constrained to a low latitude, to maximise the reduction in entry speed due to Jupiter's rotation; the latitude requirement was 1 to 6.6° (the low-latitude limit was invoked to avoid flying the probe through Jupiter's ring). Entry occurred at 6.57°N, at a speed of some 48 km s^{-1}.

The spin rate of the probe was recorded about 1 h prior to entry (from the magnetic field sensor in the lightning and radio emissions detector instrument – Lanzerotti *et al.*, 1998) at 10.4 rpm. Post-entry measurements show that the spin rate decayed from 33.5 rpm 4.9 minutes after entry to 14.2 rpm at the end of the mission 45 minutes later (interestingly within 5% of the terminal spin rate in a drop test on Earth). Evidently there was significant spin-up of the probe during entry, presumably from asymmetric ablation in the heat shield. The decay of the probe spin was lessened by the presence of three spin vanes, although the torque due to these vanes was less than the parachute swivel torque.

Frequency analysis of the Doppler shift of the telemetry signal shows motions (with an amplitude of 0.5 m s^{-1} in line-of-sight velocity) with a period of 20–25 s, and a higher frequency component with a period of 5 s or so. The latter is attributed to pendulum motion under the parachute.

Two significant anomalies occurred during the descent. First, data acquisition and transmission began at a rather greater depth than anticipated. Descent measurements began at a pressure of 0.35 bar some 53 s later than the planned altitude of 0.1 bar, 50 km above the 1 bar level. This delay has been determined to be due to a wiring error – the wires of the two *g*-switches were crossed: specifically G1 was to go 'high' at 6 *g* and down at 4.5 *g*, while G2 would trigger at 25 *g* and reset at 20 *g*. The expected sequence would have G1 switching on first, then G2, then G2 off, then G1 off. Telemetry from the probe determined that in fact G2 had triggered first.

A second issue which caused considerable difficulty in the scientific interpretation of the telemetered experiment data is that the probe temperatures became uncomfortably high. This appears in part to be due to more rapid heat

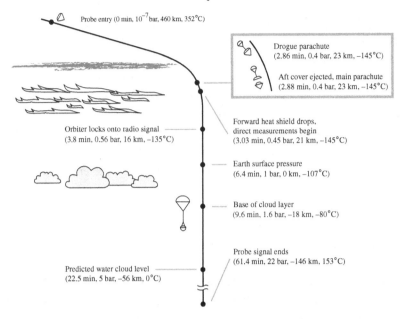

Probe entry (0 min, 10^{-7} bar, 460 km, 352°C)

Drogue parachute
(2.86 min, 0.4 bar, 23 km, −145°C)

Aft cover ejected, main parachute
(2.88 min, 0.4 bar, 23 km, −145°C)

Orbiter locks onto radio signal
(3.8 min, 0.56 bar, 16 km, −135°C)

Forward heat shield drops,
direct measurements begin
(3.03 min, 0.45 bar, 21 km, −145°C)

Earth surface pressure
(6.4 min, 1 bar, 0 km, −107°C)

Base of cloud layer
(9.6 min, 1.6 bar, −18 km, −80°C)

Probe signal ends
(61.4 min, 22 bar, −146 km, 153°C)

Predicted water cloud level
(22.5 min, 5 bar, −56 km, 0°C)

Figure 22.4. Schematic of the Galileo Probe's mission timeline.

transfer in the probe than was expected; hot gas appears to have circulated inside the probe. The experiments therefore were operating at temperatures at which they had not been calibrated (for example, the atmospheric structure instrument recorded temperatures from 35 K colder to 65 K warmer than the calibration range, and rapid temperature changes of 7.3 K min^{-1}). That said, while the probe specification called for its operation to a depth of 10 bar after nominally 33 minutes of descent, the last transmissions were received from some 23 bar (at an ambient temperature of 425 K), 57 minutes after the start of descent (in fact 61.4 minutes after the entry interface). A schematic overview of the probe's entry and descent is shown in Figure 22.4.

Scientifically, another challenge was that the probe entered in an 'atypical' region, specifically a '5 μm hotspot', a region of meteorological downwelling, where the air had been dried. This, however, can hardly be attributable to the engineering design of the probe itself but is rather a matter of mission goals – any single probe is likely to suffer this problem on a complex planet. The multiple vehicle approach of the 1970s has advantages other than simple reliability through redundancy.

Many pre-mission analyses of the aerothermodynamic environment of Jovian entry and the design of appropriate thermal protection are discussed in two volumes of the series *Progress in Astronautics* and Aeronautics, namely vol. 56 *Thermophysics of Spacecraft and Outer Planet Entry Probes* (1976) and vol. 64 *Outer Planet Entry Heating and Thermal Protection* (1979), both published by AIAA.

23

Huygens

Among many early concepts for a Titan probe (e.g. Murphy *et al.*, 1981b) it is not surprising that a Galileo-like architecture was envisaged. As initially proposed in 1982, the concept of the Cassini–Huygens mission was to be a joint effort between NASA and ESA, and NASA was to supply the Galileo flight spare probe, and ESA would provide an orbiter delivery vehicle. However, in many respects the Titan probe grew in scope and complexity, in part because of the international nature of the mission.

As the joint study progressed, the roles were reversed, and ESA studied designs for an entry and descent probe (Scoon, 1985). These studies led to some quite novel ideas (e.g. Sainct and Clausen, 1993), which in all probability would not have been explored had the probe development remained in the USA.

The probe changed from an initially spherical shell (the shape adopted by the Galileo probe) to a flatter design. This also opened up novel heat shield architectures, with options such as a beryllium nose cap and a jetisonnable carbon–carbon decelerator (although in the end, neither of these concepts was adopted and a more technologically conservative heat-shield design was used – a prudent measure given the novelty of this mission for ESA).

The mass budget (Table 23.1) deserves some brief comment. In broad terms the mass breakdown is typical (e.g. with 15% of the mass devoted to power systems), although the front shield is rather conservative. Note the formidable harness mass, balance mass, and the need for significant mass for separation hardware.

The complexity of the atmospheric photochemistry on Titan required more sophisticated instrumentation (Table 23.2) than Galileo – specifically a gas chromatograph–mass spectrometer for *in situ* chemistry measurements during descent, rather than the simpler mass spectrometer used on Galileo. Additionally, the prospect of getting at least close to the surface of Titan, and perhaps surviving contact with it, invited surface science instrumentation, and some kind of surface imager. The Huygens payload allocation defined in the Phase A study was some 40 kg.

Table 23.1. *Huygens mass budget (kg)*

Item	Probe	PSE[1]
Subsystems		
Front-shield subsystem	78.75	
Back-cover subsystem	16.13	
Separation subsystem	11.40	10.29
Descent-control subsystem	12.13	
Inner structure subsystem	41.41	
Thermal subsystem	20.60	1.50
Electrical-power subsystem	44.73	
Probe harness subsystem	12.61	
Command and data-management subsystem	23.10	
Probe-data relay subsystem	6.04	16.30
Experiments		
Doppler wind experiment (DWE)	1.90	1.90
Surface-science package (SSP)	4.87	
Gas chromatograph/mass spectrometer (GCMS)	17.2	
Huygens atmospheric-structure instrument (HASI)	5.77	
Descent imager/spectral radiometer (DISR)	8.07	
DISR cover	3.63	
Aerosol collector/pyrolyser (ACP)	6.18	
Fasteners, etc.	0.95	
Balance mass	2.95	
Total	**318.32**	**29.99**

[1] Probe support equipment on the Cassini orbiter.

The dense atmosphere, and the desire to perform scientific measurements from an altitude as high as possible, and certainly above 150 km, meant that a realistic mission would need to last some 2–3 h.

For a 2–3 h mission, primary batteries are the obvious energy source: $LiSO_2$ batteries – indeed using the same cells as flown on Galileo – were selected. The energy budget incorporated a healthy margin, in part because of reasonable conservatism (this being Europe's first planetary probe, and Titan being an almost unknown object) and in part because the energy budget had to be coarsely-quantized – there could only be an integral number of batteries – which was reduced from an initial 6 to 5 in Phase B. Each battery comprises two strings of 13 cells in series.

The command and data handling requirements in some respects are fairly trivial – the CDMU acts as a bent pipe for experiment data, formatting the experiment packets (and some housekeeping information) into transfer frames broadcast by the telecommunications system.

As far as the sequencing of operations is concerned, events are tied to occur an interval after a deceleration threshold is encountered; in a sense, a clockwork

Table 23.2. *Huygens payload experiments*

Instrument	Allocated mass (kg)	Power (W) (typical/peak)	Energy (Wh)	Typical data rate (bit s^{-1})
DWE	1.9	10/18	28	10[1]
SSP	3.9[2]	10/11	30	704
GCMS	17.3	28/79	115	960
HASI	6.3	15/85	38	896
DISR	8.1[3]	13/70	42	4800
ACP	6.3	3/85	78	128

[1] Housekeeping only (i.e. not collected as packets from the experiment, but voltages, temperatures, etc., as are recorded in other probe subsystems).

[2] Note that this is lower than the value recorded in the system mass budget, suggesting this experiment exceeded at least its initial allocation.

[3] This does not include the cover, which was added at a later stage.

timer like those on the earliest probes could perform this function. On Huygens the function is implemented by a pair of computers (using MAS 281 silicon-on-sapphire radiation-hard processors). In the latter part of the descent, events are referenced to an altitude determined by two redundant radar altimeters; should they fail, the sequence reverts to a 'time–altitude table' based on a model descent profile. The CDMS also acts as a conduit for reprogramming experiment software and operating cruise checkouts – these functions became of critical importance when the probe mission had to be redesigned following the discovery of poor receiver performance.

The descent control system (Neal and Wellings, 1993; Underwood, 1993) comprises three separate parachutes – a 2.6 m pilot chute, deployed through a breakout patch in the aft cover of the probe by a mortar. This inflates and stabilises the probe, and then pulls off the aft cover and the bag enclosing the main parachute, some 8.3 m in diameter. This slows the probe to a speed of around 50 m s^{-1}. The ballistic coefficient of the probe plus main parachute is significantly lower than that of the front shield, which is then allowed to fall away after firing explosive attachment bolts.

The descent would take some 8 h under the main parachute, so this is released after approximately 15 minutes, the bridle holding the main chute being released and a third (3 m) stabiliser chute being inflated. This parachute remains attached until and after surface impact at 4.67 m s^{-1}.

All the parachutes are of the disk–gap–band design, owing to that design's strong space heritage and relatively good damping performance. The riser, bridle, etc. are made of Kevlar, and the canopy itself of nylon.

Consideration was given to metallizing these components to prevent differential charging, but the thermomechanical difficulties in doing so, the possible degradation of communication performance, and the possibility of enhancing ambient electric fields to discharge levels ('probe-induced lightning') argued against doing so.

The thermal design of the probe reflects the several different environments it encounters. First, the 22-day coast in the Saturnian system after release from the orbiter would allow the probe, in a totally dormant state apart from the operation of three redundant clocks, to become unacceptably cold. The probe therefore includes 35 radioisotope heater units. These, and an envelope of multilayer insulation, assure an acceptable radiative equilibrium in free space at 10 AU.

The probe would also get cold during the atmospheric descent, where the thick and cold atmosphere would quickly remove heat from the probe. Conventional multilayer insulation does not insulate well in the presence of an atmosphere, thus a layer (some 10–15 cm thick) of a closed-cell foam (Basotect) retards heat leak from the probe. Note that the probe is dissipating several hundred watts during its descent.

Initially, the foam was applied in discrete panels, each wrapped in a plastic coating, with only millimetre gaps between the panels. It was found with some surprise during testing that substantial convective heat transfer could take place, even in these small gaps, compromising the performance of the insulation. A new packaging technique eliminated the problem, but this episode yet again stresses the importance of testing.

A final, perhaps unexpected, driver on the thermal design arose somewhat late in the project, when the mission design for the interplanetary trajectory to Saturn was revised to incorporate two Venus flybys, such that Cassini and the Huygens probe it carried would be exposed to a solar flux of some 3800 W m^{-2}.

The prime strategy here was to use the Cassini orbiter's 4 m high-gain antenna as a sunshade, to shield both the probe (whose battery performance would be degraded by high temperatures during cruise) and the orbiter's sensitive instrumentation from the solar flux. Even then the equilibrium temperatures would be undesirably high, and thus the MLI coating of the probe incorporates a 'radiative window' – a hole on the antisun side which allows an extra radiative loss of heat to lower the temperature at this point rather more.

The telecommunications system was required to return a modest data rate during the descent (initially the data rate was expected to increase from about 1 kbps at the start of the descent, when the orbiter would be some 100 000 km away, to about 8 kbps at the end of the mission). Two data links were included, to eliminate single-point failures, and the original intent was that data would be sent redundantly on both channels, with one channel delayed by several seconds.

The rationale here is that swinging under the parachute, or some other break in the datastream, would be transient (less than those few seconds), thus the second staggered channel would allow recovery of any data lost during a transient on the first channel.

In 1992, the Cassini mission suffered a heavy descope, as a result of budget pressure from Congress, which essentially pitted the science missions CRAF and Cassini against the International Space Station. In the end, Cassini and the station survived, but CRAF was deleted and Cassini seriously descoped – the most prominent effects being a deferment of software development and the deletion of the scan platforms supporting the science instruments. Also deleted was the probe relay antenna, a dedicated dish that would track the probe. Instead, the whole spacecraft would be slewed, and the probe relay receiver would use the body-fixed 4 m high-gain antenna.

This change required the mission (and in particular the orbiter delay time) to be re-optimised, with the ODT being raised to some 5.2 h. Remarkably, this permitted a data rate (largely due to the size of the HGA) of some 8 kbps per channel, for the whole mission.

The structure of Huygens is moderately simple (see Figure 23.1) – almost all units are bolted onto a large honeycomb disc, the 'experiment platform' (73 mm thick). This is attached via a number of insulating fibreglass brackets to an exterior metal ring, to which the thin metal shell is attached. Also mounted on this ring is the front shield (since clearly the aerodynamic loads from the shield must be transmitted to the bulk of the probe mass, which is attached to the

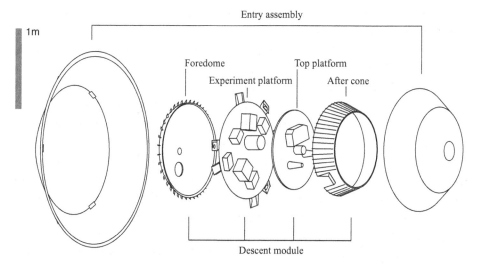

Figure 23.1. Probe exploded view. Note the small spin vanes ringing the descent module's foredome.

experiment platform). This ring also attaches, via explosive bolts, to the spin–eject device. This is the load-bearing structure transmitting ground support and launch loads to the experiment platform; at probe separation a set of springs and rollers push the probe away at about $30\,\mathrm{cm\,s^{-1}}$ (pulling three 19-pin umbilical connectors apart) and set it spinning gently ($\sim 7\,\mathrm{rpm}$) for attitude stability during the coast phase.

Some elements, notably the parachutes and the probe antennae, are attached to a second, smaller honeycomb platform that forms the upper surface of the probe in its descent configuration. This top platform is connected to the experiment platform via a set of titanium struts whose principal function is to carry the parachute inflation loads.

The descent module's shell does not carry significant mechanical loads – it attaches to the experiment sampling ports on the bottom of the probe, and attaches to the foam insulation. In principle it also acts as a shield against electromagnetic interference, notably possible lightning or electrostatic discharge on Titan. It is made from a pressed aluminium alloy about $2\,\mathrm{mm}$ thick, with stiffening plates added.

One extra feature on this structure is the presence of 36 spin vanes, small wings protruding radially and a few degrees from vertical. As the probe descends, these vanes exert a slight torque rotating the probe in a horizontal plane. The probe's rotation is decoupled from that of the parachute by means of a swivel in the parachute riser.

In equilibrium, the vanes at an angle T ($\sim\tan T$) would exert no net torque if the probe were rotating at a rate $\omega = TV/R$, where V is the descent rate and R is the radius around which the vanes are mounted. At this rotation rate, the airflow on the vanes would have zero incidence and thus there is no torque. In reality, since the swivel exerts a small retarding torque for non-zero rotation rates, the steady-state spin rate will be slightly lower than this.

However (as for the thermal equilibrium of a satellite in a low orbit) dynamic effects are important. The spin-up time of the probe, defined by the moment of inertia of the probe divided by the derivative of vane torque with spin rate, is not negligibly small compared with the descent time.

Thus from some initial (and unknown, since nondeterministic spin torques may occur due to uneven ablation during the entry phase) spin rate, the spin will slowly tend to a value given by the expression above, but will take some tens of minutes to reach that rate. That equilibrium rate is itself changing, as the probe descent rate drops with time.

The entry protection system's most obvious feature is a $2.7\,\mathrm{m}$ diameter front shield. This decreases the ballistic coefficient to a level that reduces the peak heating rate to levels that are tolerable by the thermal-protection material on the front surface of the shield. It also ensures that the probe decelerates to a Mach number low enough to permit deployment of the parachute at an altitude consistent with the scientific requirements. (For mass reasons, the shield was reduced from an initial 3 m diameter during phase B; while a smaller front shield gives a higher ballistic coefficient, the incremental area is relatively 'expensive' since rather higher structural rigidity is required – not only the mass of the outboard 0.3 m is saved, but also the additional stiffness needed inboard to support that mass.)

The relatively modest entry heat loads afforded by the large scale height in Titan's atmosphere allow lighter thermal protection to be used than for Galileo. The material used is AQ60, a French resin-doped silica fibre tile.

The peak deceleration during entry (with the relatively steep angle of $-64°$, and a speed of about 6 km s^{-1} at the 1270 km entry interface) was expected to be around 12 g. Stagnation point heat loads peak at around 600 kW m^{-2}, divided approximately equally between radiative and convective fluxes. An attempt was made to observe the radiated emission of entry from the ground (Lorenz *et al.*, 2006) but this was unsuccessful.

In fact, during the early phases of the project, radiative heat loads were re-evaluated in model studies and found to be rather ($\times2$!) larger than originally anticipated. The radiative heat loads are also sensitive to the composition of Titan's atmosphere, which was not exactly known. The radiative emission depends on both the argon and methane abundance. The variation with methane mole fraction is nonmonotonic – initially it increases, as the availability of CN-radiating molecules increases (e.g. by 20% between 2 and 3% CH_4); above some amount (3–6%, depending on the argon abundance) the endothermicity of CH_4 dissociation takes over and lowers the temperature of the shock layer. Increasing argon abundance increases the electron-number density in the shock layer, which results in a more efficient population of the excited CN states and hence an increase in the radiative flux – for 0–10% argon, the effect is again a 20% increase in flux. These variations underscore that sophisticated entry aero-thermodynamic calculations involving nonequilibrium chemistry and radiative heat transfer, together with as narrow a range of compositions as scientifically justified, are needed to obtain a robust and efficient entry protection design.

The shock layer would radiate onto the back side of the probe (the rear face of the front shield, and the aft cover of the probe itself). These surfaces could be protected with a lighter (and cheaper) thermal protection system. The material used, Prosial, is a resin foam of silica bubbles. Its expense is considerably reduced because it can be sprayed onto the protected surfaces, while AQ60 must be carefully machined into tiles that can be precisely mounted on the front shield.

A significant design flaw in the Italian-built probe relay radio receiver (part of the ESA-supplied support equipment on the orbiter) surfaced some 3 years after launch, shortly after Cassini swung by the Earth. An end-to-end telemetry test was performed, with a DSN antenna performing the role of the probe; although the link behaved nominally, no data was recovered. The problem was eventually traced to inadequate bandwidth on the bit synchronizer in the receiver.

The problem and its solution were subtle and complex; in principle either a lower Doppler shift, or higher signal-to-noise, would improve matters, but the automatic gain-control switches in the receiver would switch in above preset signal levels and in fact degrade performance! These and several other aspects were hard-coded in firmware – straightforward solutions could have been easily implemented by telecommand had these parameters been left flexible.

An option that has not been necessary to implement is to substitute science telemetry packets with dummy 'zero' packets; the bit transition density in these packets would allow the Viterbi lock-state machine in the synchronizer to 'catch up'. Although this clearly results in loss of the effective telemetry bandwidth, this would be better than leaving which packets would be corrupted to chance.

It must be stressed that Huygens is not a lander (the original study called it a Titan Atmosphere Probe), although it has always been recognized that the probe may continue to transmit after surface impact (which occurs at the very modest speed of $5\,\mathrm{m\,s^{-1}}$). No explicit design features were introduced to permit or enhance surface operations, beyond the mission energy, link and thermal budgets including margin to allow at least 3 minutes of surface operation.

Among environmental hazards that were considered during the development phase were the possibility of lightning discharges in the Titan atmosphere (recall that all four Pioneer Venus probes suffered sensor failures during descent which have been attributed to electrical interactions with the atmosphere). Significant test effort was devoted to demonstrating tolerance of nearby strikes, and the probe incorporates discharge rods to alleviate any triboelectric charge buildup.

Parachute performance is always an area of concern on probe projects. Some scientific effort was expended in order to try to understand the likely constraints on wind gust amplitudes (it is impossible to guarantee parachute dynamic performance – getting the probe back to vertical within a few seconds so that the staggered radio links are not broken longer than the overlap period).

The mission took place on 14 January 2005, and can be judged to have been a great success. After parachute deployment, the presence of a transmission from Huygens was detected by radio telescopes on the ground (although a transmitter in which the carrier is suppressed would be more energy efficient in terms of data transmission, the existence of the unsuppressed carrier made this sort of radio science much more feasible). Analysis of the directly detected radio signals includes Doppler wind measurement, and very long baseline interferometry (VLBI), measuring the position of the spacecraft in the sky.

The parachute descent (Figure 23.2) took 2.5 hours, right at the maximum end of the predicted range of descent times (presumably due to parachute drag performance and/or deployment altitude – the atmospheric models seem to have predicted the actual pressure-density profile rather accurately). After impact, the probe continued working normally, with some 72 minutes of data received by Cassini before it passed below the horizon. The surface data included the impact (with a peak deceleration of ~15 *g*, indicating a soft solid surface) and images showing a cobble-littered plain, suggestive of past fluvial activity.

Some performances of the probe systems deserve comment. Two radio channels A and B, corresponding to entirely independent data handling systems, were carried, in part for simple redundancy, and in particular to guard against data loss from swinging under the parachute by having one stream delayed by several seconds with respect to the other. Channel A was equipped with an

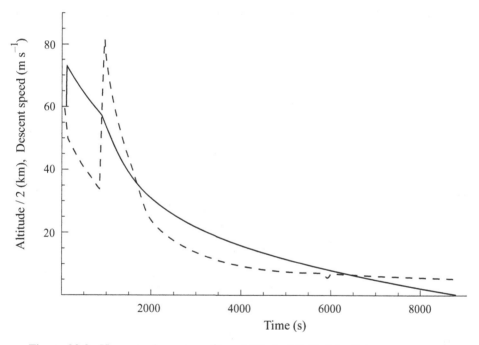

Figure 23.2. Huygens descent profiles. Altitude (divided by 2 to use common scale) is shown by the solid line. After initial deceleration and front-shield release in the first ~100 s of descent, the probe was at a terminal velocity (dashed line) that steadily decreased with time as the probe descended into denser air. At 900 s (~115 km altitude) the main chute was released and the probe accelerated under gravity to a new terminal descent speed, which declines steadily until surface impact at 4.6 m s^{-1} at 8969 s.

ultrastable oscillator (which allowed the direct radio science from the ground), and the corresponding receiver on Cassini was also equipped with such an oscillator. However, although both transmitting chains operated nominally (with essentially no data loss from chain B), and both receivers were powered on, the oscillator for Cassini's chain A receiver was not, and so data modulated on that channel, and the on-board measurement of Doppler frequency, was lost.

The two radar altimeters initially generated false altitudes, roughly half the true value, until the altitude became low enough for the echo to suppress the false lock. There was minimal impact to the mission, but this was a salutary lesson that comprehensive testing is needed of such systems.

Another unexpected behaviour was strong short-period ($\sim$1 Hz) motion of the probe – buffeting. Relatively little parachute swinging was noted, but the more rapid motions dominated accelerometer and tilt sensor data, making it difficult to retrieve information on wind gusts. Again, analysis of drop-tests on Earth would have helped anticipate (or alleviate) such effects. Additionally, the spin history of the probe (Figure 23.3) appears not to have been as anticipated.

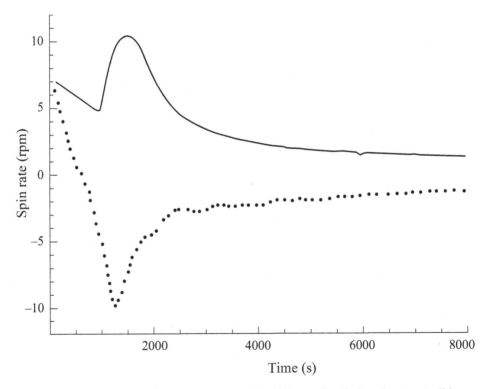

Figure 23.3. The expected spin-rate profile of the probe during descent (solid line) and that reconstructed from various datasets (Lebreton *et al.*, 2005, dots). The reason for the reversal in spin direction is not yet understood.

The location of the landing site was determined by combining the knowledge of the entry state with Doppler measurements and on-board altitude data. This iterative process yielded a location within 5 km of the location determined by correlating surface features seen by the descent imager with a map made by an imaging radar on Cassini about 10 months later. Landing co-ordinates were estimated as 10.2° S, 192.4° W.

24

Mars Pathfinder and Sojourner

The Mars Pathfinder mission began as MESUR (Mars Environmental Survey), a 1991 proposal for a network of as many as 16 Mars landers to perform network science (meteorology and seismology on distributed sites) using nominally inexpensive landers. One prominent approach to reducing the unit cost of the landers was to use a semi-hard landing approach with airbags rather than a retrorocket system. The landing system proposed was sufficiently radical that a technology demonstration/flight validation was designed, originally MESUR Pathfinder, on which work formally began in 1993.

With the loss of Mars Observer and the onset of the Discovery programme in NASA, the Pathfinder concept was 'adopted' by the Discovery programme, and became the most widely cited example of the 'faster, better, cheaper' (FBC) approach (see McCurdy, 2001). NEAR technically was the first selected Discovery mission, but took rather longer to be built and reach its target. Note also that there were other FBC programmes within NASA, including the Small Explorer Earth orbiters, and the New Millenium technology validation programme. The success of some non-NASA projects like the Clementine moon orbiter, which came out of the Strategic Defense Initiative (the 'Star Wars' programme) also set the stage for the FBC era.

As an aside, one viewpoint of the background to the development of Pathfinder is described in Donna Shirley's book *Managing Martians* (1998). Andrew Mishkin's *Sojourner* (2004) gives a more detailed but narrower view, of the rover engineering development specifically. Since Pathfinder was fundamentally an engineering mission, the scientific payload was in fact rather modest, the principal aim being to demonstrate the successful deployment of the lander.

The stereo camera's capabilities were leveraged by some ancillary fixtures on the lander; conical metal windsocks were mounted at three positions on the ASI/MET mast – these freely suspended structures hung at an equilibrium position determined by the windspeed and gravity. The onerous calibration of this experiment meant only a modest scientific return. A particularly fruitful investigation was the magnetic target that was imaged by IMP – over the duration of the mission, airborne dust particles progressively adhered to the target.

The entry protection system of Pathfinder (Wilcockson *et al.*, 1999) comprised a Viking-heritage sphere–cone front shield geometry (2.65 m diameter 70° half-angle cone and a spherical nose cap of 1/4 radius to diameter ratio). Entry mass was 585.3 kg, cross-sectional area 5.51 m^2. The ballistic coefficient quoted for entry conditions is 63.2 kg m^{-2}. As with Galileo, sensors embedded in the heat shield allowed its performance to be evaluated (Milos *et al.*, 1999b).

The same entry protection material as Viking was used, a Martin Marietta (later Lockheed Martin) superlightweight ablator SLA-561. This is a mix of ground cork with silica and phenolic microspheres in a silicone binder. This mix is packed into a phenolic honeycomb structure. The material had to be requalified, since peak heating on the direct-entry trajectory for Pathfinder would approach 100 W cm^{-2}, compared with the Viking heating rate (from orbit) of 30 W cm^{-2}. A 1.9 cm thick layer of the ablator was applied to the front shield; the backshell received a 0.48 cm spray-on layer of a similar material (without the honeycomb); 1.5 h prior to entry, coolant was vented from the electronics boxes. Separation of the cruise stage occurred 30 minutes prior to crossing the entry interface at 130 km.

The entry speed (e.g. Braun *et al.*, 1999a) was 7.26 km s^{-1} (inertial frame) or 7.48 km s^{-1} (relative to the rotating planet), with a flight path angle of $-13.6°$. Peak deceleration was about 16 *g*. Parachute deployment was programmed to occur as close as possible to a dynamic pressure of 600 N m^{-2}; the deceleration was monitored and a time offset computed based on the deceleration sensed 12 s after a deceleration threshold of 5 *g* was exceeded. The deceleration record shows the actuation of the parachute mortar at 171.4 s after entry interface crossing; the parachute inflated in about 1.25 s, with snatch loads of about 6.5 *g*. The probe was at an altitude of 7.9 km, flying at Mach 1.8.

Ground proximity (e.g. Spencer *et al.*, 1999) was sensed by a radar altimeter from 1.6 km down, which triggered the operation of braking rockets at an altitude of 98 m, 6.1 s prior to impact at a descent rate of 61 m s^{-1}. The parachute bridle was cut 3.8 s prior to impact (while the rockets were still burning, ensuring that the parachutes would be carried away and not interfere with the deployment of the lander). Impact occurred with a vertical velocity of 14 m s^{-1} and an estimated horizontal velocity of 6 m s^{-1}; the loads on the first bounce were 18.7 *g*. It is estimated that the lander travelled about 1 km in the subsequent 2 minutes of bouncing – several initial bounces were captured by the accelerometers.

After the vehicle came to rest, the airbags were allowed to depressurize by opening vent patches in the bags by retraction of lanyards by electrical motor.

The vents were covered with a mesh scrim to retain particulates from the inflation pyros. The airbags were then retracted by operating electrical motors for some 72 minutes. Three sides of the tetrahedral lander were opened by high-torque electric motors, such that regardless of the landing attitude of the lander, its base would be placed on the ground and the three circumferential petals would be splayed flat around it, exposing the solar arrays to the sky.

One of the petals carried the rover vehicle, initially referred to as the Microrover Flight Experiment (MFEX), subsequently named 'Sojourner'. The Pathfinder lander itself received the designation 'Sagan Memorial Station'. The rover had a size of $62 \times 47 \times 32 \, cm^3$, roughly the size of a typical mid 1990s laser printer. Its mass was 10.5 kg.

Sojourner featured a 'rocker bogie' design, wherein the three wheels on each side were arranged in a 'tree', with two wheels on a bogie that could articulate at one end of an arm that held the other wheel (see the entry that follows on MER and Figure 27.3). This arrangement permitted the vehicle to traverse larger obstacles than a simpler configuration with similar-sized wheels. Wheels were 13 cm diameter, 7 cm wide, driven by independent motors with maximum torque, via 2000:1 reduction gearing, of 3–4.5 N m. The motors were driven with 15.5 V, having a 10 mA no-load current and a torque slope (from which soil mechanical properties were deduced by sensing motor current) of approximately $4 \, N \, m \, mA^{-1}$. The vehicle could move forward at a maximum speed of 0.4 m per minute, and turn at $7° \, s^{-1}$ (the front and rear wheels could be independently steered). Steering position was sensed by potentiometer, while wheel rotation positions were sensed by optical encoder.

Sojourner navigation was accomplished principally by ground command using the Pathfinder camera to determine Sojourner's location and status. The vehicle itself had a number of small cameras, and a structured-light obstacle-detection system.

Manoeuvres were in general commanded directly, although some 20 high-level commands (to move to a specified location in X–Y space around the lander) were sent, and two higher-level (move up to the rock at a specified X–Y position) were also used. Operations were complicated by the difference between the day lengths on Earth and Mars, such that the lander schedule shifted by 37 minutes every day.

Power was supplied by a $0.22 \, m^2$ (16.5 W Mars Noon – 45 W AM0 at Earth) solar array (13 strings of 18 GaAs on Ge cells), and a lithium thionyl chloride ($LiSOCl_2$) primary battery. The latter, comprising nine D-cells providing 150 W-hr, was important in that the principal scientific instrument on Sojourner, an alpha-X-ray backscatter spectrometer, required long integration times (5–16 h, i.e. overnight) to obtain statistically useful numbers of counts. Deposition of dust from the atmosphere resulted in the partial obscuration of the solar array and a steady drop in output power of 0.2% per day.

The small Sojourner vehicle was more susceptible to diurnal temperature changes than the large lander. One measure adopted to mitigate the night-time ambient temperature drop (to $-110\,°C$) was the incorporation of silica aerogel insulation. The rover also had three 1 W radioisotope heater units to limit the low temperatures reached at night in the warm electronics box (WEB) at its core.

Sojourner was controlled by an 80C85 processor running at 10^5 instructions per second, accessing 576 KB of RAM and 176 KB of PROM. The computer controlled or read some 70 sensor and actuator channels. Half-duplex communications between the lander and the rover were conducted via an off-the-shelf UHF radio modem.

Experiments were conducted with the rover to determine the adhesion properties of the soil and the abrasion resistance of surface materials. These tests also indicated some triboelectric charging of the surface dust.

Downlink of data from Pathfinder was at up to 9 kbps to the 70 m DSN stations. After September 27, communications became unreliable – the last data were received on October 7; subsequent attempts to regain contact persisted until March 1998. It is believed that component failure in the communication system due to the deep thermal cycling was responsible for the end of the mission.

Sojourner travelled some 104 m in total, always within 12 m of the lander. It is possible that Sojourner may have episodically operated after communications between Earth and Pathfinder were lost, but of course there is no evidence of this.

A large number of images were acquired by the Imager for Mars Pathfinder (IMP) experiment, a stereo CCD camera mounted in a turret on a mast that was extended vertically soon after landing. The individual camera frames were quite small, but could be mosaiced together by tilting and panning the turret to form large composite panoramic images. The generation of stereo image products greatly facilitated scientific interpretation and public appreciation of the scene observed by the lander.

Among the main scientific results of Pathfinder, the site was confirmed by the lander imagery to be what had been suspected from orbit, namely an outwash plain created by fluid flow that transported rocks across the surface. The imager also detected several dust devils. A couple of anomalously bright spots (1–3 pixels across) in distant images are interpreted to be the entry shell and back cover 1–2 km away.

Significant effort was devoted to attempts to identify the minerals present in local rocks, using imaging through various filters to build up a reflectance spectrum. These analyses were somewhat impaired in the sense that most rocks were covered to a greater or lesser extent with surface dust. An additional complication was that the diffuse illumination from the sky had an intensity and colour distribution that depended significantly on the viewing geometry and time of day. The APXS instrument on the rover also suffered challenges due to poor collimation – the particles were somewhat scattered in the Martian atmosphere. Nonetheless some worthwhile mineralogical studies were made by both instruments.

The lander's meteorology package also recorded winds, temperatures and pressures; some signatures of dust devils were also noted in the latter data. Measurements of windspeed were made in two ways – by a hot-wire anemometer, and by a windsock experiment wherein weighted conical vanes were suspended from a mast such that their orientation depended on the windspeed.

The total data return was some 2.3 Gbit, including 16 500 lander images, 564 rover images, 16 APXS measurements and 8.5 million pressure, temperature and windspeed records.

25

Deep Space 2 Mars Microprobes

The DS-2 mission was the second 'Deep Space' mission in NASA's New Millennium technology validation programme (Smrekar *et al.*, 1999). It was to demonstrate miniaturized penetrators to enable subsurface and network science. The spacecraft that flew were radically smaller – by two orders of magnitude – than anything NASA had previously flown to the planets. The project cost a remarkably modest $29.6 million.

The original concept anticipated deployment at low latitude on Mars, and a payload including a microseismometer. As the mission evolved, and the delivery opportunity as a 'piggyback' payload on the Mars Polar Lander emerged, the mission concept had to change. In particular, the low-temperature environment at high latitudes on Mars reduced the expected energy capacity of the batteries (and thus the penetrators' lifetime) to the point where it was no longer likely that worthwhile seismic data would be acquired.

The new payload therefore centred on measuring the volatile content of the high-latitude soil. The same thermal environment that eroded the energy capability of the mission also made it likely that water might be trapped as ice in the soil.

Entry performance was driven by the entry conditions (at 6.9 km s^{-1} with a flight path angle of $-13.1°$, as for MPL) and the allowed flight parameters (velocity, angle of incidence) at impact (Braun *et al.* 1999b). A significant and unusual aspect was that since the probes were delivered as a secondary payload, no orientation or spin-up was provided to ensure any given orientation at entry (see Figure 25.1 for a view of the mounting structure/separation mechanism). (It may be noted in this context that the Viking entry bodies were hypersonically stable flying backwards.) The result was first the use of a 45° half-angle cone, instead of the more usual 70°, which offers higher drag ($C_d \sim 1.7$, versus 1.05 for the 45° cone). The second aspect was the very tight requirement on having the centre of mass as far forward as possible. These two aspects gave the system a

Figure 25.1. A DS-2 Microprobe mounted in the structure that attached it to Mars Polar Lander (mounted in turn on mechanical ground support equipment). The probes were ejected with a relative speed of <0.3 m s^{-1} and no spin. Note the very sturdy-looking 'spider' fitting that holds the microprobe.

strong 'weathercock' stability. The forward CoM requirement necessitated a novel two-part design, wherein the penetrating forebody was surrounded by, rather than mounted in front of, an aftbody that would remain on the surface to perform communications.

To satisfy the permitted range of impact speed (140–210 m s^{-1}) the ballistic coefficient β was required to be in the range 18–49 kg m^{-2}, while the penetration incidence angle constraint (i.e. the velocity relative to vertical) of $<30°$ required $\beta < 44.5$ kg m^{-2}. The angle of attack (i.e. the orientation of the vehicle relative to the velocity vector) was to be less than $10°$ – exceeding the incidence or angle of attack limits would lead to 'skip', the aftbody not remaining embedded securely in the ground. Nominally, the 0.35 m diameter shell and as-built mass of 3.6 kg gave $\beta = 36.5$ kg m^{-2}, leading to a peak deceleration of 12.4 g at an altitude of 44 km and a peak stagnation point heating of 175 W cm^{-2}, after about 100 and 80 s respectively; integrated stagnation point heating would be 8085 J cm^{-2} (Micheltree *et al.*, 1998). Impact (at a nominal altitude of 6 km – the southern high latitudes are elevated terrain, well above the 6 mbar Martian datum) would be at 191 m s^{-1} (around Mach 0.8) and 20° incidence, around 270 s after crossing the entry threshold (defined as a radius of 3522.2 km, around 142 km above the surface). The aftbody had to tolerate decelerations of some 60 000 g. The fore-body, penetrating further than the aftbody, therefore had a longer stroke over which to decelerate, so its impact loads were specified at 30 000 g. The 3-σ landing ellipse determined from Monte-Carlo simulations was about 180 × 20 km in extent: these simulations gave a probability of 76% for each probe (or 94% for either) to satisfy the impact conditions.

Tests to achieve reliable sub-surface soil sampling passively with holes or blades on the forebody were unsuccessful, and an auger drill had to be included in the design. This 9 mm diameter drill had an 8.5 mm stroke, driven by a 1 W motor modified for high-g impact loads. The sample drizzled during drilling into a small cup; after drill operation the sample was sealed inside the forebody by a small pyrotechnic door mechanism.

The soil water detector comprised a small cup (able to hold about 160 mm^3 of sample) around which a nichrome heater wire was wound. Thermistors were mounted on the edge of the cup and in the centre. The temperature rise experienced by the sensors for a given applied heating current would depend on the amount and thermal properties of the material deposited in the cup. It therefore acts as a crude form of thermal analyser. In particular, a deviation from a smooth heating curve would be observed if significant amounts of ice were present in the sample.

More sensitive detection of water was accomplished with a small absorption cell. A tunable diode laser emits light in a narrow bandwidth (nominally around 1.37 μm) that is swept across a wavelength range by modulating the current to the diode. The light from this source passes through a small volume that may be filled with gas from the sample; when the laser is at wavelengths where water vapour absorbs strongly, the light received by a photodiode is attenuated. The frequency-sweeping approach allows a more sensitive and robust detection than would a fixed wavelength.

Temperature sensors (platinum resistance thermometers) were embedded in the walls of the forebody to monitor the secular cooling of the probe after its emplacement into the Martian soil. The cooldown curve would yield information on the thermal conductivity (and indeed the temperature) of the soil (Urquhart and Smrekar, 2000; Smrekar *et al.*, 2001).

A pressure sensor, using a micromachined silicon membrane, was to be flown. An earlier concept, also robust enough to be deployed on a penetrator, used a small radioactive source (in fact the same source used in domestic smoke detectors) as an ionization gauge.

Entry and impact accelerometers were installed on the aftbody and forebody respectively. The former were off-the-shelf micromachined devices (Analog Devices ADXL250, often used in cars for airbag actuation), the latter an Endevco 7270 piezoresistive accelerometer.

As the mission development progressed, the scientific capability of the mission was eroded somewhat (science was always only a bonus – the principal goal of the mission was to demonstrate a safe delivery to the surface). In particular, the telecommunications system experienced severe development difficulties, and the original intent of fitting the entire system on a single hybrid chip was not realized. The replacement design, introduced only 12 months or so before launch, used discrete components requiring both more volume (or circuit-board space) and more electrical power. The electrical power requirement reduced the expected mission duration (again) and the energy available for sample heating, which was otherwise the dominant consumer of energy. The growth of board space required for the communications system meant that the pressure sensor could no longer be accommodated.

Unusually, mass (see Table 25.1) was not the tightest constraint on instrumentation. Volume was in general more at a premium than mass. The mass distribution of the probe was critical, however, in that the passive aerodynamic stabilisation of the entry shell required that the probe centre of mass be as far forward as possible (see Figure 25.2). This was achieved in part by the forebody-in-aftbody concentric design, and by introducing a tungsten nose to the forebody. This hemispherical nose, around 200 g in mass (about 1/3 of the total forebody mass, and many times the mass of the instruments) exploits the extremely high density of tungsten.

Table 25.1. *Mass budget (per probe) of the DS-2 Mars Microprobes*

Item	Mass (g)
Aftbody	1780 (incl. ~50 telecom system, 320 battery)
Forebody	670 (incl. 200 tungsten nose, ~10 microcontroller)
Entry system	1165
Entry mass total	3610
Spacecraft interface	2920
Total	**6530**

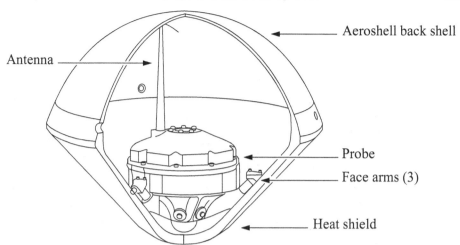

Figure 25.2. Drawing showing the layout of the probe installed in its aeroshell. The location of the probe gives a forward centre of gravity and thus 'weathercock' stability.

Little wind tunnel testing was performed, the aerodynamic performance being assessed principally by computational fluid dynamic simulation. The simulations were validated by ballistic range testing, and one hypersonic wind tunnel test at a facility in Russia.

Impact testing was a laborious aspect of the programme. Electronic components were mounted on test projectiles and shot into the ground, and taken back to the lab for health checks. The flight batteries posed a significant hazard in that they could explode if damaged by the impact. A separate series of tests was performed to evaluate impact accelerometer performance (Lorenz *et al.*, 2000). The aftbody was to remain on the surface to permit data transmission – its penetration depth was stated to be <10 cm, although in some tests on soft targets it did penetrate more than 30 cm (Lorenz *et al.*, 2000). The forebody was nominally to penetrate up to ~1 m.

The aftbody (Figure 25.3) was partially independent, in that it contained the crucial systems, namely the batteries and the communication system. Even if the forebody failed, or the umbilical cable broke, the aftbody would continue to operate and transmit data. The aftbody telecommunications system included a 6502 microprocessor, running around 8000 lines of code at 10 MHz. The receiver would operate for 1 s every minute to detect the query tones of the MGS relay spacecraft. After the first successful downlink, the sample sequence would be run.

The forebody (Figure 25.4) contained a microcontroller (Figure 25.5). This unit was based on an 8051 architecture with 64K RAM and 128K EEPROM.

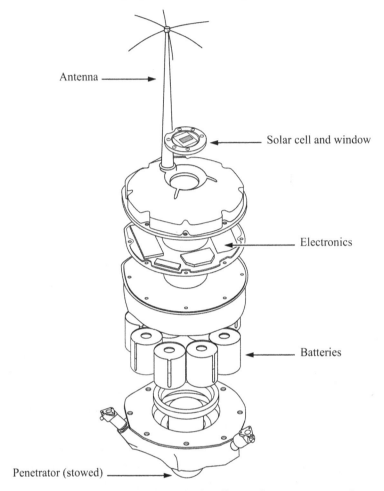

Figure 25.3. Exploded view of the aftbody. Electronics were mounted on a flat circuit board; the principal other elements were a solar-cell experiment, the antenna and the batteries.

The unit, which incorporated a 16-channel 12-bit analogue-to-digital converter, was designed for very low power (<50 mW at 1 MHz, with a 1 mW sleep mode) and low volume and mass (<8 cm^3, <90 g). The microcontroller supervised forebody operations with about 14 000 lines of code (in 8051 assembler) including the impact accelerometer sampling and the drill and heater cup. The analogue-to-digital converter on the forebody demonstrated operation at −70 °C, although its accuracy degraded somewhat at that temperature.

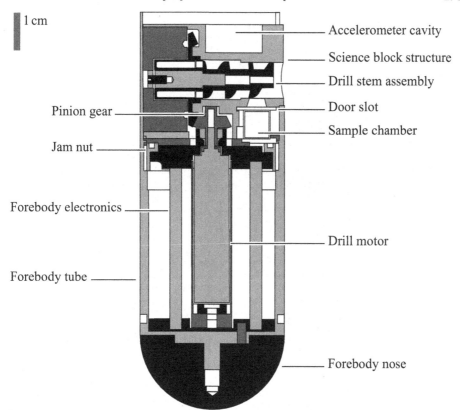

1 cm

Accelerometer cavity

Science block structure

Drill stem assembly

Pinion gear

Door slot

Jam nut

Sample chamber

Forebody electronics

Drill motor

Forebody tube

Forebody nose

Figure 25.4. Cross-section of the forebody layout. Clearly this spacecraft structure, like that of a wristwatch, is much more tightly integrated with the other subsystems. The auger drill would emerge to the right, and soil would drizzle into the sample chamber.

The forebody transmitted data along the umbilical to the aftbody twice every hour.

The DS-2 structure used some rather exotic materials. The forebody comprised a high-strength superalloy tube (MP35N) with a tungsten nose. The 'science block', into which the drill motor, thermal cup and impact accelerometer were embedded was simply aluminium alloy.

The aftbody was made from magnesium and titanium alloys. Its shape was driven by the mass distribution constraints, together with the required penetration performance. In particular, at higher angles of attack, the aftbody had a tendency to 'bounce' (more correctly, some rolling was involved) off the target.

Figure 25.5. The extreme level of miniaturization applied is evident in this photograph of the forebody microcontroller. The bevel gear for the sample drill is visible at the top left.

Introduction of tines on the front face of the aftbody helped to alleviate this tendency.

Small steel wires ('whiskers') were added to the titanium antenna to improve radiated signal performance (acting in effect as a longer antenna), while remaining robust to impact. The antenna survival was demonstrated by firing a probe backwards through a sample of the aft shield.

The heat shield, to which the probe was attached via three titanium fixtures, was of a very novel design. The structural stiffness was provided by an inner shell made from 0.8 mm thick silicon-carbide ceramic. The outer thermal-protection layer was a porous silica-rich layer called SIRCA–SPLIT (silicone impregnated reusable ceramic ablator–secondary polymer layer-impregnated technique). This thermal protection material was 1 cm thick at the nose. The backshell (hemispherical, so pressure forces during entry act through the centre of mass and do not apply torques) was made from FRCI (fibrous refractory composite insulation). While stiff, this structure was brittle, requiring careful ground handling. The brittleness was by design, in that the shell would shatter on impact without impeding the penetration.

The umbilical cable was folded as a concertinaed ribbon. The flat cable was made by depositing conductive traces on a Kapton substrate. Originally, the umbilical was to be 2 m long. Concern arose that the cable should be shielded from electrical transients, and so a deposited metallic shield was attempted. This shield layer turned out to be quite brittle and stiff (Arakaki and D'Agostino, 1999), such that the cable required a larger storage volume. This problem led to a descope of the cable to a shorter length, supported by the fact that the fore–aftbody separation in tests exceeded one metre in only two out of around fifty tests. In the event, the brittleness difficulties led the cable to not have a shield after all. The umbilical technology used in principle also allows electronic components to be installed on the same substrate (in fact many modern consumer electronic items such as CD players incorporate circuits built on flexible substrates) – a prime example being temperature sensors, so that heat flow measurements could be performed.

The batteries had very tight requirements. First was the ability to provide useful current even at temperatures of $-78\,°C$ (where many electrolytes are frozen, and the diffusion of ions in any electrolyte is slowed). The second driving requirement was the ability to tolerate 80 000 g loads at impact after a 3-year shelf life. Specially developed half-D cells by Yardney used lithium thionyl chloride. These cells presented some safety concerns in that if damaged (e.g. by short circuits induced by impact deformation) they could explode, and ordinary alkaline batteries were used during impact tests. The batteries provided for a 6 W-hr energy budget, giving an expected lifetime of 1–3 days, the low temperature and possible damage to some cells on impact being likely limiting factors.

The communications at UHF (with an RF power of only ~300 mW) permitted 7 kbps during a typical pass of MGS, which would last a little over 10 minutes. At the high latitude landing site, the polar-orbiting MGS would make several high passes per day. The receiver would detect the presence of a query tone transmitted from the orbiter (the tone frequency identifying which of the probes was to transmit) which would trigger the microprobe to transmit its data.

The DS-2 microprobes were installed on the Mars Polar Lander at KSC, and launched on 3 January 1999 on a Delta 7425. After a nominal cruise, during which there was no communication with DS-2, the commands to separate MPL from its cruise stage, and to deploy the DS-2 probes, were executed. Nothing further was heard from either MPL or the DS-2 probes.

There was never an end-to-end test of the DS-2 probes during which all systems were operated together, fired into the ground and demonstrated to operate thereafter. There was similarly no end-to-end test of the communications system, which relied on the Mars Balloon Relay on the Mars Global Surveyor orbiter. This French-supplied relay system was installed originally to support operation of a balloon to be carried on the Russian Mars-96 mission. The DS-2 communications hardware itself was delivered quite late and so was not extensively tested prior to launch.

Although the detailed failure investigation favoured separate problems, the simultaneous loss of MPL and DS-2 suggests a common failure mode. Although dual initiators are installed on the separation system, extreme temperatures or an undiscovered software error could still have prevented separation. Another common mode could be attributed to Mars itself, if the terrain were too soft or rough.

The failure mode favoured (e.g. Harland and Lorenz, 2005) for MPL is that a sensor transient when the landing legs locked into place may have been interpreted by the on-board computer as an indication of contact with the ground. This problem would have been detected in a ground test, had not an unrelated problem (subsequently fixed) occurred in that test. 'Thinking' it had landed, the vehicle would have shut down the descent engines at an altitude of above 50 m, and would have crashed onto the ground.

While perfectly plausible, this scenario does not account for the loss of both DS-2 probes. Possible failure modes include failure of the radio transmitters or batteries, perhaps by hitting rocks. Very soft terrain is another hazard, if the aftbody bearing the communications antenna were buried so deeply that the radio signals from the probes were attenuated.

Some months after the MPL/DS-2 loss, images from the Mars Global Surveyor discovered features that may indicate seepage and flow from subsurface aquifers, around 100–200 m below the ground. It was noted cynically by some science team members that the umbilical system had not been tested underwater – it would be ironic indeed if the microprobes' quest for water on Mars were fatally successful.

Whatever the technical causes of failure, the programmatic causes are all too clear – a rushed schedule, changing goals and inadequate testing.

26

Rosetta Lander Philae

ESA's Rosetta mission was launched on 2 March 2004, and is destined to reach its target comet, 67P/Churyumov–Gerasimenko, in 2014. The lander of the Rosetta mission, named Philae, is expected to be deployed around November 2014, to make the first ever controlled landing on a comet nucleus. *En route*, the mission's interplanetary trajectory takes in four gravity assists, three at Earth and one at Mars, and two asteroid flybys. Having matched the comet's orbit, Rosetta will close in to perform a comprehensive remote sensing survey of the nucleus and its environment prior to final selection of the landing site and deployment of the lander.

The finally launched mission had evolved a great deal over several iterations since the initial conception of a 'mission to the primitive bodies of the Solar System' around 1985 as a cornerstone of ESA's new Horizon 2000 science programme (this was almost a year before ESA's Giotto spacecraft had encountered comet Halley). The mission plan has always incorporated a surface element, though initially this was to obtain a sample for return to Earth. Known briefly as the Comet Nucleus Sample Return (CNSR) mission, it had by 1987 been renamed Rosetta. By the end of 1985 a joint ESA/NASA Science Definition Team had been formed to define in detail the mission's scientific objectives; NASA being envisaged as a partner for ESA on the mission. Planning began in earnest after the Giotto spacecraft's pioneering encounter with comet Halley in March 1986, which provided an important 'first look' at the type of body Rosetta was due to visit.

An ESA workshop was held in July 1986 to bring together the cometary community to look forward to the next European cometary space mission. The proceedings were published as ESA SP-249 (1986).

The report of the Science Definition Team was published in 1987 (ESA SCI (87)3). Work on the sample return mission scenario continued (see Atzei *et al.* (1989) for an overview), producing a Mission and System Definition Document (ESA SP-1125) in June 1991. This outlined the type of spacecraft and mission architecture that would be required. A large commitment from NASA was envisaged in the form of a carrier spacecraft derived from the Mariner Mark-II bus (used for Cassini). This would carry the landing stage to the comet, lifting off from the surface after about 15 days of sampling operations, to bring about 10 kg of cometary material back to Earth in an Earth Return Capsule.

Early in 1992, however, financial and programmatic difficulties within NASA (related to its own ill-fated CRAF (Comet Rendezvous and Asteroid Flyby) mission) prompted a re-examination of the original sample return concept, with a need to show that the mission could be achieved by European technology alone. As a result, Rosetta was re-oriented as a comet rendezvous and *in situ* analysis mission. A new System Definition Study (December 1993) was carried out to define the new mission. An ESA Study Report (ESA SCI(93)7) was produced. This re-examined the scientific objectives and model payload as well as outlining the new mission architecture.

The Rosetta 'comet rendezvous' concept now involved a main orbiter space-craft, carrying both a payload for remote sensing of the nucleus and *in situ* measurements of the dust, gas and plasma environment, and a ∼75 kg lander to be deployed towards the surface. Rendezvous with the target comet would occur at just over 3 AU heliocentric distance and the primary mission would last until perihelion, some two years later. The major differences between this revised design and the CNSR concept were the use of solar arrays rather than RTGs, and that the orbiter would not descend to the surface with the lander (as was the case for the CNSR scenario). Rather, it would stay in orbit around the nucleus and perform a much more extensive remote sensing investigation from rendezvous until perihelion. Rosetta proceeded along these lines, although further changes to the surface element were to occur.

Following a call for PI-led lander proposals, the initially selected configuration incorporated two ∼45 kg landers, RoLand (from a German-led consortium) and Champollion (from a French/US consortium – see Neugebauer and Bibring, 1998). In 1996 the US withdrew from Champollion (although it survived, until cancellation in July 1999, as a New Millennium mission, DS-4 (later ST-4) Champollion),[13] eventually leaving the French to team up with the RoLand consortium to provide a single, larger lander of ∼85–100 kg mass. This was called Rosetta Lander until it was given the name Philae in 2004, shortly before launch.

[13] The payload of the (DS-4 / ST-4) Champollion comet lander was as follows: CIRCLE (Champollion Infrared and Camera Lander Experiment: near-field camera, microscope, IR spectrometer) (Yelle), ISIS (stereo panoramic camera) (Bibring), CHAMPAGNE (gamma-ray spectrometer) (d'Uston), CHARGE (Chemical Analysis of Released Gas Experiment – GCMS) (Mahaffy), CPPP (Comet Physical Properties Package) (Ahrens), CONSERT (Kofman) (while still part of Rosetta), Sample return (studied briefly). The Project Scientist was Paul Weissman.

The scientific objectives of Rosetta as a whole are as follows (Schwehm and Hechler, 1994).

- Global characterisation of the nucleus, determination of dynamic properties, surface morphology and composition.
- Chemical, mineralogical and isotopic compositions of volatiles and refractories in a cometary nucleus.
- Physical properties and interrelation of volatiles and refractories in a cometary nucleus.
- Study of the development of cometary activity and the processes in the surface layer of the nucleus and in the inner coma (dust-gas interaction).
- Origin of comets; relationship between cometary and interstellar material; and implications for the origin of the Solar System.

Philae's aim is to address those objectives of Rosetta that cannot be achieved from the orbiter, including: geochemical analyses requiring sampling or close contact, surface and sub-surface physical properties, ground truth measurements for the orbiter, and the high resolution study of a single site. Philae, together with the orbiter, also provides a baseline for radio transmission tomography of the nucleus. Specifically, Philae's scientific objectives are as follows

- The determination of the composition of cometary surface matter: bulk elemental abundances, isotopes, minerals, ices, carbonaceous compounds, organic volatiles – as a function of time and insolation.
- The investigation of the structure, physical, chemical and mineralogical properties of the cometary surface: topography, texture, roughness, mechanical, electrical, optical and thermal properties.
- The investigation of the local depth structure (stratigraphy), and the global internal structure.
- Investigation of the plasma environment.

The selected payload, totalling 27.6 kg, can be seen in Section 20.3. The ten instruments include a sampling drill and two evolved gas analysers, imaging and microscopy, an alpha-X-ray spectrometer, and various sensors for studying the thermal, mechanical and electromagnetic properties of the nucleus and its near-surface environment. A mass breakdown is given in Table 26.1.

The main particular challenges of Philae's design and mission arise from the low surface gravity on the comet nucleus, the uncertain nature of the terrain (topography and mechanical properties), and the wide variation in solar flux and thermal conditions to be experienced as the comet's elliptical orbit nears the Sun. Landing will occur at about 3 AU, while perihelion is at 1.29 AU.

Power is provided initially by a primary $\sim$1 kWh $LiSOCl_2$ battery. This ensures operation for the first science sequence of 120 h. Thereafter a secondary $\sim$100 Wh Li-ion battery with body-mounted solar array is intended to provide

Table 26.1. *Mass Breakdown of Philae, the comet lander of the Rosetta Mission*

Item	Mass (kg)
APX	1.32
ÇIVA	3.39
COSAC	4.95
CONSERT	1.79
MUPUS	2.16
Ptolemy	4.53
ROMAP	0.74
ROLIS	1.36
SESAME	1.76
SD2	4.77
Payload total	**26.82**
ADS (Active Descent System)	3.69
Anchors	0.89
Landing gear	9.36
Separation structure	1.30
Non-payload common electronics	5.80
Thermal subsystem	7.42
Flywheel	2.90
Solar generator	1.72
Battery	8.50
Power Hardware	0.74
Communications subsystem	2.34
System harness	6.06
Structure	18.02
Balance mass	2.32
Lander total	**97.89**
Lander support equipment on Rosetta	13.09
Total	**110.98**

power for the long-term mission of about 3 months, until the comet reaches 2 AU heliocentric distance. After that, the extended mission will sooner or later end when the lander overheats.

The basic configuration of the lander is based on a jointed tripod carrying a baseplate, on top of which sits the lander's main body, much of the external surface of which carries solar cells. The thermal design for the lander involves a central, thermally controlled 'warm' compartment within the main body, housing the main electronic equipment and other critical subsystems, the evolved gas analysers and several of the cameras. Two solar absorbers on the lander's top panel are used to absorb heat from the Sun during the early part of surface operations.

Many of the experiments require access to the external environment, however. Most of these are mounted on the 'cold balcony', an area sharing the same baseplate as the main body of the lander. Other sensors are mounted on the landing gear. Several mechanisms are required by the payload, in addition to those serving mainly the landing and anchoring aspects. The sampling drill is due to obtain samples from a depth of up to ~20 cm and feed them to the microscope and evolved gas analysers via a carousel-based sample handling and distribution system. The APXS instrument needs close contact with the surface material and is thus lowered down through the lander baseplate. The magnetometer is deployed by a hinged boom, and the MUPUS thermal and mechanical properties experiment includes a thermal probe deployed by an arm comprising two parallel booms. An electromechanical hammering mechanism then drives the probe into the surface, to a depth of around 35 cm.

The lander will be ejected from the orbiter with a speed adjustable from 5 to 52 cm s^{-1}, by means of a screw-mechanism. This speed will cancel out enough of the lander's orbital speed for it to fall towards the surface, its attitude stabilised by an internal momentum wheel.

On landing, rebound must be avoided since this could lead to overturn of the lander. As the comet approaches closer to the Sun, outgassing of water and other volatiles from the nucleus may increase to such an extent that the lander could be blown off the surface. For these reasons the lander is equipped with a redundant system of cold-gas hold-down thruster, anchoring harpoons (2) and 'foot screws' on each of the three feet.

For most of the development and construction of the Rosetta mission, the target comet had been 46P/Wirtanen, with a launch scheduled for early 2003. However, problems with the Ariane 5 launch vehicle led to a change in target and launch date. The new comet, 67P/Churyumov–Gerasimenko, is larger (~4 km diameter versus 1.2 km) and so Philae is expected to have a higher landing speed: up to 1.2 m s^{-1} as opposed to 0.5 m s^{-1} for Wirtanen (the landing gear's original design limit being 1 m s^{-1}). The deployment manoeuvre can mitigate this to some degree, and a 'tilt-limiter' structure was added to the landing gear to prevent the main body of the lander tilting too far (>5°) on landing.

Two-way relay communications are achieved via the Rosetta orbiter. Two S-band antennas are located on the lander's upper surface. The first few hours of operation are planned to take the form of a pre-programmed sequence, after which ground controllers can modify the operations plan based on the data received.

27

Mars Exploration Rovers: Spirit and Opportunity

Following the success of the Mars Pathfinder project in 1997, there was a resurgence of interest in the deployment of an untethered rover on the surface of Mars. The concept of a semi-autonomous and freely roving vehicle was mooted as a follow-on to the Viking missions of the late 1970s. Almost twenty years were to pass before a rover was to be operated on Mars. After the Mars Pathfinder mission, NASA had proposed to send a rover equipped with a geology/chemistry payload, dubbed the 'Athena' suite, to Mars in 2001. Various constraints led to the redesign of the mission for a 2003 launch, although experiments of the payload were carried on the ill-fated Mars Polar Lander. In 2000 the Mars Exploration Rover mission was selected, with a launch-date flight three years later. This time, the Athena payload was to be duplicated, carried on two identical 174 kg rovers. Designated MER-A and MER-B, the spacecraft carrying the rovers were launched to Mars on separate Delta 2 boosters, making use of the favourable 2003 window for low-energy trajectories. The rovers on each craft were targeted to different regions of Mars. The MER-A craft, carrying the 'Spirit' rover, arrived on 4 January 2004 and was directed toward Gusev crater (14.5°S, 175.5°E) in the Aeolis region of Mars. This crater is the terminus of the fluid-cut Ma'adim Vallis, and Gusev was thought to host geological clues to the presence of water on Mars. The second craft to be launched, MER-B, arrived 21 days later and carried the 'Opportunity' rover to the Meridiani Planum area, landing at 2°S, 6°W.

27.1 The spacecraft

Each rover was carried to Mars in an entry capsule that was in turn attached to a cruise stage; disc-like structures over 2.6 m in diameter and 1.6 m deep. The identical cruise stages each carried photovoltaic arrays delivering 600 W of power at beginning of life (BOL) and had a dry mass of 870 kg. The propulsion system of the cruise stages used 31 kg of pressurized hydrazine and could give a ΔV capability of around $50 \, \mathrm{m \, s^{-1}}$, along with attitude and spin manoeuvres. The cruise stages provided power, attitude control via a pressurized hydrazine reaction system, communication support and thermal control for the encapsulated rovers. This last support function took the form of a pumped closed-loop refrigeration system using a chlorofluorocarbon that took excess heat from the cores of the rovers and disposed of it via radiators mounted on the cruise stage exterior. This system was needed because each rover carried around 15 g of ^{238}Pu oxide dispersed among eight radioisotope heater units (RHUs) to allow the rovers to survive the Martian night without impractically sized solar arrays and batteries. Consequently, the unwanted heat from the electronics boxes in the rover and from the RHUs (around 1 W per RHU) had to be dumped during cruise.

Upon arrival at Mars the cruise stage was separated from each lander capsule approximately 15 minutes prior to the nominal contact with the atmosphere. An ablating material (SLA-561), used first in the Viking mission, covered the entry shield and protects the 827 kg entry capsule and its contents. In response to the failure of the Mars Polar Lander mission during its entry phase, the MER craft both implemented a low-bandwidth telemetry system that operated throughout atmospheric entry. Set events during entry, such as airbag deployment, were signalled by the broadcast of one of 128 ten-second audio-frequency tones. The aim was to provide a clearer indication of which subsystem had failed, in the event of a catastrophe. A timeline of the entry, descent and landing (EDL) sequence is shown in Figure 27.1.

The mass of the MER rovers meant that a simple parachute decelerator, as used by the lighter Mars Pathfinder mission, could not provide a safe degree of speed reduction. Thus, along with a polyester/nylon parachute, a rocket-based decelerator system was built into the backshell of the landing capsule. The rocket-assisted descent (RAD) system, consisting of three solid-fuel motors each developing around 10 kN for a little over two seconds, was used to bring the craft to a halt at a height of some tens of metres above the Martian surface. A modified solid-state camera, added late in the design sequence of the lander, allowed a smaller reaction system (three motors giving 750 Ns of impulse) to be fired to reduce the horizontal speed of the craft, and further reduce the risk of ripping the airbags. The airbags used in the MER missions were similar to, but tougher than, those of the Mars Pathfinder design. Protecting a lander twice as heavy as the Mars Pathfinder required the airbags' double inner pressure bladders to be wrapped with up to six, rather than four, layers of woven Vectran®, a synthetic polyester fibre having high tensile strength and flexibility at low temperatures. Pressurization of the airbags was performed by three solid-state gas generators delivering gas to a final pre-impact pressure of around 7 kPa; the cooling effect of

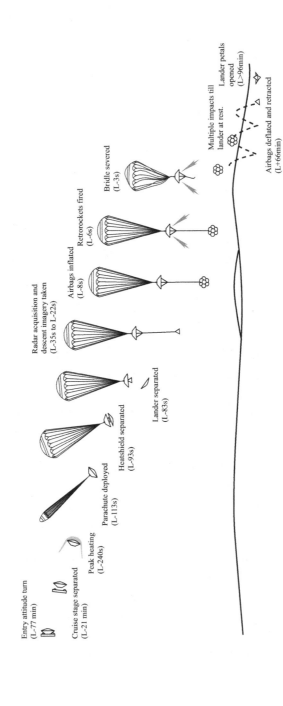

Figure 27.1. Significant events in the arrival sequences of the Mars Exploration Rovers.

the cold Martian atmosphere on the large airbag surface area necessitated the gas generators to continue to fire as the lander bounced and rolled to a halt.

The airbag system offered a low degree of dissipation for the impact energy, and thus the rover and its associated enclosure were exposed to around 30 consecutive impacts, with the initial impact speed being in the region of $10\,\mathrm{m\,s}^{-1}$. Each rover was fixed to the interior of a tetrahedral panelled enclosure formed from a space-frame carbon-composite material. The three panels around the lander's hold-down point were independently powered by hinge-motors, allowing the lander to turn itself to give the rover an upright orientation.

27.2 The rovers

Each rover has a mass of 174 kg, and when fully deployed from their cruise configuration, each stands around 1.5 m tall with a width of 2.3 m and a length of 1.6 m. Once free of its carrier, each rover communicates with terrestrial ground-stations of the Deep Space Network (DSN) directly via their low-gain and high-gain antenna at X-band frequencies of 7.2 GHz for the uplink, and 8.4 GHz for the downlink. The omnidirectional UHF antenna can communicate with orbiting spacecraft (Mars Odyssey, MGS, MEx) at rates up to 128 kbps, while the 28 cm diameter HGA allows data rates of 2 kbps with the much more distant DSN.

A notable feature of the MER project is the presence of an arm on the forward face of each rover. This arm, having five degrees of freedom, is able to position the instrument platform at its tip up to 0.8m in front of the rover within a volume of around 0.1m³. Clearly, exceptional science can be performed by locating payload sensors, and not necessarily whole instruments, on the end of an articulated arm. The inevitable tradeoff between mass and actuator size for the arm versus the reach and carried mass has been solved for the MER project by having the arm of each rover carry a four-way turret which allows one of four devices to be offered up to a surface presented to the rover. This arrangement is shown in Figure 27.2 along with the positions of the other notable rover items.

In Table 27.1 the data products from the payloads of Spirit and Opportunity are summarized and the payload instruments are described in detail thereafter.

27.2.1 Panoramic camera (Pancam)

From their vantage point 1.5 m above the Martian surface, the Pancam is able to make multi-spectral exposures of up to 30 seconds. Their angular resolution of 0.0164° rivals the acuity of the human eye and each camera is able to focus from 1.5 m to infinity over a field of view over 16° by 16° in angular size. Each camera

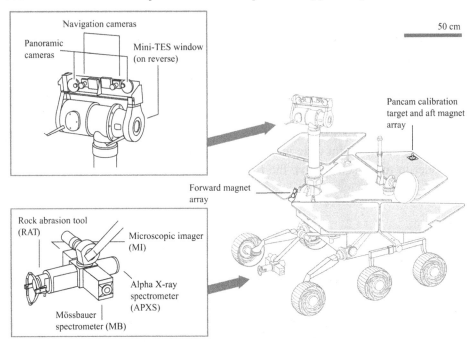

Figure 27.2. The location of the Athena payload instruments on the Mars Exploration Rovers.

Table 27.1. *A summary of the MER 'Athena' payloads and their data products*

Payload	Data products	Sensor mass and power
Panoramic camera (Pancam)	8 or 12-bit deep images of 1024 × 1024 pixels	Dual cameras, each of 270 g mass, drawing 3 W in operation
Miniature thermal emission spectrometer (mini-TES)	1024 samples at 16 bit	Mast-mounted optics feeding internally accommodated package. Total mass 2.4 kg, drawing 5.6 W in operation
Mössbauer spectrometer (MB)	512 data points per spectrum, each of 3 bytes	Arm-mounted sensor of 500 g, drawing an average of around 2 W
Alpha particle X-ray spectrometer (APXS)	256 channels for α-particle spectra, 512 channels for X-ray sectra	Arm-mounted sensor of 640 g, drawing around 1.5 W
Microscopic imager (MI)	8 or 12-bit deep images of 1024 × 1024 pixels	Arm-mounted sensor of 210 g, drawing 2.15 W
Magnetic adhesion arrays	None, analysed by other sensors	Body and arm mounted items, total mass of ∼50 g, no power use
Rock abrasion tool (RAT)	Motor current	Arm-mounted device of 685 g, drawing 11 W

is equipped with a different filter wheel, each with eight filter-positions that cover the near-UV (430 nm) through to IR (980 nm). The platform carrying the Pancam can slew through 360° and pitch by ± 90° with respect to the horizontal axis of the MER. The Pancam was developed at Cornell University in the USA.

27.2.2 Mini-TES

Built at Arizona State University in the USA, the mini-TES (e.g. Christensen *et al.*, 2003) is a Fourier transform spectrometer that obtains spectra in the wavelength range of 5 to 29.5 microns. Different minerals can be distinguished by their infrared emission spectra in this, with particular ability to discriminate between anion groups (CO_3, SO_4, SiO_4, etc.). The imaging capability is provided by having the mini-TES input optics mounted on the Pancam platform. Accordingly, mini-TES has the same azimuthal angular range as the optical imagers, but space restrictions on the optics give a restricted elevation span: 30° above, and 50° below the nominal horizon. The mast supporting the Pancam is hollow and internal mirrors direct the light gathered from the 6.3 cm diameter aperture into the mini-TES which is physically located in the MER body. The mini-TES is able to make spectral images with two levels of angular resolution, 20 mrad and 8 mrad, and the Pancam and mini-TES boresights are essentially parallel, allowing images from these systems to be overlaid.

27.2.3 Magnet array

Attached to the exterior of each rover are three arrays of magnets, provided to the Athena payload by the University of Copenhagen. The arrays are positioned on the front of each rover, on the RAT of the arm, and on the exposed deck of the rover. It is expected, on the basis of similar measurements by the Viking and Sojourner craft, that much of the Martian regolith contains a magnetic component. Dust-sized mineral particles continually rain out from the Martian atmosphere, after being lofted by winds. Only the forward-facing array can be reached by the arm, so that the trapped material may be analysed by the arm microscope, MB, and APXS sensors. The deck and RAT-located arrays are imaged by the Pancam.

27.2.4 Mössbauer spectrometer (MB)

The MER Mössbauer device, built at the Johannes Gutenberg University in Mainz, Germany, is an improved version of the device flown by that group on the Mars Pathfinder mission (Klingelhöfer *et al.*, 2003). Two cobalt-rhenium sources provide the gamma rays used to irradiate, and detect the presence of, iron nuclei and its distribution among the various oxidation states (Fe^{2+}, Fe^{3+}, Fe^{6+}). This instrument consists of two parts: the head (sources, collimation structures, and silicon detectors) which is carried on the rotating turret at the end of the rover's arm, and an electronics box (power supply, processing, memory) held within the MER. Limitations on the size and strength of sources available for use on the mission result in integration times typically lasting several hours, mitigated somewhat by the ability of the head to be manoeuvred to within a few cm of a target. The MB has a working field of view of around $1.5\,cm^2$.

27.2.5 Alpha-particle X-ray spectrometer (APXS)

Successful operation of their APXS on the Mars Pathfinder project led the team at the Max-Planck-Institut für Chemie, Mainz, Germany to develop a modified version for the Athena payload of the MERs. Like the MB, the APXS is arm-mounted and irradiates targets with alpha particles and X-rays from a curium source. The backscattered alpha particles are detected by an annular detector array, providing concentration measurements of light elements (C, O). X-rays emitted from heavier rock-forming elements (Mg, Al, Si, Ca, Fe, etc.) are detected by a high-resolution silicon drift detector, and the whole instrument has a roughly circular field of view a little over $12\,cm^2$ in area.

27.2.6 Microscopic imager (MI)

The MI is positioned 180° around from the Mössbauer spectrometer on the arm of each MER. This instrument uses a charge-coupled device CCD imager with a resolution of 1024×1024 pixels to cover an area of $30 \times 30\,mm$. The optical depth of field is $\pm 3\,mm$ and the monochromatic CCD is covered with a bandpass filter which transmits light in the spectral range between 400 nm and 680 nm. The MI relies on ambient lighting to illuminate the target and a transparent shield is used to cover the MI optics when the device is not imaging. The rod that extends from the face of the MI is a contact sensor which allows the MI to be accurately positioned and also prevents accidental impacts with target rocks. The MI is provided by Cornell University and uses the same camera body as the Pancam, giving the MI the same radiometric performance (signal to noise and sensitivity).

27.2.7 Rock abrasion tool (RAT)

Opposite the APXS on the platform carried by each rover's arm, is the RAT. This unique device is a combined grinding and coring tool that is able to remove up to

5 mm of rock from a circular region 45 mm in diameter. By exposing interior portions of rocks the RAT offers the other tools on the MER arm unprecedented access to material that will have suffered no exposure to the present Martian surface environment. Strictly, the RAT itself generates relatively little data; by monitoring the current drawn by the motors in the tool it is possible to infer the relative toughness of the rock (e.g. Bartlett *et al.*, 2005). The MER arm provides a down-force of up to 80 N for the tool to engage with the target rock, and cutting is performed by rotating an armature carrying two grinding points made of industrial diamond matrices at 3000 rpm.

The mobility system used by the two rovers borrows significantly from the design used for the Sojourner rover in the Mars Pathfinder mission. Like the Sojourner, the MER rovers are equipped with two sets of three 23 cm diameter wheels, each of which houses a bidirectional motor and 1500:1 gearbox mounted in the wheel hub. Additionally, the front and aft wheels of each rover can be yawed around a near-vertical axis, allowing the rovers to turn on the spot. On hard, level ground this drive system allows the rovers to move at a speed of 50 mm s^{-1}, and in doing so the six motors expend a total power of over 100 W. Articulation for the wheels is achieved with a so-called 'rocker-bogie' system that enables all three wheels on each side of a rover to maintain good contact with the ground in the event of encountering obstacles. Packaging constraints on the rover delivery capsule meant that the suspension was delivered to Mars in a geometry different from that used when the rover was being driven. Specifically, the forward rocker was rotated with the front wheel tucked in front of the rover, and the joint between the aft and forward rockers (indicated by a black mark in Figure 27.3 (b)) was locked flat to permit the rover to squat closer to the floor of its lander package.

27.3 Problems encountered

Although neither rover has yet to match the distance record set by Lunokhod 2, the smaller Martian rovers encounter significantly more problems arising from the dusty nature of Mars. Firstly, the progressive obscuration of the solar panels by airborne dust leads to a fall-off in the available power, mitigated somewhat by the beneficial but random action of transient aeolian phenomena, which appear to clear some of the material from the rovers. Secondly, the Lunokhod wheels were not exposed to blown dust grains, whereas the wheel assemblies on each MER are subject to both saltating and suspended particles, which place greater loads on the motor and gear shaft seals. As of late 2005, only the Spirit rover has shown signs of anomalous wheel friction, with both craft working after exceeding

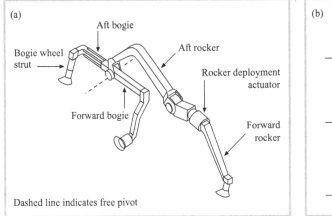
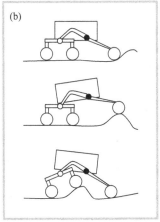

Figure 27.3. The layout and articulation of the rocker–bogie suspension used on the Mars Exploration Rovers.

their design benchmark of 90 days by almost a factor of ten. Appropriate driving rules allow that rover to manoeuvre and drive albeit with reduced efficiency. At this same stage the Opportunity rover has a jammed front wheel actuator, and has difficulties in unstowing its arm but after more than one Martian year all payloads are operating well.

Both craft have been exemplary models of the judicious mixture of caution and design innovation needed to produce successful planetary craft. In terms of risk, the project has been regarded as being challenging with successful missions of comparable complexity being developed in historically much longer periods (Dornheim, 2003). It remains to be seen whether follow-on missions will display a similar level of robustness. Indeed, the fact that the continued survival of the MER hardware is unexpected, albeit very welcome, reveals much about the uncertainties surrounding risk assessment in planetary missions and the danger of attempting to characterize unique procedures and systems.

Appendix

Some key parameters for bodies in the Solar System

Atmosphere models

Clearly the design of heatshields and parachute systems requires assumptions on the density structure of the atmosphere to be encountered. Thus atmospheric models must be constructed as a design basis – these models must provide the extreme range of conditions likely to be encountered, since extremes in any direction may drive the design.

Where *in situ* data from prior missions is available (e.g. at Mars and Venus) this of course adds considerable confidence to the model. More generally, as for the first missions to Mars, Titan, Jupiter, etc., the major source of guidance is an atmospheric refractivity profile derived from radio-occultations by prior flyby or orbiter missions. The refractivity may be converted into a mass–density profile with some assumptions on composition. However, the altitudes probed by radio occultations are generally lower than those at which peak aerodynamic heating and deceleration occur, so some assumptions must be made in propagating those measurements upward. Some of these assumptions are rather robust, such as hydrostatic equilibrium, while others are less so.

There is in model development an inherent tension, just as in the development of a mission as a whole. The engineer designing the heat shield will just want a definitive answer to the question 'what is the density at 500 km?' (or whatever), while the scientist developing a model will wish to acknowledge the widest range of uncertainty – there may be intrinsic measurement errors in a refractivity profile, there are uncertainties in the assumed composition or other factors, there may be diurnal and seasonal variations, and variations with solar activity. Thus the range between 'minimum' and 'maximum' atmospheres may be rather large, which is problematic for engineering design. Furthermore, this range alone offers little insight into the probability of the extremes or the values between them.

These uncertainties can be accommodated better with improved models, wherein the deterministic variability (e.g. the seasonal variation of pressure on Mars) are modelled explicitly, whereas stochastic variations such as 'weather' and measurement errors are modelled statistically. These models, such as the Titan-GRAM model or the Mars Climate Database, can be used in Monte-Carlo studies to determine engineering parameters such as the 3-σ loads. The 3-σ is a typical design criterion, with the expectation that the chances of exceeding such loads is less than 0.5%, a modest risk compared with the probability of other failures such as launch failure. The association of 'sigma' (i.e. a standard deviation) with a certain probability assumes, strictly speaking, Gaussian statistics, which are not always appropriate; for example, windspeeds have a much more skewed distribution, with a long high-end tail. The 'sigma' terminology is widespread, however.

These statistical models may provide more comfort than they should, in that real data is rarely available to give meaningful confidence in the statistical parameterizations used, but inasmuch as available information is folded into the estimates, unnecessary margins can be reduced. It is sobering that even with two spacecraft in orbit around Mars to provide fresh information, day-to-day variations in dust opacity and thus the temperature and density structure in the atmosphere exceeded 'worst-case' model predictions, and were nearly enough to cause failure in the Mars Exploration Rover entry and descent (and may have caused the loss of Beagle 2). The prudent designer will therefore apply as much margin as he or she can.

For Mars, the atmosphere is subject to dramatic seasonal changes (the surface pressure dropping by 30%, for example) and there are more data available. Accordingly models are more sophisticated, being based on output from numerical models as in the Martian Climate Database developed under ESA sponsorship by Laboratoire de Meteorologie Dynamique in Paris and Oxford University, http://www-mars.lmd.jussieu.fr/ or being derived from statistical descriptions of the variations in a more empirical way, as in the Mars Global Reference Atmospheric Model (MARS-GRAM) developed at NASA Marshall Research Center: Justus, C.G. and Johnson, D.L., Mars global reference atmospheric model 2001 version (Mars-GRAM 2001) User's Guide NASA/TM-210961.

For Venus and Titan at present only limited engineering models are published from the handful of measurements available. One may expect this situation to change for at least the latter two as more data arrive, and statistical models (e.g. Titan-GRAM) are under development. Papers describing the limited knowledge of the wind fields can usually be found adjacent to relevant publications in the temperature/density papers.

A.J. Kliore *et al.* (1985) Venus international reference atmosphere *Advances in Space Research*, **5**, 1–305.

R. V. Yelle, D. F. Strobell, E. Lellouch and D. Gautier (1997) Engineering models for Titan's atmosphere ESA SP-1177, 243–256.

For Jupiter, *in-situ* measurements are available from Galileo:

A. Seiff *et al.* (1998) Thermal structure of Jupiter's atmosphere near the edge of a 5-μm hot spot in the north equatorial belt. *Journal of Geophysical Research*, **103**, 22857–22890.

For the other outer planets and Triton, no *in-situ* measurements are available, and in the absence of an imminent mission, no engineering models have been developed. A literature search (e.g. ads.harvard.edu) for results from the Voyager radio-occultation experiment will yield the relevant information.

Bibliography

In addition to the sources cited in Part I for particular engineering aspects of planetary landers, and in Parts II and III for particular missions, we present below a selection of books and other sources that the reader might find useful.

Engineering

For detailed presentation of spacecraft and mission design, there are various texts available. All are good, but none are completely satisfactory in terms of covering all aspects in an up-to-date manner.

Andreyanov, V. V., Artamonov, V. V., Atmanov, I. T. *et al.*, *Avtomaticheskiye Planetnyye Stantsii (Automatic Planetary Stations)*. Nauka, Moscow, 1973. English translation available: NASA TT-F-15666, 1974.
Although out of date scientifically and to some extent technologically, this book does cover some of the subsystems aspects of planetary landers, including thermal and structural, and in considerable mathematical detail.

Bazhenov, V. I. and Osin, M. I. *Posadka Kosmicheskikh Apparatov na Planety (The Landing of Spacecraft on the Planets,* or *Space-vehicle Landings on Planets)*. Mashinostroenie, Moscow, 1978.
Focuses on atmospheric entry dynamics and, in greater detail, landing dynamics, with both mathematical detail and discussion of sub-scale drop tests.

Brown, C. D. *Spacecraft Mission Design*. 2nd edn, AIAA, Washington DC, 1998.
This text offers clear information and design guidance on the sub-systems of generic spacecraft, with many examples from actual missions and extant hardware. Its scope avoids discussion of entry and landing systems. Contains set problems after each chapter.

Brown, C. D. *Elements of Spacecraft Design*, AIAA, Washington DC, 2002.
Good overall, and excellent on astrodynamics and propulsion (including interplanetary). Has example problems. Does not cover aerothermodynamics.

Corliss, W. R. *Space Probes and Planetary Exploration*. Van Nostrand, 1965.
Written in the early years of the Space Race, and somewhat dated as a result, this wonderfully detailed book gives an overview of the state-of-the-art of planetary

exploration technology, as it was in 1965. It covers American projects almost exclusively, but does so with impressive breadth and is still of interest to this day. Contains many hard-to-find photographs and diagrams.

Cruise, A. M., Bowles, J. A., Patrick, T. J. and Goodall, C. V. *Principles of Space Instrument Design. Cambridge Aerospace Series no. 9.* Cambridge University Press, 1998.
As the title suggests, not a text on spacecraft overall but devoted to instrumentation engineering. Has excellent detail on thermal, mechanical and electronic design, and mechanisms for space.

Fortescue, P., Stark, J. and Swinerd, G. *Spacecraft Systems Engineering.* 3rd edn, Wiley, 2003.
Written in Europe, offers a usefully different perspective from most texts, citing ESA engineering standards and example spacecraft. Covers re-entry aerodynamics and aerothermodynamics.

Gilmore, D. G. *Spacecraft Thermal Control Handbook: Fundamental Technologies.* 2nd edn, Aerospace Press, 2002.
This book covers spacecraft thermal design, technologies and analysis, including some material on interplanetary spacecraft.

Griffin, M. D. and French, J. R. *Space Vehicle Design.* 2nd edn, AIAA Education Series, AIAA, Washington DC, 2004.
A good overall spacecraft design text (originally written over a decade before the first author became the NASA Administrator!), shorter than most of the others, but includes discussion on configuration design and re-entry.

Gurzadayan, G. A. *Theory of Interplanetary Flights.* Gordon and Breach, 1996.
This text covers orbital dynamics including transfer trajectories to the Moon and planets.

Hankey, W. L. *Re-entry Aerodynamics.* AIAA Education Series, AIAA, Washington DC, 1988.
Covers much of the background theory for atmospheric entry well, but lacks recent (post 1975) information for TPS systems and non-terrestrial atmosphere chemistries. Provides set problems but uses Imperial ('English') units exclusively, which for aerothermodynamic problems are particularly ugly.

Kemble, S. (2006). *Interplanetary Mission Analysis and Design.* Springer-Praxis, Chichester. This book describes current mission analysis and design techniques that may be applied to a very wide range of interplanetary missions.

Kemurdzhian, A. L., Gromov, V. V., Kazhukalo, I. F., *et al. Planetokhody (Planet Rovers) (in Russian).* 2nd edn, Mashinostroenie, Moscow, 1993.
One of the very few books on planetary rovers, from the father of space robotics, Aleksandr Kemurdzhian.

Kemurdzhian, A. L. (ed.). *Peredvizhenie po Gruntam Luny i Planet (Transport over Lunar and Planetary Soils) (in Russian.)* Mashinostroenie, Moscow, 1986.
This books covers the soil mechanics aspects of planetary rovers.

Labunsky, A. V., Papkov, O. V. and Sukhanov, K. G. *Multiple Gravity Assist Interplanetary Trajectories*. Gordon and Breach, 1998.
Addresses the important but complex topic of interplanetary transfers using multiple gravity assists.

Martínez Pillet, V., Aparicio, A. and Sánchez, F. (eds) (2005). *Payload and Mission Definition in Space Sciences*. Cambridge University Press, Cambridge.
Includes chapters on ESA mission programmatics, interplanetary trajectories and planetary instrumentation.

Meyer, R. X. *Elements of Space Technology*. Academic Press, 1999.
A general textbook on space technology.

Patel, M. R. *Spacecraft Power Systems*. CRC Press, 2004.
One of the few books covering spacecraft power.

Popov, E. I. *Spuskaemyie Apparaty (Descent Apparatus)* (in Russian). Znanie, Moscow, 1985.
A short, non-mathematical book on entry and descent probes with particular attention to Soviet/Russian examples. Available online at http://epizodsspace.testpilot.ru/bibl/biblioteka.htm.

Pisacane, V. L. (ed.) *Fundamentals of Space Systems*. 2nd edn, Oxford University Press, 2005.
Covers a wide range of general spacecraft engineering topics.

Regan, F. J. and Anandakrishnan, S. M. *Dynamics of Atmospheric Re-Entry*. AIAA Education Series, AIAA, Washington DC, 1993.
Covers many aspects of atmospheric entry, although focussing on the terrestrial case. Some four times larger than Hankey's book on re-entry in the same series.

Roy, A. E. *Orbital Motion*. 4th edn, Institute of Physics, London, 2004.
A solid introduction to orbital theory, written to a graduate level but introduces topics such as numerical integration methods and Hamiltonian techniques. Each chapter contains worked problems.

Wertz, J. R. and Larson, W. J., *Space Mission Analysis and Design*. 3rd edn, Microcosm/Kluwer, 1999.
Very strong on system engineering aspects such as defining requirements and budgets (including those for often-ignored topics such as operations, attitude errors and computing). No discussion of interplanetary trajectories or aerothermodynamics.

Missions, Technologies and Design of Planetary Mobile Vehicles. Proceedings of the CNES conference, Toulouse, September 1992. Cépaduès Editions, 1993.
Somewhat out of date in terms of robotic control and autonomy, but an excellent overview of activity at the time, and of earlier work.

For communications, the book series *Deep Space Communications and Navigation*, published by John Wiley & Sons, is worth consulting. The full text of nine monographs in this series is currently available online at http://descanso.jpl.nasa.gov/.

A series of annual international workshops on planetary probes began in Lisbon in 2003. The proceedings are an excellent source of up-to-date information; the following are for the first three:

Wilson, A. (ed.). *Proceedings of the International Workshop on Planetary Probe Atmospheric Entry and Descent Trajectory Analysis and Science*, Lisbon, 6–9 October 2003. ESA SP-544, European Space Agency, 2004.

Proceedings of the 2nd International Planetary Probe Workshop, NASA Ames Research Center, Moffett Field, CA, August 2004. Edited by E. Venkatapathy *et al.* NASA /CP-2004-213456, 2005.

Proceedings of the 3rd International Planetary Probe Workshop, Attica, Greece, 27 June–1 July 2005, published as ESA SP-544, 2006.

Reference

Bakich, M. E. *The Cambridge Planetary Handbook*. Cambridge University Press, 2000.
A reference book giving basic data for bodies of the Solar System.

Lodders, K. and Fegley, B. *The Planetary Scientist's Companion*. Oxford University Press, 1998.
A compact book of facts and figures about the main bodies of the Solar System, including details such as atmosphere profiles, and providing copious references. Emphasizes geochemistry but steps around fields such as planetary magnetospheres.

Moroz, V. I., Huntress, W. T. and Shevalev, I. L. *Planetary Missions of the 20th Century. Kosmich. Issled.* 40(5), 451–481. Translated in *Cosmic Res.* 40(5), 419–445, 2002.
This journal paper is a good, accurate and concise summary of previous planetary missions, covering both failed and successful missions.

Siddiqi, A. A. *Deep Space Chronicle: A Chronology of Deep Space and Planetary Probes 1958–2000*. Monographs in Aerospace History Volume 24. NASA SP-2002-4524. NASA, Washington, 2002.
Available online from http://history.nasa.gov/, this volume is a handy reference work, containing brief information for Solar System missions.

Thompson, T. D. *Space Log 1996*. TRW, 1997.
Sadly this highly specific work is no longer in print. It details the time, location and orbital data of every known major rocket launch from the Earth; an unparalleled resource.

Wilson, A. *Solar System Log*, Janes, 1987.
An indispensable compendium of details on planetary probes up to 1986. A rich source of facts, figures and photographs, this slim and hard-to-find work is, however, frustratingly devoid of references. Nevertheless, an important and fascinating text.

Winterbottom and Perry. *DRA Table of Space Vehicles 1958–1991*. DRA, Farnborough, 1993.
Tabulated data of lunar and planetary launches.

Planetary sciences

Cherkasov, I. I. and Shvarev, V. V. *Gruntovedenie Luny (Lunar Soil Science)* (in Russian). Nauka, Moscow, 1979.
A hard-to-find book in Russian covering many aspects of the study of lunar regolith mechanical properties through to the end of the Apollo/Luna era, including both *in situ* instrumentation and laboratory analysis of samples.

Heiken, G. H., Vaniman, D. T. and French, B. M. (eds.). *Lunar Sourcebook – A User's Guide to the Moon.* Cambridge University Press, 1991.
A detailed summary of scientific results from (especially) Apollo on the lunar environment and the properties of lunar surface material, offering a good balance of explanatory material and mission data, with copious references.

Marov, M. Ya. and Grinspoon, D. H. *The Planet Venus.* Yale University Press, 1998.
A comprehensive and readable treatment of Venus exploration, emphasizing results from Soviet missions that are not as widely advertised in the West.

Surkov, Yu. A. *Exploration of Terrestrial Planets from Spacecraft: Instrumentation, Investigation, Interpretation.* 2nd. edn, Wiley-Praxis, Chichester, 1997.
A (somewhat dated and cursory) overview of planetary exploration that introduces a uniquely detailed discussion of X-ray, gamma-ray and mass spectrometer instrumentation for planetary spacecraft, with a welcome tendency to describe Soviet missions at length.

The University of Arizona Press has a long-running series of (massive) edited books on various Solar System bodies that often make a good starting point for further exploration of the literature. Continuing exploration by telescope observation and spacecraft has led to successive volumes (now on Comets II, Asteroids III, Venus II, etc.). Of particular note for Mars and Venus missions are the following:

Hunten, D. M., Colin, L., Donahue, T. M. and Moroz, V. I. (eds.) *Venus.* University of Arizona Press, 1983.
Kieffer, H. H., Jakosky, B. M., Snyder, C. W. and Matthews, M. S. (eds.) *Mars.* University of Arizona Press, 1993.

Historical

Johnson, N. L. *Handbook of Soviet Lunar and Planetary Exploration.* American Astronautical Society Science and Technology Series, 47. Univelt, 1979.
For its time, this book provided an excellent digest of the Soviet literature that still stands up well.

Nicks, O. W. *Far Travelers: The Exploring Machines.* NASA SP-480, 1985.
An engineering-oriented view of the NASA robotic lunar and planetary missions through to Viking. Available online from http://history.nasa.gov/

Pellinen, R. and Raudsepp, P. (eds.) *Towards Mars!* Oy Raud, 2000.
A colourful book looking at Martian science and missions, with good coverage of non-US missions. See also the update, *Towards Mars! Extra, 2006.*

Perminov, V. G. *The Difficult Road to Mars: A Brief History of Mars Exploration in the Soviet Union.* Monographs in Aerospace History, Number 15. NP-1999-06-251-HQ NASA, 1999.

A candid description of the spacecraft challenges faced by the Lavochkin bureau. Gives insight into the characters and methods of that time and contains many rare diagrams and photographs.

Petrov, G. I. (ed.) *Conquest of Outer Space in the USSR 1967–1970. NASA TT-F-725, Amerind, 1973*. Translation of *Osvoenie Kosmicheskogo Prostranstva v SSSR*, Nauka, Moscow, 1971.
Following on from Skuridin's book, this contains some details of missions in progress at the time.

Reeves, R., *The Superpower Space Race: An Explosive Rivalry through the Solar System*. Plenum Press, 1994.
An excellent, balanced and detailed narrative of robotic exploration of the Solar System.

Semenov, Yu. P. (ed.) *RKK Energia im. S.P. Koroleva 1946–1996*. RKK Energia, Moscow, 1996.
One of the best of the few sources of information on the activities of OKB-1. Lunar and planetary probes are only a small part of this large book.

Semenov, Yu.P. *Rocket and Space Corporation Energia: The Legacy of S.P. Korolev*. Energia, 1994. Translated edition, Apogee Books, 2001.
A source of rare photos and drawings of early lunar and planetary probes from OKB-1.

Serebrennikov, *et al*. (eds.). *NPO im. S.A. Lavochkina. Na Zemle, v Nebe i v Kosmose*. Voennyi Parad, Moscow, 1997.
In-house publication on the activities of NPO Lavochkin.

Siddiqi, A. A. *Challenge to Apollo*. NASA SP-2000-4408, 2000. Reprinted in two volumes as *Sputnik and the Soviet Space Challenge* and *The Soviet Race with Apollo*. University Press of Florida, 2003.

Skuridin, G. A. *Mastery of Outer Space in the USSR*, 1957–1967 (in Russian). Nauka, Moscow, 1971.
Overview of Soviet space activities of the period in the form of official news releases. Includes lunar and planetary efforts.

Ulivi, P. and Harland, D. M. *Lunar Exploration: Human Pioneers and Robotic Surveyors*. Springer-Praxis, Chichester, 2004.
Detailed, balanced and technically competent story of lunar exploration.

Some useful web sites

NASA Mars Programme:
http://mars.jpl.nasa.gov/

ESA Science Programme:
http://sci.esa.int/

National Space Science Data Centre Chronology of Lunar and Planetary Exploration:
http://nssdc.gsfc.nasa.gov/planetary/chronology.html

NASA Technical Reports Server:
http://ntrs.nasa.gov/

NASA History:
http://history.nasa.gov/

Russian Space Web:
http://www.russianspaceweb.com/

RKK Energia:
http://www.energia.ru/

NPO Lavochkin:
http://www.laspace.ru/

Space Research Institute, Russian Academy of Sciences
http://www.iki.rssi.ru/

Mark Wade's Encyclopedia Astronautica:
http://www.astronautix.com/

References

Aamot, H. W. C. (1967). The philberth probe for investigating polar ice caps. CRREL Special Report 119, Cold Regions Research & Engineering Laboratory, Hanover, New Hampshire.

Adamski, D. F. (1962). The lunar seismograph experiment: ranger 3, 4, 5. NASA TR 32–272, Jet Propulsion Laboratory.

A'Hearn, M., Delamere, A. and Frazier, W. (2000). The deep impact mission: opening a new chapter in cometary science. Paper IAA-00-IAA.11.2.04 at the *51st International Astronautical Congress*, Rio de Janeiro, 2–6 October 2000.

Aleksashkin, S. N., Karyagin, V. P., Pichkhadze, K. M., Targamadze, R. Ch. and Terterashvili, A. V. (1988a). Aerodynamic characteristics of the aerostatic probe of Project VeGa. *Kosmich. Issled.* **26**(3), 434–440 (in Russian). Translation in *Cosmic Res.* **26**(3), 375–381.

Aleksashkin, S. N., Zukakishvili, R. I., Pichkhadze, K. M., Targamadze, R. Ch. and Terterashvili, A. V. (1988b). Dynamic characteristics of the sensor of the vertical component of wind speed. Aerostatic Experiment of Project VeGa. *Kosmich. Issled.* **26**(3), 441–447 (in Russian). Translation in *Cosmic Res.* **26**(3), 381–387.

Angrist, S. W. (1982). *Direct Energy Conversion*, 4th edn. Boston, Allyn and Bacon.

Arakaki, G. and D'Agostino, S. (1999). New millennium DS2 electronic packaging. An advanced electronic packaging. IEEE Aerospace Conference, Snowmass, Colorado, USA.

Atzei, A., Schwehm, G., Coradini, M., Hechler, M., De Lafontaine, J. and Eiden, M. (1989). Rosetta/CNSR–ESA's Planetary Cornerstone Mission. *ESA Bulletin*, **59**, 18–29.

Avduevsky, V. S., Marov, M. Ya., Rozhdestvensky, M. K., Borodin, N. F. and Kerzhanovich, V. V. (1971). Landing of the automatic Station Venera 7 on the Venus surface and preliminary results of investigations of the Venus atmosphere. *J. Atmos. Sci.* **28**(2), 263.

Avduevskii, V. S., Godnev, A. G., Zakharov, Yu. V. *et al* (1983). An estimate of the physical and mechanical characteristics of the soil of Venus from measurements of the impact overloads during the landings of the Venera 13 and Venera 14 automatic interplanetary stations. *Kosmich. Issled.* **21**(3), 331–339 (in Russian). Translation in *Cosmic Res.* **21**(3), 260–268.

Avotin, E. V., Bolkhovitinov, I. S., Kemurdzhian, A. L., Malenkov, M. I. and Shpak, F. P. (1979) *Динамика Планетохода (Dynamics of Planet Rovers)* (in Russian). Moscow, Nauka.

Backes, P. G., Tso, K. S., Norris, J. S. *et al.* (2000). Internet-based operations for the Mars polar lander mission. *Proc. 2000 IEEE International Conference on Robotics and Automation*, San Francisco, CA.

Barmin, I. V. and Shevchenko, A. A. (1983). Soil-scooping mechanism for the Venera 13 and Venera 14 unmanned interplanetary spacecraft. *Kosmich. Issled.* **21**(2), 171–175 (in Russian). Translation in: *Cosmic Res.* **21**(2), 118–122.

Barsukov, V. L. (ed.), (1978) *Peredvizhnaya Laboratoriya na Lune Lunokhod*-1. 2. Moscow, Nauka. 1978.

Barsukov, V. L. and Surkov, Yu. A. (eds.), (1979). *Grunt iz Materikovogo Raiona Luny (Lunar Highland Soil)*. Moscow, Nauka (in Russian).

Barsukov, V. L. (ed.), (1980). *Lunnyi Grunt iz Moria Krizisov (Lunar Soil from Mare Crisium)*. Moscow, Nauka (in Russian).

Bartlette, P. W., Carlson, L. E., Chu, P. C., Davis, K. R., Gorevan, S. P., Kusack, A. G., Myrick, T. M., and Wilson, J. J. and the Athena Science Team (2005). Summary of rock abrasion tool (RAT) results pertinent to the Mars exploration Rover Science data set. *Lunar and Planetary Science Conference*, XXXVI, abstract 2292.

Bauske, R. (2004). Dependence of the Beagle 2 trajectory on the Mars atmosphere. Presented at the *18th Int. Symposium on Space Flight Dynamics*, Munich.

Bazhenov, V. I. and Osin, M. I. (1978) *Posadka Kosmicheskikh Apparatov na Planety (The Landing of Spacecraft on the Planets, or Space-vehicle Landings on Planets)*. Moscow, Mashinostroenie.

Beattie, D. A. (2001). *Taking Science to the Moon: Lunar Experiments and the Apollo Program*. Baltimore, MD, Johns Hopkins University Press.

Bekker, M. G. (1962). Land locomotion on the surface of planets. *ARS Journal*, **32**(11), 1651–1659.

Bell, J. and Mitton, J. (eds.) (2002). *Asteroid Rendezvous*. Cambridge, Cambridge University Press.

Biele, J., Ulamec, S., Feuerbacher, B. *et al.* (2002). Current status and scientific capabilities of the Rosetta lander payload. *Adv. Space Res.* **29**(8), 1199–1208.

Biele, J. (2002). The experiments onboard the Rosetta lander. *Earth, Moon and Planets*, **90**(1–4), 445–458.

Biele, J. and Ulamec, S. (2004). Implications of the new target comet on science operations for the Rosetta lander. In Colangeli, L., Mazzotta Epifani, E. and Palumbo, P. (eds.), *The New Rosetta Targets: Observations, Simulations and Instrument Performances*. Astrophysics and Space Science Library vol. 311. Dordrecht, Kluwer, pp. 281–288.

Bienstock, B. J. (2004). Pioneer Venus and Galileo entry probe heritage. In Wilson, A. (ed.), *Proc. Int. Workshop on Planetary Probe Atmospheric Entry and Descent Trajectory Analysis and Science*, Lisbon, 6–9 October 2003. ESA SP-544, pp. 37–45.

Bird, M. K., Heyl, M., Allison, M. *et al.* (1997). The Huygens Doppler Wind Experiment. In *Huygens: Science, Payload and Mission*, pp. 139–163, A. Wilson (ed.) ESA SP-1177, Noordwijk.

Bird, M. K., Dutta-Roy, R., Heyl, M., Allison, M., Asmar, S. W., Folkner, W. M., Preston, R. A., Atkinson, D. H., Edenhofer, P., Plettemeier, D., Wohlmuth, R., Iess, L., Tyler, G. L. (2002). The Huygens Doppler wind experiment – Titan winds derived from probe radio frequency measurements. *Space Science Reviews*, **104**(1), 611–638.

Blagonravov, A. A. (ed.), (1968) *USSR Achievements in Space Research (First Decade in Space, 1957–1967)*, Moscow, Nauka. (In Russian). Translation available as JPRS-47311, (1969) Washington, Joint Publications Research Service.

Blamont, J., Boloh, L., Kerzhanovich, V. *et al.* (1993). Balloons on planet Venus: final results. *Adv. Space Res.* **13**(2), 145–152.

Bogdanov, A. V., Nikolaev, A. V., Serbin, V. I., Skuridin, G. A., Khavroshkin, O. B. and Tsyplakov, V. V. (1988). Method for analysing terrestrial planets. *Kosmich. Issled.* **26**(4), 591–603 (in Russian). Translation in *Cosmic Res.* **26**(4), 505–515.

Bonitz, R., Slostad, J., Bon, B. *et al.* (2001). Mars volatiles and climate surveyor robotic arm. *J. Geophys. Res.* **106**(E8), 623–640.

Bonnefoy, R., Link, D., Casani, J. *et al.* (2004). *Beagle 2 ESA/UK Commission of Enquiry.*

Boynton, W. V. and Reinert, R. P. (1995). The cryo-penetrator: an approach to exploration of icy bodies in the solar system. *Acta Astronautica* **35**(suppl.), 59–68.

Braun, R. D., Spencer, D. A., Kallemeyn, P. H. and Vaughan, R. M. (1999a). Mars Pathfinder atmospheric entry navigation operations. *J. Spacecraft and Rockets* **36**(3), 348–356.

Braun, R. D., Micheltree, R. A. and Cheatwood, F. M. (1999b). Mars microprobe entry-to-impact analysis. *J. Spacecraft and Rockets* **36**(3), 412–420.

Brodsky, P. N., Gromov, V. V., Yudkin, E. N., Kulakova, I. B., Kuzmin, M. M., (1995). Deepening method of the device for borehole creation in soil. Patent no. 2 04 98 53. *Bulletin of the Russian Federation Committee on Patents and Trademarks*, N34.

Brodsky, R. F. (1979) *Pioneer Venus: Case Study in Spacecraft Design.* New York, AIAA Professional Study Series.

Bruch, C. W. (1964). Some biological and physical factors in dry-heat sterilization: a general review. In Floriskin, M. and Dollfus, A. (eds.) *Life Sciences and Space Research II.* Amsterdam, New Holland.

Burgess, E. (1978). *To The Red Planet.* New York, Columbia University Press.

Buslaev, S. P. (1987). Predicting the successful landing of an automatic interplanetary station on the surface of a celestial body in the presence of uncertainty. *Kosmich. Issled.* **25**(2), 186–192 (in Russian). Translation in *Cosmic Res.* **25**(2), 149–154.

Buslaev, S. P., Stulov, V. A. and Grigor'ev, E. I., (1983) Mathematical modeling and experimental investigation of the landing of the Venera 9–14 spacecraft on deformable soils. *Kosmich. Issled.* **21**(4), 540–544. (in Russian). Translation in *Cosmic Res.* **21**(4), 439–442, 1983.

Cadogan, D., Sandy, C. and Grahne, M. (2002). Development and evaluation of the Mars Pathfinder inflatable airbag landing system. *Acta Astronautica*, **50**(10), 633–640.

Carrier III, W. D., Olhoeft, G. and Mendell, W. (1991). Physical properties of the lunar surface: Section 9.1.11 – Trafficability. In G. Heiken, D. Vaniman, B. French (eds.). *Lunar Sourcebook – A User's Guide to the Moon.* Cambridge, Cambridge University Press.

Casani, J. *et al.* (JPL Special Review Board) (2000). *Report on the loss of the Mars polar lander and Deep Space 2 missions.* JPL D-18709.

Cheremukhina, Z. P., *et al.* (1974). Estimate of temperature of Venus' stratosphere from data on deceleration forces acting on the Venera 8 probe. *Kosmich. Issled.* **12**(2), 264–271 (in Russian). Translation in: *Cosmic Res.* **12**(2), 238–245, 1974.

Cherkasov, I. I., Kemurdzhian, A. L., Mikhailov, L. N. *et al.* (1967). Determination of the density and mechanical strength of the surface layer of lunar soil at the landing site of the Luna 13 Probe. *Kosmich. Issled.* **5**(5), 746–757 (in Russian). Translation in *Cosmic Res.* **4**, 636–645, 1968a.

Cherkasov, I. I., Gromov, V. V., Zobachev, N. M. *et al.* (1968a). Soil-density meter-penetrometer of the automatic lunar station Luna-13. *Doklady Akademii Nauk SSSR*, **179**(4), 829–831 (in Russian). Translation in *Soviet Physics–Doklady* **13**(4), 336–338.

Cherkasov, I. I., Vakhnin, V. M., Kemurjian, A. L. *et al.* (1968b). Determination of the Physical and Mechanical Properties of the Lunar Surface Layer by Means of Luna 13 Automatic Station. *Moon and Planets 2* (ed. A. Dollfus). Amsterdam, North-Holland 70–76.

Chertok, B. (1999). *Rockets and People*. Moscow, Mashinostroenie.

Christensen, P. R., Mehall G. L., Silverman S. H., *et al.* (2003). Miniature thermal emission spectrometer for the Mars Exploration rovers. *Journal of Geophysical Research*, **108**(E12), 8064, DOI 10.1029/2003JE002117

CNES (1993). *Missions, Technologies and Design of Planetary Mobile Vehicles. Proceedings of the CNES Conference*, Toulouse, September 1992. Cépaduès Editions.

Colombatti, G. *et al.* (2006). Reconstruction of the trajectory of the Huygens probe using the Huygens Atmospheric Structure Instrument (HASI). *Planet. Space Sci.*, submitted.

Cooley, C. G. and Lewis, J. G. (1977). Viking 75 project: Viking lander system primary mission performance report. NASA Contractor Report CR-145148, NASA/Martin Marietta.

Corliss, W. R. (1965). *Space Probes and Planetary Exploration*. Princeton, Van Nostrand.

Corliss, W. R. (1975). The Viking mission To Mars. NASA SP-334.

Cortright, E. M. (ed.) (1975). Apollo expeditions to the Moon. NASA SP-350.

Cowart, E. G. (1973). Lunar Roving Vehicle: Spacecraft on Wheels. *Proc. Inst. Mech. Engrs.* **187**(45/73), 463–481

DeVincenzi D. L. and Stabekis P. D., (1984). Revised planetary protection policy for solar system exploration. *Adv. Space Res.* **4**(12), 291–295.

Debus A., Runavot J., Rogovsky G., Bogomolov V., Khamidullina N., and Trofimov V., (2002). Landers sterile integration implementations: example of Mars 96 mission. *Acta Astronautica*, **50**(6), 385–392.

Desai, P. N., and Lyons, D. T. (2005). Entry, descent, and landing operations analysis for the Genesis re-entry capsule. 15th AAS/AIAA Space Flight Mechanics Conference, paper AAS 05–121.

Doenecke, J. and Elsner, M. (1994). Special heat transfer problems within the Huygens probe. *Proceedings, 4th European Symposium on Space Environmental Control Systems*, 279–283.

Doiron, H. H. and Zupp, G. A. (2000). Apollo Lunar Module Landing Dynamics. *4th AIAA/ASME/ASCE/AHS/ASC Structures, Structural Dynamics, and Materials Conference and Exhibit*, Atlanta GA, 3–6 April 2000. AIAA-2000–1678.

Dornheim, M. (2003), 'Can $$$ buy time?'. *Aviation Week and Space Technology*, **158** (21), 56–58.

Dunham, D. W., Farquhar, R. W., McAdams, J. V. *et al.* (2002). Implementation of the First Asteroid Landing. *Icarus*, **159**(2), 433–438.

Eisen, H. J., Wen L. C., Hickey, G. and Braun, D. F. (1998). Sojourner Mars rover thermal performance, SAE paper 981685, 28th International Conference on Environmental Systems, Danvers, MA, July, 1998.

Ellery, A. (2000). *An Introduction to Space Robotics*. Chichester, Springer-Praxis.

European Space Agency (1986). *Comet nucleus sample return. Proceedings of an ESA Workshop held at the University of Kent at Canterbury, UK*, 15–17 July, 1986. ESA SP-249, December 1986.

European Space Agency (1987). Rosetta comet nucleus sample return: report of the science definition team. ESA SCI(87)3, December 1987.

European Space Agency (1988). Vesta: a mission to the small bodies of the solar system. ESA SCI(88)6.

European Space Agency (1991). Rosetta comet-nucleus sample return: Mission and system definition document. ESA SP-1125, June 1991.

European Space Agency (1993). *Rosetta Comet Rendezvous Mission*, ESA SCI(93)7, September 1993.

European Space Agency (2000). *BepiColombo: An Interdisciplinary Cornerstone Mission to the Planet Mercury.* ESA SCI(2000)1.

Ezell, E. C. and Ezell, L. N., (1984). On Mars: Exploration of the Red Planet 1958–1978, NASA SP-4212.

Fearn, D. G. and A. R. Martin, (1995). The promise of electric propulsion for low-cost interplanetary missions. *Acta Astronautica*, **35**, 615–624.

Fimmel *et al.*, (1983). Pioneer Venus. NASA SP-461.

Fimmel *et al.*, (1995). Pioneering Venus. NASA SP-518.

Forrestal, M. J. and Luk, V. K. (1992). Penetration into soil targets. *Int. J. Impact Engng.* **12**, 427–444.

Fraser S. J., Olson R. L., Leavens W. M., (1975). Plasma sterilization technology for spacecraft applications. Seattle, WA, Boeing Co., Aerospace Group.

Galimov, E. M., Kulikov, S. D., Kremnev, R. S., Surkov, Yu. A. and Khavroshkin, O. B. (1999). The Russian lunar exploration project. *Astronomich. Vestnik.* **33**(5), 374–385 (in Russian). Translation in: *Solar System Res.* **33**(5), 327–337.

Goldblinth, S. A., (1971) The inhibition and destruction of the microbial cell by radiations. In *Inhibition and Destruction of the Microbial Cell*, W. B. Hugo (ed.) San Diego, Academic Press.

Goldstein, D. B., Austin, J. V., Barker, E. S. and Nerem, R. S. (2001). Short-time exosphere evolution following an impulsive vapor release on the Moon. *J. Geophys. Res.* **106**(E12), 32841–32845.

Gorevan, S. P., Myrick, T., Davis, K. *et al.*, (2003). Rock abrasion tool: Mars exploration rover mission. *J. Geophys. Res.* **108**(E12), 8068.

Grafov, V. E., Bulekov, V. P., Dryuchenko, D. D. *et al.*, (1971). First experimental boring on the Moon. *Kosmich. Issled.* **9**(4), 580–586 (in Russian). Translation in *Cosmic Res.* **9**(4), 530–535.

Green, M. J. and Davy, W. C. (1981). Galileo Probe Forebody Thermal Protection. *AIAA-81–1073, AIAA 16th Thermophysics Conference*, Palo Alto, CA, June 23–25, 1981.

Grigor'ev, E. I. and Ermakov, S. N., (1983). Physical modeling of the Venera 9 and Venera 14 landing probes. *Kosmich. Issled.* **21**(4), 536–539 (in Russian). Translation in *Cosmic Res.* **21**(4), 435–438, 1983.

Gromov, V. V., Misckevich, A. V., Yudkin, E. N., Kochan, H., Coste, P., and Re, E., (1997). The mobile penetrometer, a"Mole" for sub-surface soil investigation. *Proc. 7th European Space Mechanisms & Tribology Symposium*. ESTEC, Noordwijk, The Netherlands, 1–3 October 1997. ESA SP-410, pp. 151–156.

Hall, J. L., MacNeal, P. D., Salama, M. A., Jones, J. A. and Heun, M. K. (1999). Thermal and structural test results for a Venus deep-atmosphere instrument enclosure. *Journal of Spacecraft and Rockets*, **37**,(1), 142–144.

Hall, R. C., (1977). Lunar impact – a history of project ranger. NASA SP-4210.

Hanson, A. W., (1978). Antenna design for Pioneer venus probes. *IEEE International Symposium on Antennas and Propagation*, Washington DC, May 1978.

Harland, D. M. (2000). *Jupiter Odyssey*. Chichester, Springer-Praxis.

Harland, D. M. (2002). *Mission To Saturn*. Chichester, Springer-Praxis.

Harland, D. M. and Lorenz, R. D. (2005). *Space Systems Failures*. Chichester, Springer-Praxis.

Hashimoto, T., Kubota, T. and Mizuno, T. (2003). Light weight sensors for the autonomous asteroid landing of MUSES-C mission. *Acta Astronautica*, **52**(2–6), 381–388.

Hassan, H. and J. C. Jones, The Huygens probe. ESA Bulletin 92, November 1997.

Heiken, G. H., Vaniman, D. T. and French, B. M. (eds), (1991). *Lunar Sourcebook – A User's Guide to The Moon*. Cambridge, Cambridge University Press.

Hennis, L. A. and Varon, M. N. (1978). Thermal design and development of the pioneer Venus large probe. In: *Thermophysics and Thermal Control* (R. Visjanta, ed.) Vol.65 of *Progress in Astronautics and Aeronautics*, AIAA (Presented as Paper 78–916 at the *2nd AIAA/ASME Thermophysics and Heat Transfer Conference*, Palo Alto, California, May 24–26, 1978).

Hilchenbach, M., Küchemann, O. and Rosenbauer, H. (2000). Impact on a comet: Rosetta lander simulations. *Planet. Space Sci.* **48**(5), 361–369.

Hilchenbach, M., Rosenbauer, H. and Chares, B. (2004). First contact with a comet surface: Rosetta lander simulations. In: Colangeli, L., Mazzotta Epifani, E. and Palumbo, P. (eds), *The New Rosetta Targets: Observations, Simulations and Instrument Performances*. Astrophysics and Space Science Library vol. 311. Dordrecht, Kluwer, pp. 289–296.

Hirano, Y. and Miura, K. (1970). Water impact accelerations of axially symmetric bodies. *J. Spacecraft and Rockets* **7**, 762–764.

Holmberg, N. A., Faust, R. P. and Holt, H. M. (1980). *Viking* 75 spacecraft design and test summary. NASA Reference Publication RP-1027, NASA Langley Research Center.

Hope, A. S., Kaufman, B., Dasenbrock, R. and Bakeris, D. (1997). A Clementine II mission to the asteroids. In: Wytrzyszczak, I. M., Lieske, J. H. and Feldman, R. A. (eds.), *Dynamics and Astrometry of Natural and Artificial Celestial Bodies. Proc. IAU Colloquium 165*, Dordrecht, Kluwer, pp. 183–190.

Horneck, G., (1993). Responses of *Bacillus subtilis* spores to the space environment: results from experiments in space. *Origins Life Evol. Biosph.*, **23**, 37–52.

Hunten *et al.* (eds), (1983). *Venus*. Tueson, University of Arizona Press.

Hunten, D. M., Colin, L. and Hansen, J. E. (1986). Atmospheric science on the Galileo mission. *Space Sci. Rev.* **44**, 191–240.

Ivanov, N. M. (1977). *Upravlenie Dvizheniem Kosmicheskogo Apparata v Atmosfere Marsa* (in Russian). Moscow, Nauka.

Jankovsky, R. S., Jacobson, D. T., Pinero, L. R., Sarmiento C. J., Manzella, D. H., Hofer, R. R. and Peterson, P. Y. (2002). NASA's Hall Thruster Program 2002. Paper AIAA-2002–3675 at the *38th AIAA Joint Propulsion Conference*, Indianapolis, 7–10 July 2002.

Jastrow, R. and Rasool, S. I. (eds.) (1969). *The Venus Atmosphere*. New York, Gordon and Breach.

Johnson, N. L. (1979). *Handbook of Soviet Lunar and Planetary Exploration*. American Astronautical Society Science and Technology Series, vol. 47. San Dieg., Univelt.

Johnson, N. L. (1995). *The Soviet Reach for the Moon: The L-1 and L-3 Manned Lunar Programs and the Story of the N-1 "Moon Rocket"*. 2nd edn. Huntsville, Cosmos Books.

Jones, J. C. and Giovagnoli, F. (1997). The Huygens probe system Design. In: Wilson, A. (ed.), *Huygens Science, Payload and Mission*. ESA SP-1177. European Space Agency.

Jones, R. H., (1971). Lunar surface mechanical properties from surveyor data. *J. Geophys. Res.* **76**(32), 7833–7843.

Jones, R. M. (2000). The MUSES–CN rover and asteroid exploration mission. In: Arakawa, Y. (ed.), *Proc. 22nd International Symposium on Space Technology and Science*, Morioka, Japan, 28 May–4 June 2000. pp. 2403–2410.

Kawaguchi, J., Uesugi, K. and Fujiwara, A. (2003). The MUSES-C mission for the sample and return – its technology development status and readiness. *Acta Astronautica*, **52**(2–6), 117–123.

Keating, G. M. and the rest of the MGS Aero-braking Team (1998). The structure of the upper atmosphere of Mars: in-situ accelerometer measurements from Mars Global Surveyor. *Science*, **279**, 1672–1676.

Keldysh, M. V. (ed.) (1979). *Pervye Panoramy Poverkhnosti Venery* (in Russian). Moscow, Nauka.

Keldysh, M. V. (ed.) (1980). *Tvorcheskoye naslediye Akademika Sergeya Pavlovicha Koroleva: izbrannyye trudy i dokumenty*. Moscow, Nauka.

Kelley, T. J. (2001). *Moon Lander: How We Developed the Apollo Lunar Module*. Washington DC, Smithsonian.

Kemurdzhian, A. L. (ed.) (1986). *Передвижение по Грунтам Луны и Планет (Transport on Lunar and Planetary Soils)* (in Russian). Moscow, Mashinostroenie.

Kemurdzhian, A. L., Bogomolov, A. F., Brodskii, P. N. *et al.* (1988). Study of Phobos' surface with a movable robot. In: *Phobos – Scientific and Methodological Aspects of the Phobos Study. Proceedings of the International Workshop*, Moscow, 24–28 November 1986. Space Research Institute, USSR Academy of Sciences. pp. 357–367.

Kemurdzhian, A. L., Brodskii, P. N., Gromov, V. V. *et al.* (1989a). A roving vehicle for studying the surface of Phobos (PROP). In Balebanov, V. M. (ed.), *Instrumentation and Methods for Space Exploration* (in Russian). Moscow, Nauka.

Kemurdzhian, A. L., Brodskii, P. N., Gromov, V. V. *et al.* (1989b). Instruments for measuring the physical and mechanical properties of soil, evaluating its electroconductivity, and determining the inclination of angles of the PROP roving vehicle in the framework of the "Phobos" project. In Balebanov, V. M. (ed.), *Instrumentation and Methods for Space Exploration* (in Russian). Moscow, Nauka.

Kemurdzhian, A. L., Gromov, V. V., Kazhukalo, I. F. *et al.* (1993). *Planetokhody (Planet Rovers)* (in Russian). 2nd edn., Moscow, Mashinostroenie.

Kerr, R. A. (2002). Safety versus science on next trips to Mars. *Science*, **296**, 1006–1008.

Kerzhanovich, V. V. (1977). Mars 6: improved analysis of the descent module measurements. *Icarus*, **30**, 1–25.

Kieffer *et al.* (eds.) (1993). *Mars*. Tucson, University of Arizona Press.

Klingelhöfer, G., Morris, R. V., de Souza, P. A., Jr., Bernhardt, B., and the Athena Science Team (2003). The miniaturized Mössbauer spectrometer MIMOS II of the Athena payload for the 2003 MER missions. Sixth International Conference on Mars, abstract 3132.

Knacke, T., (1992). *Parachute Recovery Systems Design Guide*. Santa Barbara, CA, Para Publishing (originally published as NWC TP 6575 by the Naval Weapons Center, China Lake); see also H. W. Bixby, E. G. Ewing and T. W. Knacke. *Recovery Systems Design Guide*. USAF, December 1978. (USAF Report AFFDL-TR-78–151.)

Koon, W. S., Lo, M. W., Marsden, J. E. and Ross, S. D. (2000). Heteroclinic connections between periodic orbits and resonance transitions in celestial mechanics. *Chaos*, **10**(2), 427–469.

Koppes, C. R. (1982). *JPL and the American Space Program*. New Haven CT, Yale University Press.

Kremnev, R. S., Selivanov, A. S., Linkin, V. M. *et al.* (1986). The VeGa balloons: a tool for studying atmosphere dynamics on Venus. *Pis'ma Astronom. Zh.* **12**(1), 19–24, (in Russian). Translation in: *Sov. Astronom. Lett.* **12**(1), 7–9.

Kubota, T., Hashimoto, T., Sawai, S., *et al.* (2003). An autonomous navigation and guidance system for MUSES-C asteroid landing. *Acta Astronautica*, **52**(2–6), 125–131.

Kurt, V. G. (1994), *Per aspera . . .* to the planets. *Space Bulletin* **1**(4), 23–25.

Landis, G. A., Kerslake, T. W., Jenkins, P. P. and Scheiman, D. A. (2004). Mars solar power, AIAA-2004–5555 (NASA TM-2004–213367).

Lanzerotti, L. J., Rinnert, K., Carlock, D., Sobeck, C. K. and Dehmel, G. (1998). Spin rate of Galileo probe during descent into the atmosphere of Jupiter. *Journal of Spacecraft and Rockets*, **35**(1), 100–102.

Latham, G. V., Ewing, M., Dorman, J. *et al.* (1970). Seismic data from man-made impacts on the Moon. *Science*, **170**(3958), 620–626.

Latham, G. V., Dorman, H. J., Horvath, P., Ibrahim, A. K., Koyama, J. and Nakamura, Y. (1978). Passive seismic experiment: a summary of current status. *Proc. 9th Lunar Planet. Sci. Conf.*, Houston, pp. 3609–3613.

Le Croissette, D. H., (1969). The scientific instruments on surveyor. *IEEE Trans. Aerospace and Electronic Systems*, **5**(1), 2–21.

Lebreton, J. -P., Witasse, O., Sollazzo, C. *et al.* (2005). An overview of the descent and landing of the Huygens probe on Titan. *Nature*, **438**(7069), 758–764.

Lei X., Zhang R., Peng L., Li-Li D. and Ru-Juan Z., (2004). Sterilization of *E. coli* bacterium with an atmospheric pressure surface barrier discharge. *Chinese Phys.* **13**(6), 913–917.

Linkin, V. A. Harri, -M., Lipatov, A., *et al.* (1998). A sophisticated lander for scientific exploration of Mars: scientific objectives and implementation of the Mars-96 small station. *Planet. Space Sci.* **46**(6/7), 717–737.

Lorenz, R. D. (1994). Huygens probe impact dynamics. *ESA Journal* **18**, 93–117.

Lorenz, R. D. (2001). Scaling laws for flight power of airships, airplanes and helicopters: application to planetary exploration. *Journal of Aircraft*, **38**, 208–214.

Lorenz, R. D. (2002). An artificial meteor on Titan? *Astronomy and Geophysics*, **43**(5), 14–17.

Lorenz, R. D. (2006). *Spinning Flight: Dynamics of Frisbees, Boomerangs, Samaras and Skipping Stones*. New York, Springer.

Lorenz, R. D., Moersch, J. E., Stone, J. A., Morgan, R. and Smrekar, S. (2000). Penetration tests on the DS-2 Mars microprobes: penetration depth and impact accelerometry. *Planet. Space Sci.* **48**, 419–436.

Lorenz, R. D. and Ball, A. J. (2001). Review of impact penetration tests and theories. In Kömle, N. I., Kargl, G., Ball, A. J., Lorenz, R. D. (eds.), *Penetrometry in the Solar System*. Vienna, Austrian Academy of Sciences Press.

Lorenz, R. D. and Mitton, J. (2002). *Lifting Titan's Veil: Exploring the Giant Moon of Saturn*. Cambridge, Cambridge University Press.

Lorenz, R. D., Bienstock, B., Couzin, P., Cluzet, G. (2005). Thermal design and performance of probes in thick atmospheres: experience of Pioneer Venus, Venera, Galileo and Huygens. Submitted to *3rd International Planetary Probe Workshop*, Athens, Greece, June 2005.

Lorenz, R. D., Witasse, O., Lebreton, J. -P. *et al.* (2006). Huygens entry emission: observation campaign, results, and lessons learned. *J. Geophys. Res.* **III** (E7) E07S11. DOI 10.1029/2005JE002603.

Maksimov, G. Yu., Construction and testing of the first Soviet automatic interplanetary stations. In Hunley, J. D. (ed.) (1997). *History of Rocketry and Astronautics, AAS History Series*, vol. 20, pp. 233–246. American Astronautical Society.

Markov, Yu. (1989). *Kurs na Mars*, Moscow, (in Russian). Mashinostroenie.

Marov, M. Ya. and Petrov, G. I. (1973). Investigations of Mars from the soviet automatic stations Mars 2 and 3. *Icarus*, **19**, 163–179.

Marov, M. Ya. and Grinspoon, D. H. (1998). *The Planet Venus*. New Haven CT, Yale University Press.

Marov, M. Ya., Avduevsky, V. S., Akim, E. L. *et al.* (2004). Phobos-Grunt: Russian sample return mission. *Adv. Space Res.* **33**(12), 2276–2280.

Marraffa L. and Smith, A. (1998). Aerothermodynamic aspects of entry probe heat shield design. *Astrophys. and Space Sci.* **260**, 45–62.

Martin Marietta Corporation, (1976). *Viking Lander "As Built" Performance Capabilities*. Martin Marietta Corporation.

McCurdy, H. E. (2001). *Faster Better Cheaper, Low-Cost Innovation in the U.S. Space Program*. Baltimore MA, Johns Hopkins University Press.

McGehee R., Hathaway M. E. and Vaughan V. L., Jr. (1959). Water-landing characteristics of a reentry capsule. NASA Memorandum 5–23–59L.

Meissinger, H. F. and Greenstadt, E. W. (1971). Design and science instrumentation of an unmanned vehicle for sample return from the asteroid Eros. In Gehrels, T. (ed.), (1971). *Physical Studies of Minor Planets, Proceedings of IAU Colloq.* 12, Tucson, AZ, March, 1971. NASA SP 267, p. 543.

Micheltree, R. A., DiFulvio, M., Horvath, T. J. and Braun, R. D. (1998). Aerothermal heating predictions for Mars microprobe. AIAA 98–0170, 36th Aerospace Sciences Meeting, January 12–15, 1998, Reno NV.

Milos, F. S. (1997). Galileo probe heat shield ablation experiment. *J. of Spacecraft and Rockets*, **34**(6), 705–713.

Milos, F. S., Chen, Y.-K., Squire, T. H. and Brewer, R. A. (1999a). Analysis of Galileo heatshield ablation and temperature data. *J. of Spacecraft and Rockets*, **36**(3), 298–306.

Milos, F. S., Chen, Y.-K., Congdon, W. M. and Thornton, J. M. (1999b). Mars pathfinder entry temperature data, aerothermal heating and heatshield material response. *J. of Spacecraft and Rockets*, **36**(3), 380–391.

Mishkin, A. (2004). *Sojourner: An Insider's View of the Mars Pathfinder Mission*. New York, Berkley Publishing Group.

Mizutani, H., Fujimura, A., Hayakawa, M., Tanaka, S. and Shiraishi, H. (2001). Lunar-A penetrator: its science and instruments. In Kömle, N. I., Kargl, G., Ball, A. J. and Lorenz, R. D. (eds.), *Penetrometry in the Solar System*. Vienna. Austrian Academy of Sciences Press. pp. 125–136

Mizutani, H., Fujimura, A., Tanaka, S., Shiraishi, H. and Nakajima, T. (2003). Lunar-A mission: goals and status. *Adv. Space Res.* **31**(11), 2315–2321.

MNTK (International Scientific and Technical Committee) (1985). *Venus–Halley Mission: Experiment Description and Scientific Objectives of the International Project VEGA (1984–1986)*. MNTK, 1985.

Mogul, R., Bol'shakov, A. A., Chan, S. L., Stevens, R. M., Khare, B. N., Meyyappan, M. and Trent, J. D. (2003). Impact of low-temperature plasmas on deinococcus radiodurans and biomolecules, *Biotechnol. Prog.* **19**,776–783.

Moore, H. J., Hutton, R. E., Scott, R. F., Spitzer, C. R. and Shorthill, R. W. (1977). Surface materials of the Viking landing sites. *J. Geophys. Res.* **82**,4497–4523.

Moore, H. J., Bickler, D. B., Crisp, J. A. *et al* (1999). Soil-like deposits observed by Sojourner, the pathfinder rover. *J. Geophys. Res.* **104**(E4), 8729–8746.

Morozov, A. A., Smorodinov, M. I., Shvarev, V. V. and Cherkasov, I. I., (1968). Measurement of the lunar surface density by the automatic station "Luna-13". *Doklady Akademii Nauk SSSR*, **179**(5), 1087–1090, (in Russian). Translation in *Soviet Physics – Doklady* **13**(4), 348–350, 1968.

Mosher, T. J. and Lucey, P. (2006). Polar Night: a lunar volatiles expedition. *Acta Astronautica*, **59**(8–11), 585–592.

Murphy, J. P., Reynolds, R. T., Blanchard, M. B. and Clanton, U. S. (1981a). Surface Penetrators for planetary exploration: science rationale and development program. NASA TM-81251, Ames Research Center.

Murphy, J. P., Cuzzi, J. N., Butts, A. J. and Carroll, P. C. (1981b). Entry and landing probe for Titan. *J. Spacecraft and Rockets,* **8**, 157–163.

Murrow, H. N. and McFall, J. C., (1968). Summary of Experimental results obtained from the NASA planetary entry parachute program, AIAA 68–934, *AIAA 2nd Aerodynamic Deceleration Systems Conference*, El Centro, CA, September 1968.

Mutch, T. (ed.) (1978). The Martian landscape. NASA SP-425.

NASA (1962). Scientific experiments for Ranger 3, 4, and 5. NASA Technical Report TN 32–199 (Revised), Jet Propulsion Laboratory.

NASA (1963). Lunar rough landing capsule development program final technical report. NASA Contractor Report CR-53814, Newport Beach, CA, Aeronutronic Division, Ford Motor Company.

NASA (1968). Surveyor project final report, parts 1 and 2. NASA Technical Report TR 32–1265, Jet Propulsion Laboratory.

NASA (1969). Surveyor program results. NASA SP-184.

Neal, M. F. and Wellings, P. J., (1993). Descent control system for the Huygens probe, *12th RAeS/AIAA Aerodynamic Decelerator Systems Technology Conference*, London, May 10–13, 1993 (AIAA 93–1221).

Neugebauer, M. and Bibring, J. -P. (1998). Champollion. *Adv. Space Res.* **21**(11), 1567–1575.

Nicholson, W. L., Munakata N., Horneck G., Melosh H. J., Setlow P. (2000). Resistance of *Bacillus* endospores to extreme terrestrial and extraterrestrial environments. *Microbiol. Mol. Bio Rev.* **64**(3), 548–572.

Northey, D. (2003). The main parachute for the Beagle 2 Mars Lander. 17th AIAA Aerodynamic Decelerator Systems Technology Conference and Seminar, Monterey, California, May 19–22, 2003.

Novak, K. S., Phillips, Sunada, J. and Kinsella, G. M. (2005). Mars exploration rover surface mission flight thermal performance, SAE 2005–01–2827 International Conference on Environmental Systems, July 2005, Rome, Italy.

O'Neill, W. J. (2002). Galileo spacecraft architecture, in *'The Three Galileos'. Proceedings of a Conference*, Padova, Italy. Dordrecht, Kluwer.

Parks, R. J. (1966). *Surveyor 1 Mission Report.* Part 1: Mission description and performance. Technical Report No. 32–1023, Jet Propulsion Laboratory, Pasadena, CA, August 31, 1966.

Pellinen, R. and Raudsepp, P. (eds.) (2000). *Towards Mars!* Helsinki, Oy Raud.

Perminov, V. G., (1990). Dynamics of soft landing of spherically-shaped probes. *Kosmich. Issled.* **28**(4), 539–544, (in Russian). Translation in *Cosmic Res.* **28**(4), 460–465. 1990.

Perminov, V. G. (1999). *The Difficult Road to Mars: A Brief History of Mars Exploration in the Soviet Union*. Monographs in Aerospace History, Number 15. NASA.

Philberth, K. (1962). Une Méthode pour mesurer les températures à l'intérieur d'un Inlandsis (a method for measuring temperatures within an ice sheet). *Comptes Rendues*, **254**, 3881.

Pichkhadze, K., Vorontsov, V., Rogovsky, G. and Pellinen, R. (2002). Technical proposal on forming the expedition with landing onto the surface of Mercury. *Geophys. Res. Abstracts* **4**, EGSO2-A-06764.

Pillinger, C. T., (2003). *Beagle: From Sailing Ship to Mars Spacecraft.* XNP Productions. Republished as *Beagle: From Darwin's Epic Voyage to the British Mission To Mars.* Faber & Faber, 2003.

Pillinger, C. T., Sims, M. R., Clemmet, J. *The Guide To Beagle 2.* copyright, C. T. Pillinger, 2003, in association with Faber and Faber.

Pioneer Venus (1980). Reprinted from *J. Geophys. Res.* **85**(A13).

Pullan, D., Sims, M. R., Wright, I. P., Pillinger, C. T. and Trautner, R. (2004). Beagle 2: the exobiological lander of Mars express. In Wilson, A. (ed.), *Mars Express: The Scientific Payload.* ESA SP-1240. ESA, Noordwijk.

Richter, L. (1998). Principles for robotic mobility on minor solar system bodies. *Robot. & auton. Sys.* **23**(1/2), 117–124.

Richter, L., Coste, P., Gromov, V., Kochan, H., Pinna, S. and Richter, H.-E. (2001). Development of the "planetary underground tool" subsurface soil sampler for the Mars express "Beagle 2" lander. *Adv. Space Res.* **28**(8), 1225–1230.

Riemensnider, D. K., (1968). Quantitative aspects of shedding of micro-organisms by humans. NASA SP108, pp. 97–103.

Rodier, R. W., Thuss, R. C. and Terhune, J. E. Parachute design for the Galileo entry probe, (1981). AIAA–81–1951, AIAA 7th Aerodynamic Decelerator and Balloon Technology Conference, October 21–23, San Diego, CA, 1981.

Rohatgi, N., Schubert, W., Knight, J., *et al.*, (2001). Development of vapor phase hydrogen peroxide sterilization process for spacecraft applications. *International Conference On Environmental Systems, Soc. of Automotive Eng.*, paper 2001–01–2411.

Rummel, J. D. (2001). Planetary exploration in the time of astrobiology: protecting against biological contamination. *Proc. Natl. Acad. Sci.*, **98**(5), 2128–2131.

Russell, C. T. (ed.) (1992). *The Galileo Mission.* Reprinted from *Space Sci. Rev.* **60**(1–4). Dordrecht, Kluwer.

Russell, C. T. (ed.) (1998). *The Near Earth Asteroid Rendezvous Mission.* Reprinted from *Space Sci. Rev.* **82**(1–2), 1997. Kluwer.

Sagdeev, R. Z., Linkin, V. M., Kremnev, R. S., Blamont, J. E., Preston, R. A. and Selivanov, A. S., (1986). The VeGa balloon experiments. *Pis'ma Astronom. Zh.* **12** (1), 10–15, (in Russian). Translation in: *Sov. Astronom. Lett.* **12**(1), 3–5, 1986.

Sagdeev, R. Z., Balebanov, V. M. and Zakharov, A. V. (1988). The Phobos project: scientific objectives and experimental methods. *Sov. Sci. Rev. E: Astrophys. Space Phys. Rev.* **6**, 1–60.

Sainct, H. and Clausen, K., (1983). Technologies new to space in Huygens probe mission to Titan. IAF-93-U.4.564, Presented at *44th IAF Congress*, Graz, Austria, October 1993.

Scheeres, D. J. (2004). Close proximity operations at small bodies: orbiting, hovering, and hopping. In Belton, M. J. S., Morgan, T. H., Samarasinha, N. and Yeomans, D. K. (eds.), (2002). *Mitigation of Hazardous Comets and Asteroids. Proceedings of the Workshop on Scientific Requirements for Mitigation of Hazardous Comets and Asteroids*, Arlington, 3–6 September 2002. Cambridge, Cambridge University Press, pp. 313–336.

Schmidt, G. R., Wiley, R. L., Richardson, R. L. and Furlong, R. R. (2005). NASA's program for radioisotope power system research and development. *AIP Conference Proceedings*, **746**, 429–436.

Schurmeier, H. M., Heacock, R. L. and Wolfe, A. E. (1965). The Ranger missions to the Moon. *Scientific American*, **214**(1), 52–67.

Schwehm, G. and Hechler, M. (1994). 'Rosetta'- ESA's planetary cornerstone mission. *ESA Bulletin*, **77**, 7–18.

Scoon, G. E., (1985). Cassini – a concept for a Titan probe. *ESA Bulletin*, **41**, 12–20.

Sears, D., Franzen, M., Moore, S., Nichols, S., Kareev, M. and Benoit, P. (2004). Mission operations in low-gravity regolith and dust. In Belton, M. J. S., Morgan, T. H., Samarasinha, N. and Yeomans, D. K. (Eds.), *Mitigation of hazardous comets and asteroids. Proceedings of the Workshop on Scientific Requirements for Mitigation of Hazardous Comets and Asteroids*. Arlington, 3–6 September 2002. Cambridge, Cambridge University Press, pp. 337–352.

Seddon, C. M. and Moatamedi, M. (2006), Review of water entry with applications to aerospace structures. *Int. J. Impact Eng.* **32**(7), 1045–1067.

Seiff, A. and Kirk, D. B. (1977). Structure of the atmosphere of Mars in summer at mid-latitudes. *J. Geophys. Res.* **82**, 4363–4378.

Seiff, A., *et al.* (1980). Measurements of thermal structure and thermal contrast in the atmosphere of Venus and related dynamical observations: results from the four Pioneer Venus probes. *J. Geophys. Res.* **85**, 7903–7933.

Seiff, A., *et al.* (1997). The atmosphere structure and meteorological instrument on the Mars Pathfinder lander. *J. Geophys. Res.* **102**(E2), 4045–4056.

Seiff, A. *et al.* (1998). Thermal structure of Jupiter's atmosphere near the edge of a 5-μm hot spot in the North Equatorial Belt. *J. Geophys. Res.* **103**(E10), 22857–22889.

Seiff, A., Stoker, C. R., Young, R. E., Mihalov, J. D., McKay, C. P. and Lorenz, R. D. (2005). Determination of physical properties of a planetary surface by measuring the deceleration of a probe upon impact. *Planet. Space Sci.* **53**(5), 594–600.

Semenov, Yu. P., (1994). *Rocket and Space Corporation Energia: The Legacy of S. P. Korolev. Energia*. Translated edition, Burnington; Apogee Books, 2001.

Semenov, Yu. P. (ed.) (1996). *RKK Energia im. S. P. Koroleva 1946–1996*. Moscow, RKK Energia.

Shaneyfelt, M. R., Winokur, P. S., Meisenheimer, T. L., Sexton, F. W., Roeske, S. B., and Knoll, M. G., (1994). Hardness variability in commercial technologies. *IEEE Trans. Nucl. Sci.* **41**, pp. 2536–2543.

Sherman M. M., (1971). Entry gasdynamic heating, NASA SP-8062, Langley Research Centre, National Aeronautics and Space Administration.

Shiraishi, H., Tanaka, S., Hayakawa, M., Fujimura, A. and Mizutani, H. (2000). Dynamical characteristics of planetary penetrator: effect of incidence angle and attack angle at impact. ISAS Science Report 677, Institute of Space and Aeronautical Science.

Shirley, D., (1998). *Managing Martians*. New York, Broadway Books.

Siddiqi, A. A. (2000). *Challenge to Apollo*. NASA SP–2000–4408. Reprinted in two volumes as *Sputnik and the Soviet Space Challenge* and *The Soviet Race with Apollo*, Gainesville FLA, (2003). University Press of Florida.

Siddiqi, A. A., Hendrickx, B. and Varfolomeyev, T. (2000). The tough road travelled: a new look at the second generation lunar probes. *J. British Interplanet. Soc.* **53**(9/10), 319–356.

Siddiqi, A. A. (2002). Deep space chronicle: a chronology of deep space and planetary probes 1958–2000. *Monographs in Aerospace History*, Volume 24. NASA SP–2002–4524. Washington NASA.

Simmons, G. J. (1977). Surface penetrators – a promising new type of planetary lander. *J. British Interplanetary Soc.* **30**(7), 243–256.

Sims, M. R., Pullan, D., Fraser, G. W. *et al.* (2003). Performance characteristics of the PAW instrumentation on Beagle 2 (the astrobiology lander on ESA's Mars Express mission). In Hoover, R. B., Rozanov, A. Yu. and Paepe, R. R. (eds.), *Instruments, Methods, and Missions for Astrobiology V. Proc. SPIE*, **4859**, 32–44.

Sims, M. R. (ed.), (2004a). *Beagle 2 Mars Mission Report.* Leicester, University of Leicester.

Sims, M. R. (ed.), (2004b). *Beagle 2 Mars Lessons Learned.* Leicester, University of Leicester.

Smith, P. H. and the Phoenix Science Team (2004). The Phoenix Mission to Mars. 35th *Lunar and Planetary Science Conference*, Houston, 15–19 March 2004, 2050.

Smrekar, S., Catling, D., Lorenz, R. *et al.* (1999). Deep Space 2: the Mars microprobe mission. *J. Geophys. Res.* **104**(E11), 27013–27030.

Smrekar, S., Lorenz, R. D. and Urquhart, M. (2001). The Deep-Space-2 penetrator design and its use for accelerometry and estimation of thermal conductivity. In Kömle, N. I., Kargl, G., Ball, A. J. and Lorenz, R. D. (eds), *Penetrometry in the Solar System.* Vienna, Austrian Academy of Sciences Press. pp. 109–123.

Spencer, D. A., Blanchard, R. C., Braun, R. D., Kallemeyn, P. H. and Thurman, S. W. (1999). Mars Pathfinder entry, descent, and landing reconstruction. *J. Spacecraft and Rockets,* **36**(3), 357–366.

Sperling, F., Galba, J., (1967). Treatise on the Surveyor lunar landing Dynamics and an Evaluation of pertinent telemetry data returned by surveyor 1. NASA Technical Report TR 32–1035, Jet Propulsion Laboratory.

Spilker, L. (ed.) (1997) *Passage to a Ringed World: The Cassini-Huygens Mission to Saturn and Titan.* NASA SP-523. NASA, Washington DC.

Spitzer, C. R. (1976). Unlimbering Viking's scoop. *IEEE Spectrum,* **13**, 92–93.

Squyres, S. W. (2005). *Roving Mars: Spirit, Opportunity and the Exploration of the Red Planet.* New York, Hyperion.

Steltzner, A., Desai, P., Lee, W., Bruno, R. (2003). The Mars exploration rovers entry descent and landing and the use of aerodynamic decelerators, 17th AIAA Aerodynamic Decelerator Systems Technology Conference and Seminar, Monterey, CA. AIAA–2003–2125.

Stooke, P. J. (2005). Lunar laser ranging and the location of Lunokhod 1. *36th Lunar and Planetary Science Conference*, Houston, 14–18 March 2005.

Stubbs, S. M. (1967), Dynamic Model Investigation of water pressures and accelerations encountered during landings of the Apollo spacecraft. NASA TN D-3980.

Surkov, Yu. A. (1997). *Exploration of Terrestrial Planets from Spacecraft: Instrumentation, Investigation, Interpretation.* 2nd. edn. Chichester, Wiley-Praxis.

Surkov, Yu. A. and Kremnev, R. S. (1998). Mars-96 mission: Mars exploration with the use of penetrators. *Planet. Space Sci.* **46**(11/12), 1689–1696.

Surkov, Yu. A., Moskaleva, L. P., Shcheglov, O. P. *et al.* (1999). Lander and scientific equipment for exploring of volatiles on the Moon. *Planet. Space Sci.* **47**(8/9), 1051–1060.

Surkov, Yu. A., Kremnev, R. S., Pichkhadze, K. M. and Akulov, Yu. P. (2001). Penetrators for exploring solar system bodies. In Kömle, N. I., Kargl, G., Ball, A. J. and Lorenz, R. D. (eds.), *Penetrometry in the Solar System.* Vienna, Austrian Academy of Sciences Press, pp. 185–196.

Thiel, M., Stöcker, J., Rohe, C., Hillenmaier, O., Kömle, N. I. and Kargl, G. (2001). The Rosetta lander anchoring harpoon: subsystem and scientific instrument. In Kömle, N. I., Kargl, G., Ball, A. J. and Lorenz, R. D. (eds.). *Penetrometry in the Solar System.* Vienna, Austrian Academy of Sciences Press, pp. 137–149.

Trainor, J H, (1994). Instrument and spacecraft faults associated with nuclear radiation in space. *Advances in Space Research,* **14**(10), 685–693.

TsUP (1985). *VeGa* (in Russian). TsUP (Spaceflight Control Centre), Moscow.

TsUP (1988). *Phobos* (in Russian). TsUP (Spaceflight Control Centre)/Informelektro, Moscow.

Tunstel, E., Maimone, M., Trebi-Ollennu, A., Yen, J., Petras, R., Wilson, R., (2005). Mars Exploration Rover mobility and robotic arm operational performance, 2005 *IEEE International Conference on Systems, Man, and Cybernetics*, Waikoloa, HI, October 10–12, 2005.

Ulamec, S., Espinasse S., Feuerbacher, B. *et al.* (2006) Rosetta Lander–Philae: implications of an alternative mission. *Acta Astronautica*, **58**(8), 435–441.

Ulrich J. A., (1966). Spacecraft sterilization techniques, NASA SP-108, p. 93.

Underwood, J C, (1993). A 12–degree of freedom Parachute/Payload Simulation of the Huygens Probe. 12th RAeS/AIAA Aerodynamic Decelerator Systems Technology Conference, London, May 10–13, 1993 (AIAA 93–1251).

Urquhart, M. L. and Smrekar, S. E. (2000). Estimation of soil thermal conductivity from a Mars microprobe-type penetrator. *31st Lunar and Planetary Science Conference*, Houston, 13–17 March 2000, 1781.

Varfolomeyev, T. (1998). Soviet rocketry that conquered space. Part 5: the first planetary probe attempts, 1960–1964. *Spaceflight*, **40**(3), 85–88.

Vaughan, V. L. (1961). Landing characteristics and flotation properties of a reentry capsule, NASA TN D-655.

Vergnolle, J.-F. (1995). Soft landing impact attenuation technologies review. *14th AIAA Aerodynamic Decelerator Systems Technology Conference*. AIAA–95–1535-CP.

Vesley D., Ruschmeyer O. R., and Bond R. G., (1966). Spacecraft contamination resulting from human contact. NASA SP108, pp. 275–283.

Vinogradov, A. P. (ed.) (1966). *Pervye Panoramy Lunnoi Poverkhnosti (First Panoramas of the Lunar Surface)*, Moscow, Nauka.

Vinogradov, A. P. (ed.) (1969). *Pervye Panoramy Lunnoi Poverkhnosti Tom 2 (First Panoramas of the Lunar Surface Vol. 2)*. Moscow, Nauka.

Vinogradov, A. P. (ed.), (1971). *Peredvizhnaya Laboratoriya na Lune Lunokhod-1. Tom 1*. Moscow, Nauka.

Vinogradov, A. P. (ed.), (1974). *Lunnyy Grunt iz Morya Izobiliya (Lunar Soil from the Sea of Fertility)*. Moscow, Nauka, (in Russian). Translated as NASA TT-F-15881, 1974.

Vojvodich, N. S., Drean, R. J., Schaupp, R. W. and Farless, D. L. (1983). Galileo atmospheric entry probe mission description, AIAA–83–0100, AIAA 21st Aerospace Sciences Meeting, Reno Nevada, January 10–13, 1983.

Von Karman, T., (1929), The impact of seaplane floats during landing, NACA TN-321, October 1929.

Vorontsov, V. A., Deryugin, V. A., Karyagin, V. P., *et al.* (1988). Method of investigation of the planet Venus using floating aerostatic stations. Mathematical Model. *Kosmich. Issled.* **26**(3), 430–433, (in Russian). Translation in *Cosmic Res.* **26**(3), 371–374, 1988.

Warwick, R. W. (2003). A low-cost, light-weight Mars landing system. *IEEE Aerospace Conference*, Big Sky, MT.

Wertz, J. R. and Larson, W. J., (1999). *Space Mission Analysis and Design.* 3rd edn., Torrence CA, Microcosm/Kluwer.

Wierzbicki T. and Yue D. Y., 1986. Impact damage of the Challenger crew compartment. *J. Spacecraft and Rockets*, **32**, pp. 646–654.

Wilcockson, W. H. (1999). Mars pathfinder heatshield design and flight experience. J. of *Spacecraft and Rockets*, **36**(3), 374–379.

Wilson, A. (ed), (1997). Huygens Spacecraft, payload and mission. ESA SP-1177.

Wilson, J. W., Shinn, J. L., Tripathi, R. K. *et al.* (2001). Issues in deep space radiation protection. *Acta Astronautica*, **49**, (3–10), 289–312.

Wilson, K. T., (1982). Rangers 3–5: America's first lunar landing attempts. *JBIS*, **36**, 265–274.

Withers, P. Towner, M. C., Hathi, B. and Zarnecki, J. C. (2003). Analysis of entry accelerometer data: a case study of Mars Pathfinder. *Planet. Space Sci.* **51**(9–10), 541–561.

Wright, I. P., Sims, M. R. and Pillinger, C. T. (2003). Scientific objectives of the Beagle 2 lander. *Acta Astronautica*, **52**(2–6), 219–225.

Yamada, T., Inatani, Y., and Honda, M., and Hirai, K. (2002). Development of thermal protection system of the Muses-c/BASH Reentry capsule. *Acta Astronautica* **51**(1–9), 63–72.

Yano, H., Hasegawa, S., Abe, M. and Fujiwara, A. (2002). Asteroidal Surface Sampling by the MUSES-C Spacecraft. In: Warmbein, B. (ed.) *Proc. Asteroids, Comets, Meteors ACM 2002*, 29 July–2 August 2002, Berlin. ESA SP-500, pp. 103–106.

Yew, C. H. and Stirbis, P. P. (1978). Penetration of projectile into terrestrial target. *J. Eng. Mech. Am. Soc. Civ. Engrs.* **104**(EM2), 273–286.

Yoshida, M., Tanaka T., Watanabe S., Takagi T., Shinohara M., and Fuji S. (2003). Experimental study on a new sterilization process using plasma source ion implantation with N_2 gas. *Journal of Vacuum Science Technology*, **21**, 4, 1230–1236.

Yoshimitsu, T., Kubota, T., Nakatani, I. and Kawaguchi, J. (2001). Robotic lander MINERVA, its mobility and surface exploration. In *Spaceflight Mechanics 2001, Advances in the Astronautical Sciences.* **108**(1), 491–501.

Yoshimitsu, T., Kubota, T., Nakatani, I., Adachi, T. and Saito, H. (2003). Micro-hopping robot for asteroid exploration. *Acta Astronautica*, **52**(2–6), 441–446.

Young, C. W. (1969). Depth prediction for earth-penetrating projectiles. *J. Soil Mech. Found. Div. Proc. Am. Soc. Civ. Engrs.* **95**(SM3), 803–817.

Young, C. W. (1997). Penetration equations. SAND97–2426, Sandia National Laboratories.

Young, R. E., Smith, M. A. and Sobeck, C. K. (1996). Galileo probe: in-situ observations of Jupiter's atmosphere, *Science*, **272**(5263), 837–838.

Young, R. E. (1998). The Galileo probe mission to Jupiter: science overview. *J. Geophys. Res.* **103**(E10), 22775–22790.

Zarnecki, J. C., Leese, M. R., Hathi, B. *et al.* (2005). A soft solid surface on Titan as revealed by the Huygens surface science package. *Nature*, **438**(7069), 792–795.

Zelenov I. A., Klishin, A. F., Kovtunenko, V. M., and Nikitin, M. D., (1988a). Characteristics of heat exchange and heat shielding of the Venera automatic interplanetary stations' descent vehicle. *Kosmicheskie Issledovaniya*, **26**(1), 28–32.

Zelenov, I. A., Klishin, A. F., Kovtunenko, V. M. and Shabarchin, A. F. (1988b). Methods of providing for thermal conditions in the Venera automatic interplanetary stations when in the atmosphere of Venus. *Kosmicheskie Issledovaniya*, **26**, 33–36.

Zimmerman, W. F., Bonitz, R. and Feldman, J. (2001). Cryobot: an ice penetrating robotic vehicle for Mars and Europa. IEEE 2001 Aerospace Conference, Big Sky, Montana.

Zupp, G. A. and Doiron, H. H. (2001). A mathematical procedure for predicting the touchdown dynamics of a soft-landing vehicle. NASA Technical Note TN D-7045. Houston, Manned Spaceflight Center.

Index